Wicking in Porous Materials

Traditional and Modern Modeling Approaches

Wicking in Porous Materials

Traditional and Modern Modeling Approaches

Edited by

Reza Masoodi

Krishna M. Pillai

CRC Press
Taylor & Francis Group
Boca Raton London New York

CRC Press is an imprint of the
Taylor & Francis Group, an **informa** business

CRC Press
Taylor & Francis Group
6000 Broken Sound Parkway NW, Suite 300
Boca Raton, FL 33487-2742

First issued in paperback 2017

Version Date: 2012920

ISBN 13: 978-1-138-07610-5 (pbk)
ISBN 13: 978-1-4398-7432-5 (hbk)

Visit the Taylor & Francis Web site at
http://www.taylorandfrancis.com

and the CRC Press Web site at
http://www.crcpress.com

Contents

Preface

Wicking or spontaneous imbibition of liquids into dry porous media is an industrial problem of great relevance—it finds application in a range of fields from textile processing to food processing to the processing of composite materials. A leading consumer product company funded our research on the wicking performance of one of their products over the last decade. These series of projects brought us to the amazing world of wicking and absorbent technology. While working in this area and conducting an exhaustive literature survey, we realized that different mathematical models based on varied flow physics are available for modeling liquid flow in porous materials during wicking. However, there was no book or publication that could present *all* the different modeling approaches in one volume in order to compare and disseminate the varied approaches. Hence, in 2010, we decided to collect these approaches and present them in the form of a book such that the basic assumptions and mathematical details associated with each of these methodologies are described. This project required about 2 years of hard work to accomplish our dream and publish this book.

The book contains some of the most important methods and philosophies for modeling wicking, from the traditional models to the latest approaches developed during the last few years. Although we tried to be as inclusive as possible, some important developments in the field may have escaped our dragnet. For example, application of the Ising model to predict the migration of moisture in a network of fibers could not be included due to a lack of contributing authors in the area. Similarly, the statistical approach to model wicking after accounting for randomness in porous media could not be included due to a shortage of time. Despite these shortcomings, we hope this book will be useful to our intended users. Although we have tried to be as meticulous and exact as possible in our work, some mistakes and oversights might have slipped in—we apologize for these misses.

In general, the book is intended for graduate students, professors, scientists, and engineers who are engaged in research and development on wicking and absorbency. In particular, the book is aimed at mathematical modelers who want to predict wicking with the help of computer programs—our goal is to provide a sound conceptual framework for learning the science behind different mathematical models, while at the same time being aware of the practical issues of model validation as well as measurement of important properties and parameters associated with various models. The layout of the book is designed to help in this task. Chapter 1 begins with an introduction on wicking, while Chapter 2 introduces the science behind wetting of surfaces. Chapter 3 introduces the reader to the basic derivation of the capillary model, and extension of the model after including gravity, inertia, and other effects. Since many of the recent models treat wicking as a flow-in-porous-media-type problem, Chapter 4 is dedicated to the science of single- and multi-phase flows in porous media, followed by a brief description of the measurement techniques and theoretical models available for the associated porous-media properties. The remaining chapters provide details of the individual models developed for wicking. Chapter 5 describes how the single-phase flow model accompanied by a sharp-front approximation is effective in modeling liquid imbibition into porous wicks. Chapter 6 describes the application of the unsaturated-flow (Richard's equation) model to predict wicking in anisotropic fibrous media. Chapter 7 discusses the application of the same unsaturated flow approach to model wicking-type liquid movements in absorbent swelling porous materials. Chapter 8 presents an application of the network models to model wicking accompanied by evaporation in porous materials. Chapter 9 explains advanced two-phase flow physics in network models used to study microscopic aspects of the wicking phenomenon. Chapter 10 describes a fractal-based approach to model wicking in porous media. In Chapter 11, wicking in deformable sponge is predicted after using the mixture theory to set up the governing equations. Finally, the use of the Lattice Boltzmann method to conduct a direct numerical simulation of wicking flows is presented in Chapter 12.

Our special thanks go to the individual authors who contributed chapters in this book—this project would not have been possible without their hardwork, support, and faith in this project. We are sure they will share our happiness and pride in publishing this book. We would like to express our appreciation to CRC Press for providing us with an opportunity to bring this work to fruition. In particular, we would like to thank the resourceful Jonathan Plant, executive editor at CRC Press, who encouraged and helped us at all stages of the project while doing a bit of firefighting for us to keep the project on track. We also thank Amber Donley, project coordinator at CRC Press, for her help. We also thank Syed Mohamad Shajahan, project manager at Techset Composition who oversaw the production of this book. A special thanks is reserved for Michelle Schoenecker and Dr. Marjorie Piechowski for helping us with the editing of our own chapters. We would also like to acknowledge Milad Masoodi's help in designing the book cover. Finally, we take this opportunity to express our deep appreciation to our families for their understanding and support during the course of this project that required many hours of activity during holidays, evenings, and weekends.

Reza Masoodi
Philadelphia, Pennsylvania

Krishna M. Pillai
Milwaukee, Wisconsin

Editors

Dr. Reza Masoodi is an assistant professor at Philadelphia University. He holds a PhD in mechanical engineering from the University of Wisconsin-Milwaukee (UWM). His PhD dissertation was about modeling imbibition of liquids into rigid and swelling porous media. His research areas of interests are flow in porous media, convection heat transfer, CFD, nano-composites, heat exchangers, and renewable energy. Dr. Masoodi has completed several research projects on single-phase flow in rigid and swelling porous media, capillary pressure, and permeability changes due to swelling. He has published several research papers, technical reports, and book chapters in the area of wicking in porous media. Dr. Masoodi is also a reviewer for more than 20 journals and served as technical committee member or peer reviewer for several scientific conferences.

Dr. Krishna M. Pillai is an associate professor at University of Wisconsin-Milwaukee (UWM) and is also the director of Laboratory for Flow and Transport Studies in Porous Media at UWM. He holds a BTech and MTech from IIT Kanpur (India), PhD from the University of Delaware, and post-doctoral fellowship at the University of Illinois, Urbana-Champaign. His research interests lie in several areas of porous media transport including flow and transport in fibrous media, wicking in rigid and swelling porous media, processing of polymer–matrix and metal–matrix composites, and evaporation modeling using network and continuum models. He has published extensively in reputed journals and presented his work in numerous international conferences and workshops. He is also a coauthor of several book chapters. He received copyright protections (sponsored by UWM Research Foundation) for his simulation codes PORE-FLOW and MIMPS. He was awarded the prestigious CAREER grant in 2004 by the National Science Foundation of the United States to model and simulate flow processes during mold filling in liquid molding processes used for manufacturing polymer composites. During his sabbatical in 2007–2008, he was a visiting researcher at CNRS at Institut de Mécanique des Fluides de Toulouse, France.

Contributors

Daniel M. Anderson
Department of Mathematical
  Sciences
George Mason University
Fairfax, Virginia

Holger Beruda
Procter&Gamble Service GmbH
Schwalbach am Taunus, Germany

Thomas M. Bucher
Department of Mechanical and
  Nuclear Engineering
Virginia Commonwealth
  University
Richmond, Virginia

Jianchao Cai
Institute of Geophysics and
  Geomatics
China University of Geosciences
Wuhan, People's Republic
  of China

Volker Clausnitzer
Groundwater Modeling Centre
DHI-Wasy GmbH
Berlin, Germany

Hans-Jörg G. Diersch
Groundwater Modeling Centre
DHI-Wasy GmbH
Berlin, Germany

Sandrine Geoffroy
Institut Clément Ader
Université de Toulouse
Toulouse, France

S. Majid Hassanizadeh
Department of Earth Sciences
Utrecht University
Utrecht, the Netherlands

Kamel Hooman
School of Mechanical and Mining
  Engineering
The University of Queensland
Queensland, Australia

Xiangyun Hu
Institute of Geophysics and
  Geomatics
China University of Geosciences
Wuhan, People's Republic of China

Vahid Joekar-Niasar
Innovation and R&D Department
Shell Global Solutions International
Rijswijk, the Netherlands

Manuel Marcoux
Institut de Mécanique des Fluides
  de Toulouse
Université de Toulouse
and
CNRS
Toulouse, France

Reza Masoodi
School of Design and Engineering
Philadelphia University
Philadelphia, Pennsylvania

Vladimir Mirnyy
Groundwater Modeling Centre
DHI-Wasy GmbH
Berlin, Germany

Krishna M. Pillai
Department of Mechanical
  Engineering
University of Wisconsin—Milwaukee
Milwaukee, Wisconsin

Mark L. Porter
Earth and Environmental Sciences
  (EES-14)
Los Alamos National Laboratory
Los Alamos, New Mexico

Marc Prat
Institut de Mécanique des Fluides
  de Toulouse
Université de Toulouse
and
CNRS
Toulouse, France

Rodrigo Rosati
Procter & Gamble Service GmbH
Schwalbach am Taunus, Germany

Marcel G. Schaap
Department of Soil, Water and
  Environmental Science
The University of Arizona
Tucson, Arizona

Mattias Schmidt
Procter & Gamble Service GmbH
Schwalbach am Taunus, Germany

Javed I. Siddique
Department of Mathematics
Pennsylvania State University
York, Pennsylvania

Hooman Vahedi Tafreshi
Department of Mechanical and
  Nuclear Engineering
Virginia Commonwealth University
Richmond, Virginia

Stéphanie Veran-Tissoires
Institut de Mécanique des Fluides
  de Toulouse
Université de Toulouse
and
CNRS
Toulouse, France

Boming Yu
School of Physics
Huazhong University of Science and
  Technology
Wuhan, People's Republic of
  China

Jijie Zhou
Shanghai Institute of Applied
  Mathematics and Mechanics
Shanghai University
Shanghai, People's Republic of
  China

1

Introduction to Wicking in Porous Media

Reza Masoodi and Krishna M. Pillai

CONTENTS

Wicking or spontaneous imbibition is the suction of a liquid into a porous medium due to the negative capillary pressure created at the liquid–air interfaces. The capillary pressure arises as a result of wetting the surface of the solid particles by the invading liquid that causes a change in the surface energy of the solid phase. In this chapter, an estimation of the capillary pressure through the Young–Laplace equation is presented, and the capillary and hydraulic radii, important parameters in the equation, are discussed. Some important applications of wicking in industry as well as in nature are briefly discussed next. Finally, an overview is presented of the different modeling approaches possible for wicking, which include the traditional capillary models, the porous-continuum models, the discrete models, and the statistical approaches.

1.1 Introduction

Imbibition is defined as the displacement of one fluid by another fluid, immiscible and more viscous than the first, in a porous medium (Alava et al., 2004). Examples of imbibition are displacement of oil by another immiscible liquid during oil recovery (Sorbie et al., 1995), invasion of a dry fiber-preform by a resin in resin transfer molding (Pillai, 2004), penetration of ink into paper (Schoelkopf, 2002), and movement of a liquid into a porous surface during the wiping-off of the liquid from the surface (Lockington and Parlange, 2003). If the driving force that causes imbibition is the capillary pressure, then such an imbibition is called wicking (also known as capillarity) (Masoodi et al., 2007). In other words, *wicking is the spontaneous imbibition of a liquid into a porous substance due to the action of capillary pressure.* Wicking may also occur from a combination of capillary pressure and an external pressure (Masoodi et al., 2010). Wicking plays an important role in many phenomena observed in science, industry, and everyday life. For example, in heat pipes, nano- and micro-wicks, wipes, sponges, diapers, and feminine pads, the wicking capability of the porous material involved is of high importance. Wicking also happens in nature; for example, in green plants, both capillary and osmotic pressures are responsible for drawing water and minerals from the roots up to the leaves (Ksenzhek and Volkov, 1998).

There are two proven approaches to modeling the wicking phenomenon mathematically. The older, more conventional method uses the Lucas–Washburn equation where the porous medium is assumed to be a bundle of aligned capillary tubes of the same radii (Lucas, 1918; Washburn, 1921). The newer method is based on Darcy's law where wicking is modeled as a single-phase flow through porous media (Pillai and Advani, 1996; Masoodi et al., 2007). Wicking is a function of the microstructure found inside porous media, the characteristics of the liquid involved, and time (Masoodi et al., 2007). The general relation between the wicking rate, the wicking time, and liquid characteristics is clearly described in both these wicking models that allow analytical results. The relation between the wicking rate and the microstructure of a porous medium is the most challenging, since the microstructure of porous media shows great variations (Bear, 1972).

Swelling is an important phenomenon that affects the wicking rate in many commercial products, such as papers, diapers, napkins, and wipes (Masoodi and Pillai, 2010; Masoodi et al., 2011). The swelling occurs due to the liquid absorption by constituent particles in plant-based porous materials such as paper and pulp when they come in contact with an organic liquid or water. Since the swelling changes the structure and molecular arrangement of the materials, it affects both the wettability and "wickability" of porous media (Kissa, 1996). In fact, the "swellability" of fibers is an important and useful property of paper for the paper industry since the swelling leads to liquid absorption in the paper matrix, which in turn leads to liquid retention inside the porous paper. In the paper industry, this liquid holding capacity, which is related to the moisture content after a paper sheet dries, is called the water retention value (TAPPI, 1991).

1.2 Capillary Pressure

Capillary pressure is due to a partial vacuum created on top of the liquid-front in capillary tubes during the wicking process (Berg, 1993; Masoodi and Pillai, 2012). The difference in the surface energies of the dry and wet solid matrices leads to the creation of a capillary force at the liquid–solid interface, which is responsible for pulling the invading liquid into porous materials (Masoodi and Pillai, 2012). The capillary force originates from the mutual attraction of molecules in the liquid medium and the adhesion of such molecules to molecules in the solid medium; the wicking phenomenon occurs when the adhesion is greater than the mutual attraction (Berg, 1993).

According to fluid statics, when the wetting and nonwetting fluids meet, the capillary pressure, which is the difference in pressures across the interface between the two immiscible fluids, is defined as

$$p_c = p_{nw} - p_w \tag{1.1}$$

where p_{nw} is the pressure in the nonwetting fluid while p_w is the pressure in the wetting fluid. A force balance over the meniscus in a capillary tube (Figure 1.1) leads to the equation

$$p_{nw}\pi R_c^2 = p_w\pi R_c^2 + \gamma \cos\theta(2\pi R_c) \tag{1.2}$$

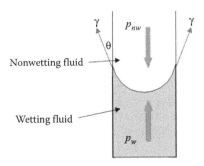

FIGURE 1.1
Meniscus (or liquid-front) in a capillary tube.

where R_c is the radius of the capillary tube. Combining Equations 1.1 and 1.2 leads to the well-known Young–Laplace equation for capillary pressure in a capillary tube:

$$p_c = \frac{2\gamma \cos\theta}{R_c} \tag{1.3}$$

Wettability is the tendency of a liquid to be attracted toward the surface of a solid phase. In a porous medium, the wettability is the main property that leads to the wicking or spontaneous imbibition (see Chapter 2). The parameter that quantifies wettability is the contact angle created by the liquid and solid phases. A lower contact angle implies a better wettability; hence, the best wettability is seen when the contact angle is zero, and the worst wettability occurs when the contact angle is 180°. (An efficient method for measuring the dynamic contact angle in porous media for a given liquid is explained in Chapter 5.)

One of the biggest challenges in using the Young–Laplace equation, Equation 1.3, is the estimation of the capillary radius. Since the real porous media with tortuous, interconnected fluid paths are not like bundles of identical, aligned capillary tubes (the "picture" used in the capillary model), some assumptions must be made to estimate the equivalent capillary radius for a real porous medium. The most accurate method of estimating the equivalent capillary radius is the capillary rise experiment, in which the upwards force due to capillary pressure is balanced by the downward force due to gravity (Dang-Vu and Hupka, 2005; Masoodi et al., 2008). The radius of the tube, employed to predict the capillary pressure given by Equation 1.3, is called the capillary radius (Masoodi et al., 2008; Masoodi and Pillai, 2012), or dry radius (Chatterjee, 1985; Chatterjee and Gupta, 2002), or static radius (Fries, 2010). The radius of the equivalent capillary tube in the wet part of the capillary tube is called the hydraulic radius (Masoodi et al., 2008; Masoodi and Pillai, 2012), wet radius (Chatterjee, 1985; Chatterjee and Gupta, 2002), or dynamic radius (Fries, 2010). Studies show that the hydraulic and capillary radii are two different variables although they are frequently used interchangeably (Chatterjee, 1985; Masoodi et al., 2008).

The analysis of micrographs of any porous medium is another way of estimating the equivalent capillary radius (Masoodi et al., 2007). Some researchers fitted circles in the pore space between particles and suggested that the circle radii should be taken as the equivalent capillary radii (Chatterjee and Gupta, 2002; Benltoufa et al., 2008). Others suggested the radius of the largest sphere (or circle, in the case of a 2-D micrograph) that could be fitted in the interparticle pore spaces should be considered the capillary radius (Lombard et al., 1989). Dodson and Sampson (1997) defined the capillary radius as the radius of a circle whose perimeter is the same as that of the pores. Masoodi and Pillai (2012) balanced the release of interfacial energy with the viscous energy dissipation at the liquid–air interface during wicking to derive the following expressions for capillary pressure (see Chapter 5 for details) that are amenable to be used with micrographs:

$$R_c = 2\frac{A_{int,s}}{C_{int,s}}\frac{\varepsilon}{1-\varepsilon} \tag{1.4a}$$

$$R_c = 2\frac{A_{int,v}}{C_{int,v}} \tag{1.4b}$$

Here, $A_{int,s}$ and $A_{int,v}$ are the cross-sectional areas of solid particles and void space, respectively, at the interface; $C_{int,s}$ and $C_{int,v}$ are the perimeters of solid particles and void space, respectively, at the interface (note that $C_{int,s} = C_{int,v}$); and ε is the porosity of porous media at the interface.

Techniques also exist to measure the hydraulic radius. An indirect estimation of the hydraulic radius involves first measuring the pressure drop along a porous sample and then employing the Hagen–Poiseuille law to calculate the hydraulic radius of the imaginary capillary tubes (Dullien, 1992; Rajagopalan et al., 2001; Masoodi et al., 2008).

1.3 Examples of Wicking

Wicking occurs in nature and in everyday life. For example, wicking is responsible for transport of water and minerals in plants (Ksenzhek and Volkov, 1998). Capillarity (another name for wicking) is considered a major performance index for industrial absorbing materials, such as wipes, diapers, and commercial wicks (Chatterjee and Gupta, 2002; Masoodi et al., 2007, 2011). Figure 1.2 shows an example of wicking in toilet paper where the liquid is being pulled up by capillary pressure at the liquid-front. An example of wicking due to the combination of an external pressure and capillary pressure occurs during the wiping of a surface by a wipe, paper napkin, or sponge (Mao and Russell, 2002, 2003; Lockington and Parlange, 2003; Gane et al., 2004; Masoodi et al., 2010). In the liquid composite molding (LCM) process for making polymer composites (see Figure 1.3), the capillary pressure is imposed at the micro-fronts inside fiber bundles of a woven or stitched fabric to model delayed wetting of bundles in dual-length-scale porous media (Pillai, 2002). The capillary pressure plays an important role in the formation of voids during the LCM mold-filling process; at low capillary numbers, the wicking flow inside fiber bundles in a fabric leads to trapping of macro-voids between the fiber bundles (Pillai, 2004).

Wicking and liquid absorption are important characteristics of textile materials (Patniak et al., 2006). One of the most important applications of wicking is in the printing industry where ink has to be absorbed efficiently

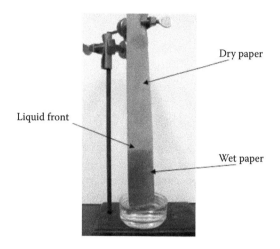

FIGURE 1.2
Wicking in toilet paper.

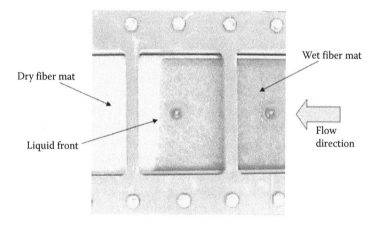

FIGURE 1.3
Injection of liquid in a mold filled with compressed fiber mats.

by paper (Gane et al., 2000; Schoelkopf et al., 2000; Schoelkopf, 2002). Using wicks made from polymers, ceramics, or textiles to dispense air fresheners or insecticides to room air is another industrial application of wicking (Masoodi et al., 2007, 2008). The astronauts use a device called spacecraft propellant management device (PMD) for movement in space, where the movement of liquid fuel through capillarity in low-gravity environments is of high importance (Fries, 2010). Heat pipes are very important in the field of microelectronic cooling (Faghri, 1995). The driving force in heat pipes is wicking or capillarity, used for moving liquid from the cold end to the hot end of the heat pipe (Figure 1.4). Researchers have made nanowick devices,

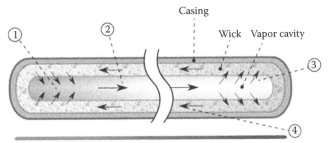

Heat pipe thermal cycle:
1. Working fluid evaporates to vapor absorbing thermal energy.
2. Vapor migrates along cavity to lower temperature end.
3. Vapor condenses back to fluid and is absorbed by the wick, releasing thermal energy.
4. Working fluid flows back to higher temperature end.

FIGURE 1.4
A schematic diagram describing fluid flow inside a heat pipe. (From Zootalures. 2006. Heat pipe mechanism, Wikipedia, The Free Encyclopedia, http://en.wikipedia.org/wiki/File:Heat_Pipe_Mechanism.png#filelinks (accessed January 4, 2012).)

using tiny copper spheres and carbon nanotubes, to passively wick a coolant toward hot electronics (Venere, 2010). Using such nanowicks, scientists have developed an advanced cooling technology for high-power electronics in military and automotive systems that is capable of handling up to 10 times the heat generated by conventional computer chips (Venere, 2010).

1.4 Modeling of Wicking in Porous Materials

There are several approaches to modeling the wicking process in porous media. To model wicking or spontaneous imbibition in a porous medium, a mathematical theory is needed to predict the all-important wicking speed in terms of the capillary pressure and the characteristics of the porous medium and the invading liquid. The following modeling approaches for wicking in porous substrates and incorporating different mathematical theories are introduced in the forthcoming chapters.

1.4.1 Wicking Models Based on the Capillary Model

In some of the traditional wicking models, the porous medium is modeled as a bundle of aligned capillary tubes. The first mathematical theory of capillary rise in a tube was published by Lucas (1918) and Washburn (1921) independently. They used momentum balance and neglected gravity and inertial effects to derive an analytical solution for meniscus height as a function of time in a vertical capillary tube. The capillary force was balanced with the viscous force in the tube to arrive at the well-known Lucas–Washburn equation. Later, others tried to include the neglected or missed force terms in the momentum equation and improved the accuracy of the capillary model (Chapter 3). In a recent development, the fractal theory was employed to include the statistical information about porous media in order to improve the accuracy of the capillary model for complex and irregular geometries (Chapter 10).

1.4.2 Porous-Continuum Models

Under the porous-continuum models, the flow in a porous medium is typically averaged over an averaging volume much larger than the individual pores or particles/fibers, but much smaller than the overall size of the medium. Thus, the averaged flow-variables are employed to predict flows in porous media after incorporating the properties of the liquid and solid phases through averaged medium properties (Chapter 4). The governing equations for wicking flows under isothermal conditions are Darcy's law and the continuity equation. Darcy's law is an empirical formula published in 1856 by Henry Darcy based on the results of his experiments on the flow of water through beds of sand (Bear, 1972). There are two general approaches to model fluid flow using the porous-continuum models: (1) the single-phase flow approach and (2) the two-phase flow approach. In the single-phase flow approach, a clearly defined liquid-front in a porous medium is postulated behind which the pore space is assumed to be completely occupied by the invading liquid.* Such a model has been used for modeling wicking in both rigid and swelling porous media (Chapter 5). In the two-phase flow approach, both the wetting and nonwetting fluids are modeled using Darcy's law, which leads to Richards' equation for predicting saturation migration in partially saturated porous media (Chapters 4 through 6). Hence, Richards' equation has been employed frequently to model wicking in rigid porous media (Chapters 4 through 6), and in nonrigid, swelling porous media (Chapter 7). The mixture theory can also be used to set up the averaged governing equations for flow in a deforming porous media (Chapter 11).

1.4.3 Discrete Models

One prominent example of the discrete models is the pore network model that uses pore-level physics to simulate fluid motion and transport within porous media. The geometry of such a pore network is designed to represent the structure of a real porous medium through a "reconstruction" process. The pore network is modeled as a network of pores (also called nodes) that are interconnected to each other through throats (also called bonds). The

---

* See some examples of sharp liquid-fronts in Figures 1.2 and 1.3.

modeling approach can be either single-phase or two-phase depending on whether just the liquid phase is being considered or both the liquid and gas phases are being considered within the pore space of the network in the mathematical model (Chapters 8 and 9).

1.4.4 Statistical Models

In statistical models, the statistical information about the microstructure of a porous medium is used to determine variability in the macroscopic (or porous-continuum) properties of a porous medium. The Lattice–Boltzmann method (LBM) is an approach based on statistical mechanics for modeling wicking in porous media. In LBM, instead of solving the Navier–Stokes equations, the discrete Boltzmann equation is solved to simulate the flow of multiple fluids within the complicated pore geometries (Chapter 12).

1.5 Summary

Wicking or spontaneous imbibition is a very common natural phenomenon where a liquid moves spontaneously into dry porous materials due to the release of surface energies. For example, the transportation of water and minerals from roots to the upper parts (leaves and branches) of trees involves wicking. Wicking is also very important in many consumer products, including wipes, diapers, feminine pads, and dispensers of insecticides and air fresheners. Wicking also has important industrial applications, such as heat pipes and nano-wicks, which are used for cooling microelectronic devices. This chapter introduces the capillary (suction) pressure, the main cause of the wicking phenomenon. The different available approaches for the estimation of the capillary and hydraulic radii, the two important wicking parameters, are explained. Finally, an overview of different modeling approaches for wicking is provided where the traditional capillary models, the porous-continuum models, the discrete models, and the statistics-based models are introduced; these models will be explored in greater detail in Chapters 3 through 12.

Nomenclature

Roman Letters

A Area (m^2)
C Perimeter (m)

p	Pressure (Pa)
R	Radius (m)

Greek Letters

θ	Contact angle (degree)
γ	Surface tension (N/m^2)
ε	Porosity (dimensionless)

Subscripts

c	Capillary
int	Interface surface of dry and wet matrix
nw	Nonwetting
s	Solid
v	Void space
w	Wetting

References

Alava, M., Dubé, M., and Rost, R. 2004. Imbibition in disordered media. *Advances in Physics* 53: 83–175.

Bear, J. 1972. *Dynamics of Fluids in Porous Media*. New York, Elsevier Science.

Benltoufa, S., Fayala, F., and BenNasrallah, S. 2008. Capillary rise in macro and micro pores of Jersey knitting structure. *Journal of Engineered Fibers and Fabrics* 3(3): 47–54.

Berg, J.C. 1993. *Wettability*. New York, Marcel Dekker.

Chatterjee, P.K. and Gupta, B.S. 2002. *Absorbent Technology*. Amsterdam, Elsevier.

Chatterjee, P.K. 1985. *Absorbency*, Amsterdam/New York, Elsevier.

Dullien, F.A.L. 1992. *Porous Media: Fluid Transport and Pore Structure*. San Diego, Academic Press.

Dodson, C.T.J. and Sampson, W.W. 1997. Modeling a class of stochastic porous media. *Applied Mathematics Letters* 10(2): 87–89.

Dang-Vu, T. and Hupka, J. 2005. Characterization of porous materials by capillary rise method. *Physicochemical Problems of Mineral Processing* 39: 47–65.

Fries, N. 2010. *Capillary Transport Processes in Porous Materials—Experiment and Model*. Göttingen, CuvillierVerlag.

Faghri, A. 1995. *Heat Pipe Science and Technology*. New York, Taylor & Francis Group.

Gane, P.A.C., Ridgway, C.J., and Schoelkopf, J. 2004. Absorption rate and volume dependency on the complexity of porous network structures. *Transport in Porous Media* 54: 79–106.

Gane, P.A.C., Schoelkopf, J., Spielmann, D.C., Matthews, G.P., and Ridgway, C.J. 2000. Fluid transport into porous coating structures: Some novel findings. *Tappi Journal* 83: 77.

Ksenzhek, O.S. and Volkov, A.G. 1998. *Plant Energetics*. San Diego, Academic Press. 1st edition.

Kissa, E. 1996. Wetting and wicking. *Textile Research Journal* 66(10): 660–668.

Lockington, D.A. and Parlange, J.Y. 2003. Anomalous water absorption in porous materials. *Journal of Physics D: Applied Physics* 36: 760–767.

Lucas, R. 1918. Rate of capillary ascension of liquids. *Kollid Z* 23: 15–22.

Lombard, G., Rollin, A., and Wolff, C. 1989. Theoretical and experimental opening sizes of heat-bonded geotextiles. *Textile Research Journal* 59(4): 208–217.

Masoodi, R. and Pillai, K.M. 2010. Darcy's law-based model for wicking in paper-like swelling porous media, *AIChE Journal* 56(9): 2257–2267.

Masoodi, R. and Pillai, K.M. 2012. A general formula for capillary suction pressure in porous media. *Journal of Porous Media* 15(8): 775–783.

Masoodi, R., Pillai, K.M., and Varanasi, P.P. 2007. Darcy's law based models for liquid absorption in polymer wicks. *AIChE Journal* 53(11): 2769–2782.

Masoodi, R., Pillai K.M., and Varanasi P.P. 2008. Role of hydraulic and capillary radii in improving the effectiveness of capillary model in wicking. *ASME Summer Conference*, Jacksonville, FL, USA, August 10–14.

Masoodi, R., Pillai, K.M., and Varanasi, P.P. 2010. The effect of hydraulic pressure on the wicking rate into the wipes. *Journal of Engineered Fibers and Fabrics* 5(3): 49–66.

Masoodi, R., Tan, H., and Pillai, K.M. 2011. Numerical simulation of liquid absorption in paper-like swelling porous media. *AIChE Journal*. DOI 10.1002/aic.12759.

Mao, M. and Russell, S.J. 2002. Prediction of liquid absorption in homogeneous three dimensional nonwoven structures. *International Nonwoven Technological Conference*, Atlanta.

Mao, N. and Russell, S.J. 2003. Anisotropic liquid absorption in homogeneous two dimensional nonwoven structures. *Journal of Applied Physics* 94(6): 4135–4138.

Pillai, K.M. 2002. Governing equations for unsaturated flow in woven fiber mats: Part 1 Isothermal flows. *Composites Part A: Applied Science and Manufacturing* 33:1007–1019.

Pillai, K.M. 2004. Unsaturated flow in liquid composite molding processes: A review and some thoughts. *Journal of Composite Materials* 38(23): 2097–2118.

Pillai, K.M. and Advani, S.G. 1996. Wicking across a fiber-bank. *Journal of Colloid and Interface Science* 183: 100–110.

Patniak, A., Rengasamy, R.S., Kothari, V.K., and Ghosh, A. 2006. Wetting and wicking in fibrous materials. *Textile Progress Woodhead* 38(1): 1–105.

Rajagopalan, D., Aneja, A.P., and Marchal, J.M. 2001. Modeling capillary flow in complex geometries. *Textile Research Journal* 71(91): 813–821.

Schoelkopf, J. 2002. Observation and modelling of fluid transport into porous paper coating structures. Ph.D. Thesis, University of Plymouth, U.K.

Schoelkopf, J., Gane, P.A.C., Ridgway, C.J., and Matthews, G.P. 2000. Influence of inertia on liquid absorption into paper coating structures. *Nordic Pulp Paper Research Journal* 15: 422.

Sorbie, K.S., Wu, Y.Z., and McDougall, S.R. 1995. The extended washburn equation and its application to the oil/water pore doublet problem. *Journal of Colloid Interface Science* 174: 289.

TAPPI Useful Methods. 1991. *Water Retention Value (WRV), UM-254, 54-56*. Atlanta, TAPPI Press.

Venere, V. 2010. Nanowick at heart of new system to cool power electronics. http://www.eurekalert.org/pub_releases/2010–07/pu-nah072210.php.

Washburn, E.V. 1921. The dynamics of capillary flow. *Physical Review* 17: 273–283.
Zootalures. 2006. Heat pipe mechanism, Wikipedia, The Free Encyclopedia, http://en.wikipedia.org/wiki/File:Heat_Pipe_Mechanism.png#filelinks (accessed January 4, 2012).

2

Wettability and Its Role in Wicking

Jijie Zhou

CONTENTS

Wicking plays a significant role in many separation processes, for example, chromatography. Porous and fabric networks can guide small amounts of liquid flow with wetting and spreading forces, and are themselves treated by chemicals in the liquid solution in the meantime. The first impression on such a spontaneous liquid-delivery process is often analogous to a capillary rise in small tubes. Similar to the chemical potential, liquids tend to minimize the summation of interfacial energies by changing their shapes. In this chapter, we interpret the deformable interfaces between immiscible fluids in terms of wetting parameters, such as surface energy, spreading parameter, and contact angle. Various concepts are brought in and connected to an illustrative and relevant role in wicking. A generalized view in film transport and hydrodynamics is outlined, and measurement methods for wetting parameters are listed.

2.1 Introduction

Wicking is referred to as the action of the spreading of a liquid through a textile network with capillary-like motions. The process of wicking involves physical chemistry of a surface, liquid properties, concave menisci, interfacial topography, ambient, and operating conditions. Application of basic sciences and engineering principles to the development, design, and operation of the process may control wicking toward a more valuable form; engineering improvement involves more intrinsic material designs than auxiliary instrumentation.

In the early nineteenth century, an English physician, Thomas Young, observed the liquid contact angle on a ridge surface and discovered the phenomena of capillary action on the principle of surface tension; a French scientist, Simon-Pierre Laplace, discovered the significance of meniscus radii with respect to capillary action. These discoveries led to the hydrostatic Young–Laplace pressure (ΔP) at a curved interface as a landmark of the field, where the pressure difference traversing the boundary is proportional to the mean curvature, that is, the average of two principle radii of curvature at the interface (or average of the maximum κ_1 and minimum κ_2 curvature values of this interface). The Young–Laplace pressure can explain shape factors such as the reason for small bubbles to merge themselves into connected large ones.

Surface-active agents (surfactants) practically act as detergents, wetting agents, emulsifiers, foaming agents, and dispersants. The effects of surfactants on wetting are a direct evidence of the intermolecular nature of a wetting force. The Marangoni effect, named after an Italian physicist, accumulates solvent molecules in opposition to the surfactant concentration gradient. Owing to concentration or temperature variations, liquid is conveyed along the consequential surface tension gradient to accommodate the Marangoni traction, to advance the liquid front, and to enlarge the wetted surface area (namely, to spread).

As an interfacial nature, surfaces carry a specific energy (surface energies) named surface tension. This energy reflects the cohesion of the condensed phase. Its quantity, from Zisman's empirical criterion, predicts wetting performance. Interfacial morphologies and energies coexist as we model wettability, capillarity, and related phenomena; in addition, the meniscus structure at the liquid advancing front counts as well as density, concentration, and pressure.

Wetting parameters play a fundamental role in wicking, similar to the Navier–Stokes equations with respect to turbulence. Hierarchical structures account for most complexity. Since 2000, the Clay Mathematics Institute has offered a million-dollar prize to the conjectures of existence and smoothness of the Navier–Stokes solutions. Wicking, or specifically imbibition, possesses complicated boundary conditions, absorbent mechanisms, and consequentially different assumptions in different models.

2.2 Wetting and Its Intermolecular Origin

Soap, a premier surfactant, was produced in 600 BC from tallow and beech ashes by the Phoenicians. The amphiphilic nature of soap molecules makes them to invade an immiscible water–oil interface, so that they become the compound of detergents and the main component of biological membranes. Modification of interfacial structures increases the affinity of water molecules to the interface. The degree of wetting or wettability is determined by intermolecular interactions when two materials are brought together, or when a liquid solution comes into contact with matter. Molecular rearrangements at interfaces lead to surface-state changes. The surface composition changes at a much faster rate than the bulk population. For instance, the interface quickly becomes more solute-rich than the solution.

A water wettable surface is termed hydrophilic; otherwise hydrophobic. Hydrophilic and lipophilic concepts are extended to interaction between molecules as well. Since disrupting the existing water structure at hydrogen-bonding sites is entropically unfavorable, water molecules barely reorient or restructure themselves around nonpolar solute or hydrophobic surfaces. As a consequence, nonpolar solute molecules deform their size and shape to become accommodated in an aqueous medium, which results in the hydrophobic effect. When a transition to wetting occurs, for entropic reasons, the liquid surface tension is so small that the free energy cost in forming a thick film is sufficiently compensated by liquid–solid interaction.

There are two theoretical approaches to formulate the foundation of wetting: the equation of state approach and Lifshitz–van der Waals/acid-base (LW/AB) approach. The first one assumes that an equation of state-type relation exists and employs the Gibbs–Duhem equations at interfaces; the other sums all attraction forces bearing on the surface molecules. Both exist as an immediate consequence of hypotheses and breaks down upon experimental checks of internal consistency. For example, surface energies are calculated on the continuum basis of the Lifshitz theory, or in terms of pairwise additivity of individual atoms or molecules. The theoretical surface energies $\gamma \approx A/24\pi(0.165 \text{ nm})^2$, where A is the Hamaker constant, are in good agreement with the experimental measurements for a variety of liquids and solids, except for strong hydrogen-bonding liquids where H-bonded network is nonpairwise additive. Perturbation of the hydrogen-bonding network is the origin for the enhancement of excess surface free energy.

Intermolecular and surface forces typically involve van der Waals, electrostatic, steric, depletion, hydrophobic, and salvation interactions. They are either essentially electromagnetic or steric in origin. In some cases, oscillatory period and fluctuation account for these forces; aggregation often raises geometric packing considerations due to solute concentration. The strength and decay length of each interaction characterize its force. The average translational kinetic of a molecule is $\frac{3}{2} kT$. Thermal energy kT is used to gauge the strength of an interaction.

As an ambiguous intermolecular force, van der Waals interaction is essentially electrostatic; it is a weak charge-fluctuation force that gradually dies away as the sixth power of molecular separation, which is very small when compared with the thermal energy kT. It may sometimes loosely refer to all intermolecular forces. The van der Waal's equation of state was first derived in 1873, assuming that a fluid composed of particles that occupy a finite volume and exert forces on each other. Taking these effects into account, the van der Waal's equation of state is cast into the form:

$$\left(P + \frac{a}{V^2}\right)(V - b) = n \cdot RT \tag{2.1}$$

where $R = 8.3143 \, \text{J mol}^{-1} \, \text{K}^{-1}$ is the universal gas constant. The interaction decays so fast that the sum of interactions between the molecules is considered two at a time, without significant influence of other nearby molecules. Van der Waals interaction defines the chemical character of many organic compounds. The typical surface tension of a van der Waals liquid or solid is $\gamma = 33 \, \text{mJ m}^{-2}$.

To elucidate the interplay between intermolecular and surface forces and the resulting wetting properties of a liquid, molecular dynamics (MD) simulations have been employed to understand liquid behaviors confined in a nano-scaled system. Despite its intermolecular origin, wetting is a macroscopic observation with many characteristic parameters to describe its physics and phenomena. Surface tension is equivalently a binary interfacial tension, suspending along a few molecular layers, where molecules attract each other the most. Surface tension is a well-defined material property on an atomically smooth and chemically homogeneous surface, when its value is irrelevant to a change in thickness. The hydrostatic Young–Laplace pressure (ΔP) traversing a curved membrane is proportional to the surface tension (γ) of the membrane which leads to

$$\Delta P = \gamma(\kappa_1 + \kappa_2) \tag{2.2}$$

where κ_1 and κ_2 are principle curvatures of the membrane (Figure 2.1).

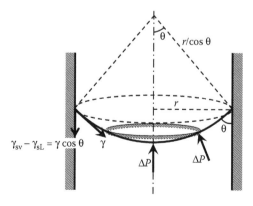

FIGURE 2.1
Liquid rise in a capillary.

2.2.1 Heuristic Derivation: How a Partial Wetting Liquid Rises into a Capillary with Wetting Parameters

When the partial wetting liquid comes into contact with the inner surface of the capillary, the immersed inner surface witnesses a solid–liquid interfacial energy (γ_{sL}) smaller than before (solid–vapor, γ_{sv}). This gives the liquid an advancing force $2\pi r(\gamma_{sv} - \gamma_{sL})$ to wet more capillaries. The advancing force is initially balanced with liquid momentum. As the liquid rises higher into the capillary, the viscous drag as well as gravity, slows down the liquid. Statically, the liquid reaches its balance when the deformable interface adapts Young's relation

$$\gamma \cos\theta = \gamma_{sv} - \gamma_{sL} \tag{2.3}$$

at the triple line, and when gravity nullifies the effect of Laplace pressure (ΔP) that originates from the liquid–vapor interfacial energy (γ). With θ angle at the circular triple line, the liquid–vapor interface has to conform into a bowl shape. Pressure along the vapor side of the interface is the same everywhere as the ambient pressure; transverse pressure is induced by radii of curvature. By neglecting the weight of the liquid in the depth of the bowl shape, transverse pressure will become constant on the interface. Consequently, the radii of curvature in the Laplace pressure formula $\Delta P = \gamma(\kappa_1 + \kappa_2)$ become a constant. Applying the triple-line boundary condition of θ, we obtain $\kappa_1 = \kappa_2 = \cos\theta/r$.

2.3 Contact Angle and Triple Line

Contact angle of a liquid to a substrate is the geometrical angle of a liquid/vapor interface meeting a solid surface, which is formed at the three-phase boundary. It is a quantitative measure of wetting affinity. The line of contact where three phases intersect is termed triple line or contact line when advancing.

Contact angle (θ) of one liquid has a narrow value of measurement on an ideal, flat, homogeneous, ridged surface, for example, on a semiconductor silicon wafer. Contact angle of mercury with glass is about 140°. Realistically, the contact angle on a rough surface may vary by 30° or 40°. Upon volume change, a liquid droplet may maintain its root area on the substrate invariant by adjusting its height and contact angle. Such an observation of pinning established hysteresis mechanisms of the triple line. While the triple line is pinned, the contact angle exists between two limiting values—advancing contact angle (θ_a) and receding contact angle (θ_r), which are the maximum and minimum values, respectively. Beyond these values, the line of contact moves.

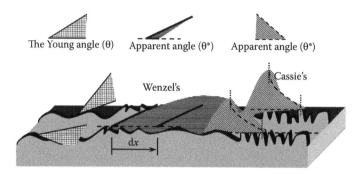

FIGURE 2.2
Apparent contact angles.

The liquid, as described by Shuttleworth and Bailey (1948), leaves a bit of "wonder" in its wake, that is, the receding contact angle may not have a unique value. Experimental range of observable contact angles, that is, the measurable and reproducible values of the contact angle, is in between the advancing and the receding contact angles. There is another contact angle, associated with the thermodynamic equilibrium configuration of the three-phase boundary/ intersection system. This equilibrium contact angle is coupled with a global energy minimum, sometimes named as the Young angle linked to solid surface energies, and commonly referred to as the contact angle.

Contact angle on a chemically heterogeneous surface (a mixture of θ_1 and θ_2 with f_1 and f_2 representing the fractional surface areas) is studied in Cassie and Baxter (1944) relation

$$\cos \theta^* = f_1 \cos \theta_1 + f_2 \cos \theta_2 \qquad (2.4)$$

The apparent angle (θ^*) is restricted between θ_1 and θ_2. Particularly, if the heterogeneous surface consists of air, that is, the liquid is lifted up by air pockets on roughness features, the apparent contact angle (θ^*) on this composite surface can be written as

$$\cos \theta^* = \varphi_s \cos \theta - (1 - \varphi_s) \qquad (2.5)$$

where φ_s is the pillar density on the solid surface. The defects and contaminants make many values of the contact angle possible, and thus leading to pinning of the triple line (Figure 2.2).

2.3.1 Roughness

Similar to a chemically heterogeneous surface, the apparent contact angle is modified on a rough-textured surface. Influence of geometric roughness (*r*: ratio of a rough surface area to a smooth planar one) on the contact angle was first explained by the Wenzel (1936) model via counting in the surface energy over the whole rough surface area. Wenzel's relation gives

$$\cos \theta^* = r \cos \theta \qquad (2.6)$$

Since roughness $r > 1$, it does not alter nonwetting into a wetting surface, but makes the hydrophilic surface more hydrophilic and the hydrophobic surface more hydrophobic. Besides following contours of the rough surface as in Wenzel's relation, the liquid may suspend a composite surface of the solid material and air pockets (Cassie and Baxter relation). Owing to the small solid–liquid contact area in the Cassie and Baxter state, the hysteresis is small and a liquid drop may roll off easily. The nematic arrangement seen in rice leaves, leads to anisotropic wetting; high base radii and rounded protrusions in roughness design, as seen in lotus leaves, are advantageous to self-cleaning and superhydrophobicity.

On a rough surface, as illustrated by He et al. (2003), two distinct apparent contact angles are possible in terms of reaching the local stable energy state, modeled by either Cassie's or Wenzel's theory, although one has a global lower energy than the other local minimum; that is associated to the thermodynamic equilibrium configuration, liquid meniscus on rough surfaces can be in the bi-stable Wenzel/Cassie state, depending on the initial location of a droplet. Furthermore, the influence of roughness on the contact angle is asymmetric when comparing hydrophilic ($\cos \theta > 0$) with hydrophobic ($\cos \theta < 0$) wetting. From partial wetting to complete wetting, Wenzel's model ceases to be applicable as $\cos \theta$ approaches 1; from hydrophobic to nonwetting, roughness dramatically alters hydrophobicity into superhydrophobicity. Such observations often involve the coexistence of two scales of roughness or random roughness. Shibuichi et al. (1996) have demonstrated that hierarchical roughness enables plant surfaces not only to reach a very high degree of hydrophobicity (i.e., $\cos \theta$ close to –1), but also a low contact angle hysteresis (i.e., a small difference between the advancing and receding angle). On the other, ideal substrates with an atomically smooth surface, for example, semiconductor silicon wafers, elastomers cross-linked from a liquid film, and the Pilkington floated glass are investigated to diminish the hysteresis ($\Delta\theta = \theta_a - \theta_r$). The difference between the advancing and receding angle reflects the change in the moisture content of the substrate—as a liquid drop is about to advance, the liquid sees a rough dry solid beyond the drop; when the liquid recedes, it leaves a trail of a mixture of solid and liquid at its wake. Consequential imbibition and drainage, in lieu of wetting, are still under investigation. Pinning of a contact line on either chemical heterogeneities or roughness incorporates nonlinearity into wicking, is similar to evaluating contributions from a dynamic load or a static friction. Since neither roughness nor fractional surface area is a continuum variable, applying them into wicking becomes a model-sensitive approach.

The maximal "chemical hydrophobicity" for water is 120°. Although hierarchical roughness promotes superhydrophobicity, the involved roughness is in the order of micrometers and submicrons. However, several chromatographic wicking studies involve nanometer-sized interstices.

2.3.2 Fine Structure of the Triple Line

Exquisite surface structures have been fabricated to study both beneficial and unfavorable effects of roughness on wetting. At pinning sites where the contact line is jammed before it suddenly jumps, the local curvature makes a range of contact angle accessible at that neighborhood, and that range allows a drop to deform against an external force and to cling on a surface. Microscopically, the contact line pinned at a single nonuniform spot adopts a characteristic saddle shape. Such fine structures are studied for the advancing front of hemi-wicking. The triple line is an averaged observation on a macroscopic level. The validity of the Wenzel model is restricted to such large drops that projection of a single defect is evened out.

The contact angle becomes size dependent in the nano regime, for example, water contact angle on carbon nanotubes. Line tension on small drops becomes an additional force to the contact angle θ that is given in Young's relation $\gamma \cos \theta = \gamma_{sv} - \gamma_{sL}$. At the molecular level, the definition of triple line is practically not the merging line as liquid thickness changes to zero, for example, tetrasiloxane meets the solid in layers resembling an inclined wavy slope. For small wetting molecules, there are also visible deformations on the Young angle; there exists a point of inflection in the immediate vicinity of the triple line. Such relevant spatial observations are typically a few to tens of nanometers. When the surface roughness is not in a regular pattern, it is hard to incorporate the geometry factors into consideration on the surface with nanotopographies, not only because of the difficulty of fabrication, but also because of the lack of dedicated work on superhydrophilic surfaces. In an effort of modeling macroscopically observed viscous dynamics on nanotextured surfaces, theoretical understanding often assumes that the effect of pinning and depinning from each nanopillar is not important.

A particular note is that contact angle in dynamic wetting is speed-dependent because the dynamic contact angle contributes to the formation of newly wetted solid surface. When a liquid sheet impinges onto a moving solid substrate, the actual contact angle depends not only on the wetting speed and material properties of the contacting media, but also on the inlet velocity of the liquid and geometry near the moving contact line. In complete wetting regime, a velocity-dependent dynamic contact angle

$$\theta \propto v^{1/3} \tag{2.7}$$

gives the relationship between θ and the speed of spreading (v).

2.4 Interfacial Tension and Spreading Parameter

Interfacial width between immiscible polymer phases is in the order of 1–3 nm, where one polymer protrudes into the other domain by a random walk of its loop. The interfaces for small molecules are typically a few atomic layers thick, where surface atoms are not in the same arrangement as in the bulk. On an ideal case in solids, the so-called singular surface, each asymmetrically located atom moves toward the interior, which results in an appreciable reduction of lattice constant between the surface atoms and interior atoms. Such a perpendicular shift of the surface layer may accompany lateral shift or restructuring of the original lattice. It is common that pure material surfaces reduce its surface energy by chemical and physical adsorption. Particularly, enrichment of surfactants on the surface of a liquid effectively reduces the surface energy. Liquid molecules on a liquid–vapor interface are prone to maintain a greater space than those inside the bulk and as a result interfacial molecules experience more attractive force than the interior ones. In contrast to solids, molecule lattice or spatial pattern on liquid surface seldom shifts, reconstructs or shrinks. The liquid surface acts as a stretched membrane, characterized by a surface tension (or interfacial tension) that opposes its distortion.

As liquid deforms continuously when subjected to shear stress, capillary adhesion and stable shapes of dew is a consequence of surface tension. Surface tension is also the physical origin of capillarity and wetting phenomena. Surface tension can be viewed as energy per unit area, which is numerically similar to the surface free energy required for generating more surface or surface area. Surface energy of simple liquids and solids in their own vapor is determined by the material property of a substance. In a foreign vapor, the process of creating surface area results in lowering the value of surface energy. While supplying energy to create a surface, the work required is proportional to the number of molecules on the new surface area, that is, $dW = \gamma\, dA$. If cohesion energy inside a liquid is U per molecule, the molecules missing half of the neighbors on the surface find themselves short of $U/2$. If Θ^2 is roughly each molecular exposed area, the surface tension of this liquid becomes $U/2\Theta^2$. The intensity of surface free energy is driven by the density of atomic dislocation and defects rather than their precise arrangement.

When the amount of liquid in a sessile drop exceeds its advancing contact angle, the triple line moves and this can be explained in both energy and force angle of view. The surface free energy can be viewed as surface tension, a force per unit length, directed toward the liquid along the local radius of curvature on any intersected curve, that is, $dF = \gamma\, dL$, or $d\sigma = \gamma\, dx$. The rate at which the rim of the drop advances over the solid surface is determined by the influence of gravity viscosity and capillarity, the volume of liquid in the drop, the contact angle, and the slip at the contact line.

In the field of materials science, surface tension is also referred to as surface stress, interfacial free energy, or surface free energy. The liquid sees a substrate (or another immiscible liquid) as one of the three types of wetting, that is, there are three types of behaviors of a liquid drop with its substrate: (i) complete wetting, (ii) partial wetting, and (iii) partial nonwetting. Lenient usage of the term "wetting" is also applied to the substance other than water. These behaviors are characterized by spreading the parameter and contact angle. By force balance of surface stress at the triple line (or alternatively from surface-free-energy point of view, work done to displace the contact line), contact angle θ is given in Young's relation: $\gamma \cos \theta = \gamma_{sv} - \gamma_{sL}$. Spreading parameter ($S$) is defined as

$$S = \gamma_{sv} - \gamma_{sL} - \gamma \tag{2.8}$$

The three types of wetting correspond to sign changes: (i) $S > 0$ and $\theta = 0$—complete wetting, (ii) $S < 0$ and $\theta < 90°$—partial wetting, and (iii) $S > 90°$—partial nonwetting (Figure 2.3).

Disjoining pressure of a film is a repulsive pressure analogous to the repulsive van der Waals force across the adsorbed liquid film. It only becomes significant when the film is very thin, that is, thinner than 100 nm and it causes the liquid to spread on surfaces. When liquid is pure water, electrically charged layers spontaneously form at both liquid–gas and solid–liquid interfaces; such charged layers and their electrostatic interaction are typical within a range of 10 nm. The water surface could be slightly deformed, which allows it to relax the additional stress from confinement and artificial boundary. Disjoining pressure is rather long range, varying as $1/e^3$. For example, a film of thickness e and surface tension γ is placed on a solid wafer whose surface tension is γ_{sv} and γ_{sL}. As liquid thickness vanishes on the substrate, the energy of the bare solid is recovered, that is,

$$\gamma_{sv} = \lim_{e \to 0} \gamma_{sL} + \gamma + \varphi(e) \tag{2.9}$$

Therefore, when the film thickness tends to zero, $\varphi(e, e = 0) = \gamma_{sv} - \gamma_{sL} - \gamma = S$; when the film thickness tends to large $\varphi(e, e = \infty) = 0$. This corrective term

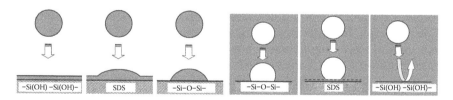

FIGURE 2.3
Wetting and nonwetting on a chemically modified surface.

$\varphi(e)$ is associated with the interfacial energies and one of formulations for the disjoining pressure produced by the two interfaces of the liquid film is

$$p(e) = -d\varphi/de > 0 \qquad (2.10)$$

Disjoining pressure is related to the chemical potential $\mu = \mu_o - v_o\, p(e)$; if the film thickness is changed by adding molecules $dN = de/v_o$, change of the energy includes

$$M\,dN = \mu_o\,dN + p(e)\,de \qquad (2.11)$$

Disjoining pressure is also a combination of various physical forces, that is, molecular, electrostatic, and structural; when the presence of intermolecular interactions is summarized into the disjoining pressure, the effective interface potential becomes the cost of free energy to maintain a homogeneous wetting film of thickness e. In case of partial wetting, for example, as described by Derjaguin and Kussakov (1939), the capillary pressure produced by meniscus at curved interfaces is in equilibrium with the connected disjoining pressure produced by the liquid-adsorbed layer.

These two aspects—energy and force—are typical approaches to study surface (interfacial) tension. The Young–Laplace pressure can explain capillary bridges, as a lateral cohesive force to consort the dissolved particles into clusters, or to maintain wet hairs adhering to bundles. Surface-to-volume ratio is a typical parameter in wicking studies. In hydrodynamic or dynamic studies, interfacial tension is also referred to as tension or wetting tension (τ) (Figure 2.4).

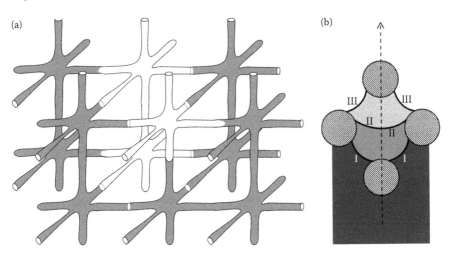

FIGURE 2.4
Wicking into porous network. (a) Partial filling of connected narrow pores; (b) sequential menisci while liquid passing obstacles in time course.

Complete wetting is often seen with light oils on high-chemical-binding-energy solids. Neinhuis and Barthlott (1997) have documented that self-cleaning superhydrophobicity is achieved by plants through epidermal roughness together with hydrophobic properties of the epicuticular wax. Langmuir (1938) found that surface treatments such as depositing a mono-layer of oleic acid on clean surfaces may dramatically modify wetting properties of the surfaces; for example, contact angle measurements, depend greatly upon both the characteristics of the underlying solid and such modifications.

2.5 Synopsis of Film Transport and Hydrodynamics

Whitaker (1964) and Whitaker and Jones (1966) recognized that surfactants have a stabilizing influence on film flow below an inclined surface, that is, transition Reynolds number, for the film flow to become unstable, is increased due to the addition of surfactants. Literally, transverse-film laminar diffusion is not directly influenced by perpendicular convection. In addition, Lenormand (1990) has experimentally studied the rich mechanisms of film transport, which involves displacement of the meniscus between immiscible fluids, hydrodynamics of nonwetting fluids, and possible flow with disconnected structures by the film along the roughness of the walls. Interdisciplinary investigation may include osmotically driven transport, thermocapillary flow, electrowetting, shear thinning, polymer rheology, coalescence of emulsions, and rupture of form bubbles.

Wetting of porous media plays a primary role in a variety of scientific endeavors, such as oil recovery, printing, textile production, and so on. One practical approach to characterize such systems is to measure the rate of penetration of the liquid into a porous structure and then to model that structure as pores and capillaries. Film dynamics, instability, capillary waves, and optical distortion of menisci have been used as principles to motivate measurement of the relationship between observation and wetting parameters. Macroscopic advancing of the contact line depends on the overall statistics of individual molecular displacements that occur near the triple line. Chapter 3 presents the important aspects of hydrodynamics such as the traditional Lucas–Washburn model and inertial effects. The precise form of the Lucas–Washburn equation is crucial for an effective characterization of porous media by capillary penetration experiments. Traditional film study at the mechanical level employs lubrication approximation. The laws of conservation that are universally applied in fluidics include

Continuity equation:

$$\frac{\partial \rho}{\partial t} + \nabla(\rho \mathbf{u}) = 0 \tag{2.12}$$

Navier–Stokes equation:

$$\frac{\partial(\rho \mathbf{u})}{\partial t} + \mathbf{u} \cdot \nabla(\rho \mathbf{u}) = -\nabla p + \mathbf{F} + \eta \nabla^2 \mathbf{u} + \left(\zeta + \frac{\eta}{3}\right) \operatorname{grad} \operatorname{div} \mathbf{u} \qquad (2.13)$$

Energy equation:

$$\frac{\partial \rho(e + u^2/2)}{\partial t} + \nabla \left[\mathbf{u}(\rho e + \rho u^2/2 + p) - \kappa \nabla T \right] = q + \mathbf{Fu} + \Psi \qquad (2.14)$$

In the initial stages of capillary penetration, the inertial effects are important as the influence of the dynamic contact angle should not be ignored. As the viscous friction force takes effect, the Lucas–Washburn equation gives

$$L^2 \propto t \qquad (2.15)$$

which relates the distance of penetration L spatio-temporally to wetting properties, such as the capillary geometry, surface tension, and contact angle. Owing to minimization of energy, existence of gravity, or Rayleigh–Taylor instability; ripples, waves, or uniformly distributed drops may form, while opposite effects act on the film.

A few trustworthy phenomena of surface-tension-driven instabilities are Rayleigh–Taylor instability, Rayleigh–Plateau instability, and Saffman–Taylor instability. A suspended film under a horizontal panel turns into an array of drops; the wavelength depends on the competition of the corrugation between gravity and surface tension, known as Rayleigh–Taylor instability. The fastest wavelength that overshadows others is $2\pi \sqrt{2}\ \kappa^{-1}$, where

$$\kappa^{-1} = L_c = \sqrt{\gamma/\rho g} \qquad (2.16)$$

is the capillary length. The liquid cylinders form wave-like dripping and even necklace-like gobbling droplets is the axial symmetric Rayleigh–Plateau instability. The wavy profile induces a Laplace pressure $\gamma/(R + e)$. Albeit the observation is often on a liquid jet or efflux, the Rayleigh–Plateau instability occurs in the cylindrical film coating inside a capillary tube. In the Saffman–Taylor instability, a viscous force initiates fingering on the advancing front and the Ohnesorge number is so small that inertia is less important. All in all, despite reversible condensation, wetting, seeping, and wicking are not symmetric observations from drying or dewetting. Because of a similar argument, it is worthy to note that imbibition and drainage mechanisms are different (Figure 2.5).

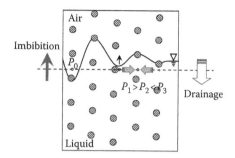

FIGURE 2.5
Interfacial displacement at the liquid advancing front.

2.6 Measurement of Wetting Parameters

Every man's memory is his private literature

Aldous Leonauard Huxley

Axisymmetric drop shape analysis is a common static method for measurement of interfacial tension. In this method, a numerical prediction for the equilibrium shapes of a pendant drop at different surface tension values is fitted to micrographic contours of the drop. Theoretical image fitting analysis of interfacial properties through configuring the drop without fitting an apex, approaches accuracy in less numerical steps. Another shape-fitting method is liquid-bridge-equilibrium fitting for the surface tension value. In microgravity, as all bubbles and drops have a spherical shape, the capillary pressure technique is the method of choice for fast data acquisition and measured capillary pressure gives access to dynamic interfacial tensions as tensiometry measurements.

Criteria and dimensional analyses are necessary to inspect assumptions and relative importance of relevant effects, and to justify experimental data. Liquid surfaces are easily contaminated; the exposed liquid surface tension may change with time when the test approaches quasi-static equilibrium. Dynamic surface tension measurement can be conducted under nearly hydrostatic conditions at each instance of recording time, that is when both bulk convection and mass transfer are insignificant. Surface tension measurement of high viscous liquids treats the viscosity effect separately through empirical approaches, to calibrate the curves of known liquids having similar surface tensions but different viscosities for viscosity correction (Figure 2.6).

An indirect method presents an interior interfacial tension through interior surface temperature distribution. Luminance measurement at spatial resolution is captured through thermochromic liquid crystals that alter their

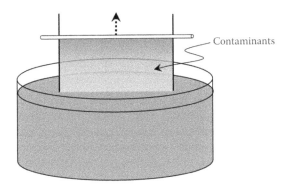

Contaminants

FIGURE 2.6
Diagrammatic drawback in force balance tensiometry.

optical properties with temperature. A calibration function transforms the measured luminance distribution into corresponding temperatures and interfacial tension values. Knowing the temperature on both inner and outer surfaces, the heat transfer can be evaluated.

The wetting principles are used on contact angle goniometers for measurement. An optimal method of choice depends upon the nature and conditions of the liquid to be measured. For measuring small contact angles, an inverted bubble method, originally proposed by McLachlan and Cox (1975), has a higher resolution. A widely employed technique for partial-wetting and superhydrophobic contact angles is to visualize a sessile drop and to couple the projected point of the triple line with a tangent line at the liquid–gas interface. The sessile-drop-tangent method also involves imaging process, which is more straightforward than the imaging process for the pendant-drop method, though it requires an accurate alignment of light passing in the direction of the liquid–solid interface. The droplet height is included in the best-fitting algorithm for superhydrophobic contact angles.

Furthermore, the pendant drop method, bubble pressure method (Jaeger's method), and the sessile drop technique, capillary rise method, and the stalagmometric method explain the surface tension values. For example, experimental measurement of surface tension on small liquid samples was devised by Ferguson 1924, 1933, based on measurement of pressure difference with optics principles across a liquid sample placed in a small diameter capillary on condition of one-flat meniscus of a liquid sample.

Traditional measurement for interfacial tension includes Du Noüy Ring method and Wilhelmy balance tensiometry (Ludwig Ferdinand Wilhelmy 1812–1864). Some variations are the spinning drop method for low interfacial tension, the interpreting-test-ink method on substrates, and drop volume method when it becomes complex to form a sufficient and stable pendant drop. A variety of techniques have been developed over the years with an optical goniometer, image analysis, and commercial computer-controlled instruments.

Owing to the triple-line-pinning phenomenon, two reproducible angles that characterize wetting, advancing (θ_a), and receding contact angles (θ_r) are approached in Wilhelmy balance tensiometry to measure the hysteresis ($\Delta\theta = \theta_a - \theta_r$) and to compare with goniometric methods. In Wilhelmy balance tensiometry, advancing and receding contact angles are calculated from immersion and emersion force measurements, with buoyancy correction from a force–balance equation. Captive-drop goniometry studies the advancing contact angle by feeding liquid into the drop through a needle connected to a syringe and recedes by removing. In comparison, tilting-plate goniometry confirms the value of advancing contact angles (θ_a). Experimental data for receding contact angles (θ_r) are not so consistent statistically as the advancing contact angles. Data agreement or disagreement among these methods reflects the substrate properties on wetting/dewetting physics.

In the subsequent chapters, we can further access the methodology for porosimetry, by directly and indirectly measuring the capillary and hydraulic radii, capillary pressure, estimation of tortuosity, permeability, relative permeability, and measuring wicking parameters.

2.7 Summary

Owing to liquid capability to deform continuously under shear, surface energy becomes the physical origin of the capillarity and wetting phenomena. Surface tension is determined by intermolecular forces, modified by surface coating and roughness, and balanced by the thermodynamic equilibrium configuration at intersection. Wetting behaviors are characterized with spreading parameter (S). A positive spreading parameter of a liquid on a substrate corresponds to a complete wetting case. Partial wetting depends on the contact angle, with obtuse angle linked to nonwetting and acute angle linked to wettable character for the liquid on substrates. In this chapter, wicking is dissected as the capillary-like action when the wettable liquid displaces a nonwetting fluid through porous and fabric networks, which involves the study of material properties, immiscible interfaces, complexity, and rate of transport.

References

Cassie, A.B.D. and S. Baxter. 1944. Wettability of porous surfaces. *Trans. Faraday Soc.* 40:546–551.

Derjaguin, B.V. and M.M. Kussakov. 1939. Anomalous properties of thin molecular films. *Acta Physicochim. USSR* 10:153–174.

Ferguson, A. 1924. VIII. A note on Mr. Ablett's paper on the angle of contact between paraffin wax and water. *Philos. Mag. Series* 6. 47(277):91–93.

Ferguson, A. 1933. Surface tension and its measurement. *J. Sci. Instrum.* 10(2):34–37.

He, B., N.A. Patankar, and J. Lee. 2003. Multiple equilibrium droplet shapes and design criterion for rough hydrophobic surfaces. *Langmuir* 19:4999–5003.

Langmuir, I. 1938. Overturning and anchoring of monolayers. *Science* 87:493–500.

Lenormand, R. 1990. Liquids in porous media. *J. Phys. Condens. Matter* 2:SA79.

McLachlan, D. and H.M. Cox. 1975. Apparatus for measuring the contact angles at crystal–solution–vapor interfaces. *Rev. Sci. Instrum.* 46:80–83.

Neinhuis, C. and W. Barthlott. 1997. Characterization and distribution of water-repellent, self-cleaning plant surfaces. *Ann. Bot.* 79:667–677.

Shibuichi, S., T. Onda, N. Satoh et al. 1996. Super water-repellent surfaces resulting from fractal structure. *J. Phys. Chem.* 100:19512–19517.

Shuttleworth, R. and G.L.J. Bailey. 1948. The spreading of a liquid over a rough solid. *Discuss. Faraday Soc.* 3:16–22.

Wenzel, R.N. 1936. Resistance of solid surfaces to wetting by water. *Ind. Eng. Chem.* 28:988–994.

Whitaker, S. 1964. Effect of surface active agents on the stability of falling liquid films. *Ind. Eng. Chem. Fundam.* 3:132–142.

Whitaker, S. and L.O. Jones. 1966. Stability of falling liquid films. Effect of interface and interfacial mass transport. *AIChE J.* 12:421–431.

3

Traditional Theories of Wicking: Capillary Models

Reza Masoodi and Krishna M. Pillai

CONTENTS

In this chapter, the capillary model as a traditional method for modeling wicking in porous media is introduced. Despite its weaknesses, including its applicability to only one-dimensional (1-D) wicking, and reducing the complex geometry of a real porous medium into a simplifying bundle-of-parallel-tubes-type geometry, the capillary model enjoys significant popularity with the wicking modeling community due to the simplicity of the mathematical model, the easy measurability of its few parameters, and the availability of vast literature on this model. The governing equation for the capillary model is derived using momentum balance in a single capillary tube. The governing equation is solved analytically for two simple cases: inclusion of only the viscous effect that leads to the classic Lucas–Washburn equation, or inclusion of both the gravity and viscous effects lead to an equation that can be solved analytically in two ways: either implicitly or in an explicit manner using the Lambert function. The governing equation after including the viscous, gravity, and inertial effects is solved numerically where inclusion of the inertial effects leads to significant deviations in the prediction of the liquid-column height during wicking. Although it is possible to make the capillary model comprehensive by adding terms pertaining to liquid acceleration and energy losses below the tube, recent research indicates that the effect of these additional terms is quite negligible. A simple variant of the Lucas–Washburn equation, after assuming a linear increase in the hydraulic diameter with time, is found to be quite accurate for predicting wicking in paper-like swelling porous media. Later, a novel method to measure the hydraulic diameter using the falling head permeameter is presented. The capillary model, after including the viscous and gravity effects, is found to be quite accurate for predicting wicking in polymer wicks if the capillary and hydraulic radii are measured independently.

3.1 Introduction

The traditional approach to model the wicking phenomenon mathematically is to use the Lucas–Washburn (L–W) equation (Lucas, 1918; Washburn, 1921), which is based on the capillary model for a porous medium. In such a model, the porous medium is assumed to consist of a bundle of aligned capillary tubes of the same radii. A mathematical explanation of the capillary rise phenomenon in a tube was first proposed by Lucas (1918) and later independently developed by Washburn (1921). They used the momentum balance equation after neglecting the gravity and inertial effects to derive an analytical solution for predicting changes in the meniscus height with time.

The L–W equation, in essence, balances the viscous pressure drop in a capillary tube with the capillary (suction) pressure at the meniscus to pull the liquid up the tube. The capillary pressure is obtained by the Young–Laplace equation, which relates the pressure difference across the interface between two static immiscible fluids in a capillary tube to the radius of the tube, the surface energies of the fluids and solid involved, and the contact angle (Masoodi and Pillai, 2012). The Young–Laplace equation is named after Thomas Young, who developed the qualitative theory of surface tension in 1805 (Young, 1805), and Pierre-Simon Laplace, who published the mathematical description later (Laplace, 1806). It may also be called the Young–Laplace–Gauss equation, as Gauss unified the work of Young and Laplace by using the Bernoulli's principle to derive this equation (Finn, 1999). The viscous energy loss during the derivation of the L–W equation is obtained by applying the Hagen–Poiseuille equation for liquid flows in tubes (Sutera and Skalak, 1993).

Several researchers included the missing or neglected physical effects during the capillary-rise phenomenon and modified the L–W equation accordingly by adding such "missing" terms (e.g., Rideal, 1922; Bosanquet, 1923; Szekely et al., 1971; Masoodi et al., 2007; Lavi et al., 2008; Fries, 2010). Such missing terms pertained to the gravity effects, the inertial effects, the viscous loss in liquid below the tube, and the viscous loss associated with the entrance effects. The experimental results showed that the modified capillary model is superior to the traditional L–W equation (e.g., Szekely et al., 1971; Stange et al., 2003; Masoodi et al., 2007; Fries, 2010).

Wettability is an initial requirement for wicking that is quantified by the contact angle, which is the angle between the tangent lines of the liquid–air and solid–liquid interfaces at the solid–air–liquid contact line (see Chapter 2 for a discussion on wettability). The dynamic contact angle is a major parameter in determining the capillary pressure, which in turn is important in determining the wicking rate. Dussan and Davis (1974) and Dussan (1979) reviewed previous work on the estimation of dynamic and static contact angles; it is generally acknowledged that the dynamic contact angle is related to the velocity-based capillary number (Jiang et al., 1979; Bracke et al., 1989).

While using the capillary models, some researchers noticed that the capillary and hydraulic radii might have different values, and suggested considering them as two independent parameters to obtain more accurate theoretical predictions (Chatterjee, 1985; Masoodi et al., 2008). The transition in flow pattern when the viscous, inertial, or gravity effects are dominant have also been studied (Stange et al., 2003; Fries and Dreyer, 2008). Dimensionless studies of capillary flow detailed the transition between different zones where the inertial, gravity, and viscous forces are dominant (Ichikawa and Satoda, 1994; Fries, 2010).

Despite having been studied and applied so extensively, the capillary models, including the L–W equation, have several weaknesses and problems when applied to model wicking or imbibition in porous media: (1) The capillary model based on the 1-D flow in straight tubes is applicable to 1-D wicking flows and cannot be extended to model two-dimensional (2-D) or three-dimensional (3-D) wicking flows in complicated geometries. (2) The real complex microstructure of a porous medium is replaced by an imaginary structure made of a bundle of capillary tubes. (3) In real porous media, the different pores are connected to each other, while in the capillary models, the capillary tubes are not interconnected. (4) In real porous media, there is a distribution of pore diameters while the capillary tube diameters are assumed to be constant in the capillary models. (5) The tortuous path of fluid flow in porous media is replaced with a straight-line fluid motion in the capillary models. (6) The assumptions about the pore structure of the porous substrate invariably lead to the use of some type of fitting parameter, such as hydraulic diameter, capillary diameter, or tortuosity, in the capillary models.

Although the above-listed deficiencies of the capillary model negatively affect the applicability and the scientific basis of the model, it is still a very popular approach and is used extensively wherever the wicking flows are 1-D. The reasons for its continuing success are its simplicity, small number of material parameters to be measured, easy measurability of the parameters, easy experimental validation, and finally, extensive published literature on the model. It also helps that many wicking phenomena can be treated as 1-D in nature.

3.2 Wicking in Rigid Porous Materials

In rigid porous media, the size of the constituent particles and the pores between them remain constant during the wicking process. As a result, the tube diameter remains constant in the capillary model.

3.2.1 Theory of Capillary Models

Since a porous medium under capillary models is treated as a bundle of aligned capillary tubes of the same size, the fluid flow in each tube is assumed to represent fluid motion inside the pore space of a porous material. Therefore, the mathematical theory of capillary models will be based on the flow of a liquid through a single capillary tube. It turns out that an *approximate* integral analysis based on average velocity through a capillary tube is sufficient for the development of the final forms of the L–W equation and other capillary models in porous media.

The equation of mass balance for a deforming control volume (CV) just below the meniscus, as shown in Figure 3.1, is (White, 2010)

$$\left(\frac{dm}{dt}\right)_{sys} = 0 = \frac{d}{dt}\int_V \rho \, dV + \int_A \rho \vec{v}_r \cdot \overline{dA} \tag{3.1}$$

where $\vec{v}_r$ is the velocity of the liquid relative to the moving control-volume boundary, $\overline{dA}$ is the inlet and outlet areas, V is volume, ρ is density, and m is mass. Assuming the height of the CV to increase as a result of the moving liquid-front, it is clear that the liquid will not leave such a CV; this means that the relative velocity at the outlet is 0. Using this fact in Equation 3.1 leads to the expression

$$\frac{d}{dt}\int_0^h \rho \pi R^2 dz - \rho w(\pi R^2) = 0 \tag{3.2}$$

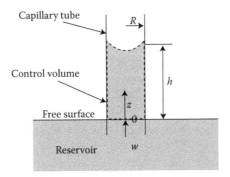

FIGURE 3.1
A schematic of liquid rise in a capillary tube.

According to Equation 3.2, the average z-direction liquid velocity at the inlet, w, is identical to the rate of increase in the meniscus (or CV) height inside the capillary tube

$$w = \dot{h} \tag{3.3}$$

The integral form of the momentum balance equation for the CV shown in Figure 3.1 can be expressed (White, 2010) as

$$\sum F_z = \frac{d}{dt}\int_V \rho w \, dV + \int_A \rho w \vec{v}_r \cdot \overline{dA} \tag{3.4}$$

where F_z is any vertical force acting on the CV, w is the z-component of velocity, ρ is the density of the liquid, $\vec{v}_r$ is the fluid velocity relative to the CV surface, V is the CV volume, and A is the cross-sectional area at the inlet/outlet of the CV. Using Equation 3.3 and the Leibnitz rule, the first integral on the right-hand side of Equation 3.4 can be developed as

$$\frac{d}{dt}\int \rho w \, dV = \frac{d}{dt}\int_0^h \rho \dot{h}\pi R^2 \, dz = \frac{d}{dt}(\rho\pi R^2 \dot{h}h) = \rho\pi R^2 \dot{h}\ddot{h} + \rho\pi R^2 \dot{h}^2 \tag{3.5}$$

To transform the second integral on the right-hand side of Equation 3.4, consider there is one inlet at $z = 0$ and no outlet at $z = h$; therefore

$$\int_A \rho w \vec{v}_r \cdot \overline{dA} = \rho \dot{h}(-\dot{h})\pi R^2 = -\rho\pi \dot{h}^2 R^2 \tag{3.6}$$

Let us now consider the left-hand side of Equation 3.4. There are three major forces that act on the liquid inside the CV, namely the gravity force (F_g), the pressure force (F_p), and the viscous force (F_v). The pressure force (F_p) can be split into two parts: the "upper" pressure force (F_{up}) that acts on the top of CV where $z = h$, and the "lower" pressure force (F_{lp}) that acts at the CV entrance where $z = 0$. As a result, the left-hand side can be expressed as

$$\sum F_z = F_{up} + F_{lp} + F_g + F_v \tag{3.7}$$

Note that the capillary force is the major suction force, created due to the capillary pressure appearing at the meniscus, which pulls the liquid upwards in the tube while the gravity and viscous forces resist this motion. By definition, the capillary force is equal to the capillary pressure multiplied by the cross-sectional area of the tube. Both the capillary and atmospheric pressure forces act on the CV top, so the upper pressure force (F_{up}) can be expressed as

$$F_{up} = -\int_A (p_a - p_c)dA = -\pi R^2 (p_a - p_c) \tag{3.8}$$

Note that since F_{up} is acting downward by definition, there is a negative sign before the integrand. To find the pressure at the CV entrance (point "0" in Figure 3.1), an energy balance equation between point "0" and the free surface outside the tube will be used (White, 2010):

$$\frac{p_{fs}}{\gamma} + \frac{(v_{fs})^2}{2g} + z_{fs} - h_L = \frac{p_0}{\gamma} + \frac{(v_0)^2}{2g} + z_0 \tag{3.9}$$

Note that $p_{fs} = p_a$, $v_{fs} = 0$, $z_{fs} = z_0$, $v_0 = w$, and $h_L \sim ((v_0)^2/2g)$ (minor loss for the inward-projecting pipe); therefore

$$p_0 = p_a - \rho w^2 \tag{3.10}$$

Since F_{lp} is acting upwards, it is positive, and hence

$$F_{lp} = \int_A p_0 \, dA = \pi R^2 (p_a - \rho w^2) = \pi R^2 (p_a - \rho \dot{h}^2) \tag{3.11}$$

where the last step involves the use of Equation 3.3. The downward gravity force can be found by estimating the weight of the liquid within the capillary tube as

$$F_g = -\int_V \rho g \ dV = -\int_0^h \rho g \pi R^2 \ dz = -\rho g \pi R^2 h \qquad (3.12)$$

The pressure drop due to viscosity can be found using the Hagen–Poiseuille law (White, 2010) as

$$\Delta p = \frac{8 \mu L Q}{\pi R^4} \qquad (3.13)$$

The pressure drop due to viscosity for the tube shown in Figure 3.1 can be reexpressed as

$$\Delta p = \frac{8 \mu}{\pi R^4} h(\pi R^2 \dot{h}) = \frac{8 \mu}{R^2} h \dot{h} \qquad (3.14)$$

Therefore, the viscous force associated with the above pressure drop, arising due to the wall shear-stress, is

$$F_v = -\frac{8 \mu}{R^2} h \dot{h} (\pi R^2) = -8 \pi \mu h \dot{h} \qquad (3.15)$$

Substituting the forces obtained through Equations 3.8, 3.11, 3.12, and 3.15 in Equation 3.7 leads to

$$\sum F_z = \pi R^2 p_c - \rho g \pi R^2 h - 8 \pi \mu h \dot{h} - \rho \pi \dot{h}^2 R^2 \qquad (3.16)$$

Further substituting Equations 3.5, 3.6, and 3.16 in Equation 3.4 results in the following governing equation for liquid-front movement in a capillary tube

$$R^2 p_c = 8 \mu h \dot{h} + \rho g R^2 h + \rho R^2 \frac{d(h \dot{h})}{dt} \qquad (3.17)$$

The left-hand side is related to the capillary pressure-induced suction force while the first, second, and third terms on the right-hand side of Equation 3.17 arise due to the viscous, gravity, and inertial forces, respectively. If we use R_h for the hydraulic radius of the tube, then using R_h in place of R in Equation 3.14 results in

$$p_c = \frac{8 \mu h \dot{h}}{R_h^2} + \rho g h + \rho \frac{d(h \dot{h})}{dt} \qquad (3.18)$$

Note that all terms in Equation 3.18 have dimensions of pressure; the equation balances the capillary pressure with a sum of "pressures" due to the viscous, gravity, and inertial forces.

3.2.2 Capillary Pressure

Using the Young–Laplace equation (Masoodi and Pillai, 2012), the capillary pressure in a tube can be expressed as

$$p_c = \frac{2\sigma \cos(\theta_d)}{R_c} \tag{3.19}$$

where σ is the surface tension of the liquid, θ_d is the dynamic contact angle, and R_c is the capillary radius. In general, the capillary radius is taken as different from the hydraulic radius in order to improve the accuracy of the capillary models (Chatterjee, 1985; Chatterjee and Gupta, 2002; Masoodi et al., 2008).

The contact angle formed at the meniscus changes with the speed of the meniscus (Dussan and Davis, 1974; Dussan, 1979). The contact angle formed when the meniscus is stationary is called the *static* contact angle, while the contact angle formed when the meniscus is moving is called the *dynamic* contact angle. The dynamic and static contact angles are related to each other through the expression (Bracke et al., 1989)

$$\frac{\cos(\theta_d) - \cos(\theta_s)}{\cos(\theta_s) + 1} = -2Ca^{0.5} \tag{3.20}$$

where the capillary number, Ca, is defined as

$$Ca = \frac{\mu \dot{h}}{\sigma} \tag{3.21}$$

Equation 3.20 states that the dynamic contact angle changes as the velocity of the meniscus changes. However, except at the beginning and end of the capillary motion, the meniscus speed does not change significantly, and so the dynamic contact angle can be assumed to remain constant during wicking.

At the start of the wicking process, it takes time to form the meniscus in a capillary tube and so the capillary pressure reaches its final value, as given by Equation 3.19, only after some time. To take into account the delay due to meniscus formation, Stange et al. (2003) suggested a meniscus formation factor as

$$S(t) = 1 - e^{-4.6t/t_r} \tag{3.22}$$

where t_r is the reorientation time of the liquid. Siegert et al. (1964) recommended the following expression for t_r:

$$t_r = 0.413\sqrt{\frac{\rho R^3}{\sigma}} \tag{3.23}$$

After including the meniscus formation factor $S(t)$, Equation 3.19 can be modified as

$$p_c = S(t)\frac{2\sigma\cos(\theta_d)}{R_c} \tag{3.24}$$

Note that the effect of the term $S(t)$ is important in the initial moments of the wicking process and can be neglected, unless one wants to study the initial stage of wicking.

3.2.3 Missing or Neglected Terms of the Capillary Model

Fries (2010) included the local acceleration of the liquid below the tube, the viscous dissipation below the tube, and the convective flow losses at tube entrance, and added four more terms to the previous expression for the capillary model:

$$p_c = \frac{8\mu h\dot{h}}{R_h^2} + \rho gh + \rho h\ddot{h} + \frac{11}{12}\rho R_h\ddot{h} + 2\frac{\mu}{R_h}\dot{h} + \frac{23}{24}\rho\dot{h}^2 + \frac{1}{2}\rho K\dot{h}^2 \tag{3.25}$$

Here, K is a function suggested by Sparrow et al. (1964) to include the entrance effects. However, Fries (2010) claimed that these new terms are negligible and Equation 3.18 is accurate enough for predicting liquid flows in capillary tubes.

3.3 Solving the Governing Equation

For predicting the rise of the liquid–air meniscus, Equation 3.18 can be solved as a second-order differential equation in h. In the different variants of the capillary model, usually one or two of the listed terms in the equation are used at a time. Four different cases of such capillary models are presented here.

3.3.1 Capillary Model with Viscous Effects: Lucas–Washburn Equation

When the gravitational and inertial terms are neglected in Equation 3.18, the capillary (suction) force is balanced by the viscous (friction) force through

$$\frac{2\sigma\cos(\theta_d)}{R_c} = \frac{8\mu h\dot{h}}{R_h^2} \tag{3.26}$$

This equation can be solved using the initial condition $h\,(t=0)=0$; the solution is the well-known *Lucas–Washburn* equation (also known as *Washburn* equation) of the form

$$h = \sqrt{\frac{\sigma R_e \cos(\theta_d)}{2\mu}} \, t \qquad (3.27)$$

in which R_e, the effective capillary radius, is obtained by the expression

$$R_e = \frac{R_h^2}{R_c} \qquad (3.28)$$

3.3.2 Capillary Model with Viscous and Gravity Effects

When the inertial term is neglected in Equation 3.18, the capillary pressure is balanced by the viscous and gravity terms:

$$p_c = \frac{8\mu h \dot{h}}{R_h^2} + \rho g h \qquad (3.29)$$

This nonlinear ordinary differential equation for the meniscus height h can be solved easily by the separation-of-variables technique and by using the initial condition $h\,(t=0)=0$ (Masoodi et al., 2008):

$$p_c \ln \left| \frac{p_c}{p_c - \rho g h} \right| - \rho g h = \frac{\rho^2 g^2 R_h^2}{8\mu} t \qquad (3.30)$$

Equation 3.30 is an implicit expression for the liquid-front height in a capillary tube. Although Equation 3.30 is usable in this implicit form, it can also be converted to a more usable, explicit form using the Lambert function. By dividing all terms in Equation 3.30 by $-p_c$ and subtracting 1 from both sides, it turns into

$$\ln \left| \frac{p_c - \rho g h}{p_c} \right| + \frac{\rho g h}{p_c} - 1 = -\frac{\rho^2 g^2 R_h^2}{8\mu p_c} t - 1 \qquad (3.31)$$

Taking exponential of both sides leads to

$$\left(\frac{\rho g h}{p_c} - 1 \right) e^{\frac{\rho g h}{p_c} - 1} = -e^{-\frac{\rho^2 g^2 R_h^2}{8\mu p_c} t - 1} \qquad (3.32)$$

which is similar to the Lambert function. The Lambert function, also called the omega function or the product logarithm, is a set of functions that are branches of the inverse relation of the function (Corless et al., 1996)

$$f(W) = We^W \qquad (3.33)$$

The general strategy is to move all instances of the unknown to one side of the equation and transform it into the form $y = W(x)e^{W(x)}$, at which point the function W provides the value of the variable x:

$$y = W(x)e^{W(x)} \Leftrightarrow x = W(y) \qquad (3.34)$$

Applying the idea of Lambert function, Equation 3.34, to Equation 3.32 leads to

$$\left(\frac{\rho g h}{p_c} - 1\right) = W\left(-e^{-\frac{\rho^2 g^2 R_h^2}{8\mu p_c}t - 1}\right) \qquad (3.35)$$

Such a solution can be easily manipulated into the following more convenient form:

$$h = \frac{p_c}{\rho g}\left[1 + W\left(-e^{-\frac{\rho^2 g^2 R_h^2}{8\pi p_c}t - 1}\right)\right] \qquad (3.36)$$

which is an explicit equation that gives liquid-front height in a capillary tube as a function of time.

To compare the predictions of Equations 3.30 and 3.36 and thus validate the proposed transformation, consider the dimensionless variables

$$H = \frac{h}{h_e} \qquad (3.37a)$$

$$T = \frac{t}{t_e} \qquad (3.37b)$$

such that h_e and t_e are obtained from the expressions

$$h_e = \frac{p_c}{\rho g} \qquad (3.38a)$$

$$t_e = \frac{8\mu p_c}{\rho^2 g^2 R_h^2} \qquad (3.38b)$$

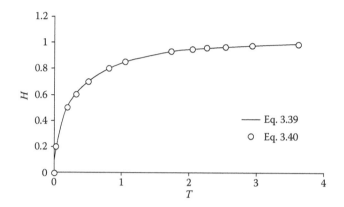

FIGURE 3.2
A comparison of predictions obtained from the implicit solution, Equation 3.39, and the explicit solution, Equation 3.40.

On employing these transformations, Equations 3.30 and 3.36 turn, respectively, to the following dimensionless forms:

$$\ln(1 - H) + H = -T \tag{3.39}$$

$$H = 1 + W(-e^{-T-1}) \tag{3.40}$$

Figure 3.2 shows that predictions obtained from the implicit expression, Equation 3.39, and the explicit expression, Equation 3.40, for the dimensionless height of the liquid column in a capillary tube are indeed identical.

To predict the height of the liquid column using Equation 3.40, it is important to estimate the value of the Lambert function. Though tabulated forms of the function are available in the literature, its more usable, analytical expressions are also possible. For example, Fries (2010) used the following approximate expression for $W(x)$ during the numerical computation of Equation 3.40:

$$W(x) \approx -1 + \frac{\sqrt{2 + 2ex}}{1 + (4.13501\sqrt{2 + 2ex}/12.7036 + \sqrt{2 + 2ex})} \tag{3.41}$$

Here, "e" is the Euler's number.

3.3.3 Capillary Model with Viscous and Inertial Effects

When the gravitational effects are not significant (i.e., when the height of the liquid column is much smaller than its final steady-state value), the inertial and viscous effects are dominant, and hence the gravity term in Equation 3.18 can be dropped

$$p_c = \rho \frac{d(h\dot{h})}{dt} + \frac{8\mu h \dot{h}}{R_h^2} \tag{3.42}$$

To solve this equation, the following two initial conditions are needed:

$$h(t = 0) = 0 \tag{3.43a}$$

$$\dot{h}(t = 0) = 0 \tag{3.43b}$$

The solution of Equation 3.42 with the above initial conditions is

$$h = \sqrt{\frac{p_c R_h^2}{4\mu}\left[t + \frac{\rho R_h^2}{8\mu}\left(e^{-\frac{8\mu}{\rho R_h^2}t} - 1\right)\right]} \tag{3.44}$$

which is identical to the expression presented by Bosanquet (1923).

3.3.4 Capillary Model with Viscous, Gravity, and Inertial Effects

In the case of including all the terms on the right-hand side of Equation 3.18, it is not possible to solve the resulting governing equation analytically. Numerical methods, such as the Runge–Kutta method, may be employed to find the numerical solution (e.g., Szekely et al., 1971). One may use Equations 3.37 and 3.38 to convert Equation 3.18 into the dimensionless form

$$H\dot{H} + H + \omega \frac{d(H\dot{H})}{dT} = 1 \tag{3.45}$$

where ω is a positive dimensionless number, obtained through the expression

$$\omega = \frac{\rho^2 g R_h^4}{64\,\mu^2 h_e} \tag{3.46}$$

Note that a small value of ω indicates a small influence of the inertial forces in Equation 3.45.

It is interesting to note that the coefficient ω can be related to the other recognized dimensionless numbers through the expression

$$\omega = \left(\frac{Fr \cdot Bo}{8\,Ca}\right)^2 \tag{3.47}$$

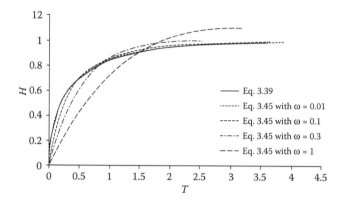

FIGURE 3.3

The numerical predictions of the dimensionless liquid-column height as a function of the dimensionless time obtained from Equation 3.45 and compared with the prediction of Equation 3.39.

where Fr is the Froude number, Bo is the Bond number, and Ca is the capillary number. These numbers are defined as

$$Fr = \frac{v}{\sqrt{gh_e}} \tag{3.48}$$

$$Bo = \frac{\rho g R_h^2}{\sigma} \tag{3.49}$$

$$Ca = \frac{v\mu}{\sigma} \tag{3.50}$$

Note that the Froude number represents a ratio of the inertial and the gravity forces, the Bond number represents a ratio of the gravity and surface tension forces, and the capillary number represents a ratio of the viscous and surface tension forces.

Figure 3.3 illustrates the numerical predictions of Equation 3.45 for different values of ω. It is clear that for small values of ω, the inertial forces become insignificant and the predictions of Equations 3.39 and 3.45 converge. However, as ω increases, the two predictions begin to differ significantly because of the increasing influence of the inertial effects.

3.4 Wicking in Swelling Porous Materials

In the porous media that swell on coming in contact with a liquid, the average pore size reduces with time as a result of the swelling of solid particles or fibers constituting the porous medium. As a result, the radius of the imaginary capillary tubes representing the interconnected pores should also vary accordingly in order for the capillary models to predict an accurate wicking rate.

3.4.1 Modified Lucas–Washburn Equation for Swelling Porous Media

Schuchardt and Berg (1990) assumed the radius of the capillary tubes to decrease linearly with time as a result of swelling. Based on this assumption, the capillary and hydraulic radii change with time as

$$R_c = R_0 \tag{3.51a}$$

$$R_h = R_0 - at \tag{3.51b}$$

where R_0 is the initial radius and a is a swelling constant. Note that the capillary radius is taken to be a constant since it is applicable near the liquid-front as it is moving into a porous medium in the completely dry state. Note, however, that the hydraulic radius, which represents the porous medium already wetted by the liquid and hence swelling, is deemed to decrease with time. The capillary and hydraulic radii are assumed to be equal and constant in the beginning to be consistent with the assumptions made by Schuchardt and Berg (1990). Substituting Equations 3.51a and 3.51b into Equation 3.26 leads to

$$\frac{2\sigma \cos(\theta_d)}{R_0} = \frac{8\mu h \dot{h}}{(R_0 - at)^2} \tag{3.52}$$

After separating the variables and rearranging, we get

$$\frac{2\sigma \cos(\theta_d)}{8\mu R_0}(R_0 - at)^2 \, dt = h \, dh \tag{3.53}$$

Integrating Equation 3.53 while using the initial condition of $h(t = 0) = 0$ results in the modified L–W equation as proposed by Schuchardt and Berg (1990):

$$h = \sqrt{\frac{\sigma R_0 \cos(\theta_d)}{2\mu}\left(t - \frac{at^2}{R_0} + \frac{a^2 t^3}{3R_0^2}\right)} \tag{3.54}$$

Note that upon neglecting the swelling effect (i.e., $a = 0$), Equation 3.54 reduces to Equation 3.27, the original L–W equation for rigid porous media.

3.5 Measuring the Wicking Parameters

To use the capillary model for wicking modeling in a porous medium, the parameters of the capillary model pertaining to the liquid and porous medium properties should be measured *independently* (Masoodi et al., 2007, 2008). These parameters include capillary radius (R_c), hydraulic radius (R_h), liquid surface tension (σ), and dynamic contact angle (θ_d). The methods of measuring liquid surface tension (σ), capillary radius (R_c), and dynamic contact angle (θ_d) are explained in Chapter 5. Here, a method to measure the hydraulic radius is introduced.

3.5.1 Hydraulic Radius (R_h)

Here, the falling head permeameter method used by Masoodi et al. (2007) to measure the hydraulic radius of polymer wicks is described (Figure 3.4). The polymer wick was assumed to be made of parallel capillary tubes. The polymer wick was connected to the nozzle of a syringe filled with a liquid (see Figure 3.4). The liquid passed through the wick due to the force of gravity. By applying the Hagen–Poiseuille law, the flow rate Q through the wick can be described by

$$Q = N\frac{\pi\rho g R_h^4 (L_{sr} + L_w)}{8\mu L_w} \tag{3.55}$$

Through the mass-conservation principle, the flow rate is also related to the rate of depletion of the liquid in the syringe:

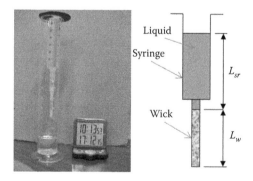

FIGURE 3.4
A photo and schematic of the falling head permeameter used for measuring the hydraulic radius of the polymer wicks. (From R. Masoodi, K.M. Pillai, and P.P. Varanasi. Darcy's law-based models for liquid absorption in polymer wicks. *AICHE J.* 53(11):2769–2782, 2007.)

$$Q = -\frac{dL_{sr}}{dt}\pi R_{sr}^2 \qquad (3.56)$$

After substituting Equation 3.55 into Equation 3.56 and integrating with the initial condition of $L_{sr}(t = 0) = L_{sr0}$, the following equation is derived:

$$\ln\left(\frac{L_{sr} + L_w}{L_{sr0} + L_w}\right) = -\frac{N\rho g R_h^4}{8R_{sr}^2 \mu L_w}t \qquad (3.57)$$

The number of capillary tubes passing through the wick cross section, N, can be related to the wick and hydraulic diameters through the relation

$$N = \frac{\varepsilon \pi R_w^2}{\pi R_h^2} \qquad (3.58)$$

Substitution of Equation 3.58 into Equation 3.57 leads to

$$\ln\left(\frac{L_{sr} + L_w}{L_{sr0} + L_w}\right) = \frac{\varepsilon R_w^2 R_h^2 \rho g}{8\mu L_w R_{sr}^2}t \qquad (3.59)$$

Therefore, by measuring the height of the fluid level, L_{sr}, at various times and plotting $\ln(L_{sr} + L_w)/(L_{sr0} + L_w)$ versus time, the slope of the trend line can be measured, which in turn should be equal to the coefficient of t in Equation 3.59. Since all the other parameters in the coefficient $(\varepsilon R_w^2 R_h^2 \rho g/8\mu L_w R_{sr}^2)$ are measurable independently, the unaccounted parameter R_h can then be estimated.

3.6 Applications of the Capillary Models in Predicting Wicking

3.6.1 Wicking in Polymer Wick

In this section, the capillary models will be applied to model liquid imbibitions in cylindrical wicks made of sintered polymer beads. If the capillary radius (or the "dry" capillary radius) is assumed to be the radius of imaginary capillary tubes constituting the polymer wicks, then, by definition, the hydraulic radius is the ratio of the fluid volume inside the capillary tubes to the wetted surface area of the tubes:

$$R_h = \frac{\varepsilon \pi R_w^2 h}{2N\pi R_c h} \qquad (3.60)$$

The following equation also holds for the wicks and is obtained by equating the pore volume in the wick with the total volume of the cylindrical tubes inside the wick:

$$N\pi R_h^2 h = \varepsilon\pi R_w^2 h \qquad (3.61)$$

From Equations 3.60 and 3.61, one can deduce that

$$R_c = 2R_h \qquad (3.62)$$

Equation 3.62 is commonly used in the capillary models for wicking; the accuracy of this relation was studied by Masoodi et al. (2008). In their approach, Equation 3.30, after including the viscous and gravity effects, was employed to model the wicking process. Nomenclature corresponding with the various capillary models that were tested is as follows (Masoodi et al., 2008):

1. "Capillary model (measured R_h)": Measure the hydraulic radius R_h first, and then use Equation 3.62 to estimate the capillary radius R_c.
2. "Capillary model (measured R_c)": Measure the capillary radius R_c first, and then use Equation 3.62 to estimate the hydraulic radius R_h.
3. "Capillary model (measured R_c and R_h)": Measure both the capillary radius R_c and the hydraulic radius R_h independently.

The mass absorbed as a function of time is plotted in Figure 3.5 for a wicking study conducted with three alkanes (hexadecane, decane, and dodecane) in polymer wicks made from polyethylene. It is clear that the capillary model, Equation. 3.30, is most accurate when the capillary and hydraulic radii are measured *independently*. Of the two "partial" measurement approaches used, the direct measurement of the capillary radius yields a better prediction compared with the direct measurement of the hydraulic radius.

3.6.2 Wicking in Paper-Like Swelling Porous Media

Schuchardt and Berg (1990) studied the wicking of liquids in a porous medium made from cellulose and superabsorbent fibers. They applied the modified L–W equation, Equation 3.54, for predicting the wicking of water in paper reinforced with fibers of a superabsorbent material. As shown in Figure 3.6, this simple model, built on the assumption that the hydraulic radius in a swelling porous medium decreases linearly with time, compared better with the experimental data than the conventional L–W model.

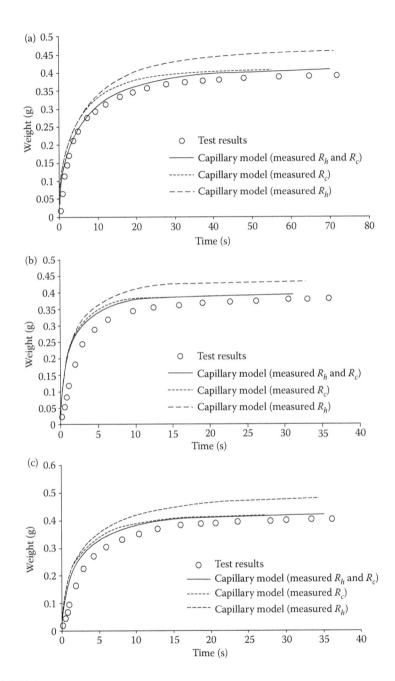

FIGURE 3.5

Mass absorption plots using polypropylene wicks with the three liquids—(a) HDEC = hexadecane, (b) DEC = decane, and (c) DDEC = dodecane—in the wicking experiments. (From R. Masoodi, K.M. Pillai, and P.P. Varanasi. Role of hydraulic and capillary radii in improving the effectiveness of capillary model in wicking, *ASME Summer Conference*, Jacksonville, FL, USA, August 10–14, 2008.)

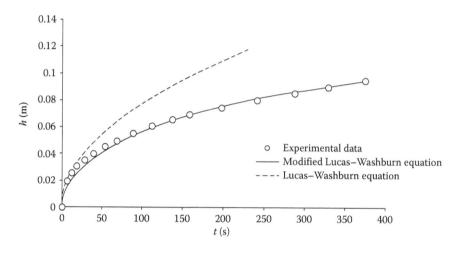

FIGURE 3.6
Wicking height versus time plot for water absorption in the 13%FC CMC/cellulose composite-paper strip. (From R. Masoodi, and K.M. Pillai. *AIChE Journal*, 56(9):2257–2267, 2010.)

3.7 Summary and Conclusions

In this chapter, we introduced the capillary model for modeling wicking in porous media. The governing equation for the capillary model was derived using the integral approach after including the viscous, gravity, and inertial forces. The capillary pressure applied at the meniscus of the liquid column rising in the capillary tube is based on the Young–Laplace equation after including the effects of fluid motion by choosing the dynamic contact angle in the formulation. It is also possible to include the effects of initial delay in meniscus formation on the capillary pressure through the meniscus formation factor. Although an extended form of the momentum balance equation can also be developed after including the local acceleration and viscous dissipation of the liquid below the tube along with the convective flow losses at tube entrance, the effect of such terms is negligible and the original equation after including only the viscous, gravity, and inertial forces is sufficiently accurate to handle most capillary tube rise phenomena. The capillary model after including only the viscous forces leads to the well-known L–W (or Washburn) equation. The capillary model after including both the viscous and gravity forces yields a wicking prediction equation that compensates for a lack of accuracy in the L–W predictions when the liquid column is close to attaining the final steady-state height. An explicit form of this formulation is developed after using the Lambert function. An analytical solution is possible for the capillary model after including the viscous and inertial effects; however, a numerical solution is needed after adding gravity along with the viscous and inertial effects to the capillary model; a parametric study of this solution with the help of a dimensionless number indicates that significant differences can be induced by the inclusion of the inertial forces. An alternate form of the L–W equation can be developed for modeling wicking in swelling porous materials after varying the hydraulic radius linearly with time.

To use the capillary model for predicting wicking in porous materials, it is important to measure the model parameters (capillary radius, hydraulic radius, surface tension, and dynamic contact angle) independently. A novel method based on the falling head permeameter is described to estimate the hydraulic radius. The measurement methods for the rest of the parameters are described in Chapter 5 of this book.

Two example applications of the introduced capillary models were described. In one application, the wicking in a polymer wick was modeled after including the viscous and gravity effects where it was shown that measuring both the capillary and hydraulic radii independently improves the accuracy of the capillary model. In the other application, the modified L–W equation was used to achieve a satisfactory prediction of wicking in a paper-like swelling porous medium.

Nomenclature

Roman Letters

A	Area (m²)
a	Swelling coefficient (Equation 3.51b)
Bo	Bond number (Equation 3.49)
Ca	Capillary number (Equation 3.50)
F	Force (N)
Fr	Froude number (Equation 3.48)
g	Acceleration due to gravity (m/s²)
h	Height of liquid in a capillary tube (m)
$\dot{h}$	$= dh/dt$
$\ddot{h}$	$= d^2h/dt^2$
H	Dimensionless height (Equation 3.37a)
$\dot{H}$	$= dH/dT$ (Equation 3.45)
K	Entrance correction factor (Equation 3.25)
L	Length (m)
m	Liquid mass (kg)
N	Number of capillary tubes passing through a wick (Equation 3.55)
p	Pressure (Pa)
Q	Volume flow rate (m³/s)
R	Radius (m)
S	Correction factor for reorientation time of meniscus (Equation 3.22)

t	Time (s)
T	Dimensionless time (Equation 3.37b)
V	Volume (m³)
v	Velocity (m/s)
w	z-component of velocity (m/s)
W	Lambert function (Equation 3.34)
z	z-coordinate

Greek Letters

θ	Contact angle (degree)
σ	Surface tension (N/m)
μ	Viscosity of liquid (kg/m.s)
ε	Porosity
ρ	Density (kg/m³)
γ	Specific weight ($\gamma = \rho g$)
ω	A dimensionless coefficient (Equation 3.46)

Subscripts

0	Initial value
a	Atmosphere
c	Capillary
d	Dynamic
e	Effective
fs	Free surface
g	Gravity
h	Hydraulic
lp	Lower pressure
p	Pressure
r	Reorientation
s	Static
sr	Syringe (Figure 3.6)
up	Upper pressure
v	Viscosity
w	Wick
z	z-component

References

C.H. Bosanquet. On the flow of liquids into capillary tubes. *Philos. Mag. Ser.* 6, 45(267):525–531, 1923.

M. Bracke, F. De Voeght, and P. Joos. The kinetics of wetting: The dynamic contact angle. *Prog. Colloid Polym. Sci.* 79:142–149, 1989.

R.M. Corless, G.H. Gonnet, D.E.G. Hare, D.J. Jeffrey, and D.E. Knuth, On the Lambert W function, *Adv. Comput. Math.* 5:329–359, 1996.

P.K Chatterjee. *Absorbency*, Amsterdam/New York, Elsevier, 1985.

P.K. Chatterjee, and B.S. Gupta. *Absorbant Technology*, Amsterdam, Elsevier, 2002.

E.B. Dussan, and S.H. Davis. On the motion of a fluid-solid interface along a solid surface. *J. Fluid Mech.* 65:71–95, 1974.

E.B. Dussan V. On the spreading of liquids on solid surfaces: Static and dynamic contact lines. *Ann. Rev. Fluid Mech.* 11:371–400, 1979.

N. Fries. *Capillary Transport Processes in Porous Materials—Experiment and Model*, CuvillierVerlag, Göttingen, 2010.

N. Ichikawa, and Y. Satoda. Interface dynamics of capillary flow in a tube under negligible gravity condition. *J. Colloid Interface Sci.* 162:350–355, 1994.

B. Lavi, A. Marmur, and J. Bachmann. Porous media characterization by the two-liquid method: Effect of dynamic contact angle and inertia. *Langmuir* 24(5):1918–1923, 2008.

T.-S. Jiang, S.-G. Oh, and J.C. Slattery. Correlation for dynamic contact angle. *J. Colloid Interf. Sci.* 69:74–77, 1979.

R. Lucas. Rate of capillary ascension of liquids. *Kollid Z* 23:15–22, 1918.

P.S. Laplace. *Celestial Mechanics*, Suppl- book 10, Vol 4, Paris, Gauthier-Villars, 1806.

R. Finn. Capillary surface interfaces. *Notices AMS* 46(7):770–781, 1999.

N. Fries, and M. Dreyer. An analytic solution of capillary rise restrained by gravity. *J. Colloid Interf. Sci.* 320(1):259–263, 2008.

R. Masoodi, and K.M. Pillai. Darcy's law-based model for wicking in paper-like swelling porous media. *AIChE J.* 56(9):2257–2267, 2010.

R. Masoodi, and K.M. Pillai, A general formula for capillary suction pressure in porous media. Accepted for publication in *J. Porous Media* 15(8):775–783, 2012.

R. Masoodi, K.M. Pillai, and P.P. Varanasi. Darcy's law-based models for liquid absorption in polymer wicks. *AICHE J.* 53(11):2769–2782, 2007.

R. Masoodi, K.M. Pillai, and P.P. Varanasi. Role of hydraulic and capillary radii in improving the effectiveness of capillary model in wicking, *ASME Summer Conference*, Jacksonville, FL, USA, August 10–14, 2008.

E.K. Rideal. On the flow of liquids under capillary pressure. *Philos. Mag. Ser.* 6(44):1152–1159, 1922.

S.P. Sutera, and R. Skalak. The history of Poiseuille's law. *Annu. Rev. Fluid Mech.* 25:1–19, 1993.

D.R. Schuchardt, and J.C. Berg. Liquid transport in composite cellulose-superabsorbent fiber network. *Wood Fiber. Sci.* 23(3):342–357, 1990.

J. Szekely, A.W. Neumann, and Y.K. Chuang. The rate of capillary penetration and the applicability of the washburn equation. *J. Colloid Interf. Sci.* 35(2):273–278, 1971.

M. Stange, M.E. Dreyer, and H.J. Rath. Capillary driven flow in circular cylindrical tubes. *Phys. Fluids* 15(9):2587–2601, 2003.

C.E. Siegert, D.A. Petrash, and E.W. Otto. Time response of liquid-vapor interface after entering weightlessness. Technical Report NASA TN D-2458, Lewis Research Center, Cleveland, Ohio, 1964.

E.M. Sparrow, S.H. Lin, and T.S. Lundgren. Flow development in the hydrodynamic entrance region of tubes and ducts. *Phys. Fluids* 7(3):338–347, 1964.

E.V. Washburn. The dynamics of capillary flow. *Phys. Rev.* 17:273–283, 1921.

F.M. White. *Fluid Mechanics with Student DVD*, 7th edition. New York, McGraw-Hill, 2010.

T. Young. An essay on the cohesion of fluids. *Philos. Trans. R. Soc. London.* 95, 65–87, 1805.

4

An Introduction to Modeling Flows in Porous Media

Krishna M. Pillai and Kamel Hooman

CONTENTS

This chapter begins with a review of the predictive tools used to analyze fluid flow through a porous medium where the governing equations for such flows will be presented. First, the well-known Darcy's law along with the continuity equation for single-phase flow through a saturated porous medium will be discussed in the context of an upscaling technique called the volume-averaging method. Then, we will shift our attention to multiphase flows through permeable media where the generalized form of Darcy's law along with the coupled mass balance equations will be presented. Later, the unsaturated flow model as a derivative of the multiphase flow model with its assumption of a stationary air phase will be taken up as an effective description of moisture transfer during the wicking process. Any flow model will require suitable values of various porous-media properties before it is applicable to practical situations. Hence finally, a description will be provided of various ways to estimate or measure various properties of interest including the permeability, relative permeability, and capillary pressure.

4.1 Introduction

A solid interspersed with interconnected pores can be defined as an effective porous medium since it allows passage of fluids and solutes through it. Examples of such porous media range from sand to cloth to sponge. Wicking or imbibition of liquids into dry porous materials forms a subset of flow and transport phenomena possible in porous media. In this chapter, a synopsis of flow models developed to predict the flow of a single fluid and multiple fluids will be provided. Special emphasis will be placed on the measurement of important parameters associated with the porous media flow where details of the measurement techniques will be presented. Such details are likely to help the reader in understanding and using the material described here as well as in the other chapters.

4.1.1 Importance of Averaging while Modeling Porous-Media Flows

Defining a porous medium as a combination of interconnected voids, which are permeable to flow of one or more fluids through the openings, one is intuitively tempted to solve the known governing equations (such as Navier–Stokes) for flow through the pores using a computational fluid dynamics (CFD) software and get the local velocity distributions. Based on these pieces of information one can then find the wall shear stresses and the total resistance to fluid flow through such porous material. This *continuum* approach is feasible for modeling flows through a cluster of a small, finite number of particles. However, it is technically impossible to implement this approach in most cases of practical applications. This is because any typical example of a porous medium (such as sand, fabric, or fibrous sheets) typically consists of a very large number of rather irregularly shaped particles or fibers arranged in often extremely complex arrangements. As a result, firstly, it is extremely tedious to create a mesh to model such an interstitial flow, and then secondly, it is extremely expensive (computationally) to solve for such flows. Hence, such a continuum approach is almost never adopted for solving flows in any reasonably sized porous domains.

Therefore, *porous-continuum* approaches are put forward to model the complex flow behavior in a porous medium. As continuum models are derived from averaging over an infinitesimal volume of the material, in a very similar fashion, porous-continuum models can be thought of as results of averaging continuum models over a representative elementary volume occupied by both solid and fluid phases [1]. It is well known that during any averaging process some information is lost and thus a porous-continuum model is less accurate, but easier to implement compared with the pore-scale analysis.

The averaging process for developing a porous-continuum description of a phenomenon in terms of governing partial differential equations (PDEs) is also called the "upscaling" process. There have been several theoretical approaches for developing such PDEs in terms of averaged flow variables, such as the continuum theory of mixtures [2,3] and the method of homogenization [4,5]. Of these, however, volume averaging remains one of the most respected methods [6] because of its emphasis on physical reasoning behind its assumptions and mathematical manipulations rather than a mere exercise in mathematical technique.

Although flow quantities in porous-media studies are typically expressed in the form of several types of averages, two are especially important: (1) phase average (also called the volume average); (2) intrinsic phase average (also called the pore average). For any flow-related quantity q associated with the fluid flowing through a porous medium, the phase and intrinsic phase averages ($\langle q \rangle$ and $\langle q \rangle^f$) are defined, respectively, using the terminology developed by the volume averaging community [6] as

$$\langle q \rangle = \frac{1}{V} \int_{V_f} q \, dV$$

and

$$\langle q \rangle^f = \frac{1}{V_f} \int_{V_f} q \, dV \tag{4.1}$$

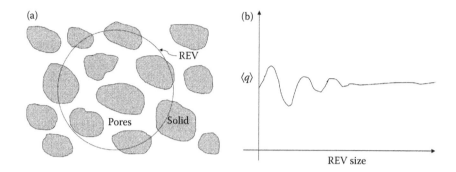

FIGURE 4.1
(a) A typical representative elementary volume (REV) used for averaging in porous media. (b) A typical variation in the value of $\langle q \rangle$ (or $\langle q \rangle^f$) with an increase in the REV size.

Here q is integrated over an averaging volume called representative elementary volume, or REV. As shown in Figure 4.1a, REV can be typically thought of as a sphere of a fixed volume. If the REV is on the order of the size of solid-phase particles, a slight change in the REV size leads to large fluctuations in the averaged quantities, Figure 4.1b. This also means that a slight movement of REV during the creation of a porous-continuum model out of a real porous medium will lead to unreasonable fluctuations in the averaged quantities. Hence, it is important that a steady average value is returned by an REV—in other words, the REV radius has to be much larger that the solid constituents (particles or fibers) of a porous medium. On the other hand, a typical REV, although being much larger than the constituent particles of a porous medium, has to be much smaller than the overall dimensions of a porous medium in order for a meaningful upscaled or averaged flow-description can be scripted through the porous-continuum equations.

The method of volume averaging for deriving flow and transport equations in porous media has been developed over last few decades by the painstaking efforts of several prominent researchers [6–10]. This method is used to create governing equations in terms of the volume-averaged flow variables. One starts with a partial differential equation for point-wise or microscopic mass, momentum, and energy balance within the fluid phase, and then averages each additive term in that equation. Through the use of two averaging theorems [8], one can transform the microscopic mass-balance equation for single-phase flow in porous media into the macroscopic mass-balance equation:

$$\text{microscopic:} \quad \nabla \cdot \vec{v} = 0 \quad \rightarrow \quad \text{macroscopic:} \quad \nabla \cdot \langle \vec{v} \rangle = 0 \qquad (4.2)$$

Here $\vec{v}$ is the point-wise fluid velocity, $\langle \vec{v} \rangle$ is the phase-averaged fluid velocity (also called Darcy velocity or specific discharge). Similarly, the momentum balance equation can be transformed from the microscopic form into the macroscopic one as

microscopic: $0 = -\nabla p + \mu \nabla^2 \vec{v} + \rho \vec{g}$ → macroscopic: $\langle \vec{v} \rangle = -\dfrac{K}{\mu} \left(\nabla \langle p \rangle^f - \rho \vec{g} \right)$

$$(4.3)$$

where the macroscopic form of momentum balance is also called Darcy's law. Here ρ is density, μ is viscosity, K is the permeability of porous medium, p is the hydrodynamic fluid pressure, $\langle p \rangle^f$ is the intrinsic phase-averaged (or pore averaged) pressure, and $\vec{g}$ is the acceleration due to gravity.

4.2 Single-Phase Flow through a Porous Medium

In this section, we will discuss the case where all of the pore space within a porous medium is occupied or saturated by a single fluid or phase. The motion of fluid under such circumstances is called the single-phase flow through a porous medium. The PDEs for a pore-continuum model under isothermal conditions will consist of a mass balance (continuity) equation and a momentum balance equation.

4.2.1 Mass Balance Equation

The most general form of mass-balance statement, also known as the continuity equation, for single-phase flows through porous media can be expressed as

$$\frac{\partial(\rho\varepsilon)}{\partial t} + \nabla \cdot (\rho \langle \vec{v} \rangle) = 0 \qquad (4.4)$$

where ρ is density while ε is porosity. One of the assumptions inherent in this relation is that the particles/fibers constituting the porous medium do not absorb any of the passing fluid. If the fluid moving through the porous medium is incompressible, this relation reduces to

$$\frac{\partial\varepsilon}{\partial t} + \nabla \cdot \langle \vec{v} \rangle = 0 \qquad (4.5)$$

If the porous medium is of a rigid, nondeforming kind, then this relation further simplifies to

$$\nabla \cdot \langle \vec{v} \rangle = 0 \qquad (4.6)$$

This is the classic form of the mass balance equation which is often seen in the diverse fields involving porous media including convection, ground-water flow, and composites processing. We will do well to reiterate that this final form is valid only for the flow of an incompressible fluid through a rigid, nondeforming porous medium consisting of nonabsorbing particles.

4.2.2 Momentum Balance Equation

Henry Darcy [11] reported his empirical formula in 1856 to relate water flow rate through a sand column to the pressure gradient by applying a "hydraulic resistance" term. Note that, in essence, Darcy's empirical equation was not initially presented as a force balance. Lage [12] offers a very interesting review on the historical development of porous media models starting from Darcy's equation to what we now refer to as Darcy's law, and which has also been called the Hazen–Darcy equation. According to Darcy's equation, the fluid velocity is linearly proportional to the pressure gradient, that is,

$$-\frac{\partial \langle p \rangle^f}{\partial x} = \frac{\mu}{K} \langle u \rangle \tag{4.7}$$

Here, flow is happening along the x direction with $\langle u \rangle$ being the x-direction volume-averaged (Darcy) velocity. It is important to reiterate that the pressure and velocities described here inside a porous medium do not refer to any point-wise pressure or velocity, but to averaged pressure and velocity inside the medium.

A more recognizable form of the Darcy's law is given as

$$\langle u \rangle = -\frac{K}{\mu} \frac{\partial \langle p \rangle^f}{\partial x} \tag{4.8}$$

such that the velocity is in the direction of declining average pressure. Using the vector notation, Darcy's law can be generalized for a general 3-D flow space as

$$\langle \vec{v} \rangle = -\frac{K}{\mu} \nabla \langle p \rangle^f \tag{4.9}$$

Note that since the point-wise velocity can be expressed in terms of its components as $\vec{v} = u\hat{i} + v\hat{j} + w\hat{k}$, therefore, the volume averaged form of velocity can be expressed as $\langle \vec{v} \rangle = \langle u \rangle \hat{i} + \langle v \rangle \hat{j} + \langle w \rangle \hat{k}$.

It has to be emphasized that the form of Darcy's law expressed through Equations 4.8 and 4.9 is valid for *isotropic* porous media, the examples of which include sand, paper, certain non-woven fabrics, and so on. However, one often comes across porous materials made of elongated pores aligned in a certain direction, thereby creating various forms of *anisotropic* porous media. The examples of such media found in wicking applications include fiber mats made from unidirectional fibers, woven or stitched fabrics made from fiber bundles, as well as paper made from oriented fibers. In such cases, the permeability is expressed in the form of a tensor, **K**, and Darcy's law takes the form

$$\langle \vec{v} \rangle = -\frac{\mathbf{K}}{\mu} \cdot \nabla \langle p \rangle^f \tag{4.10}$$

For modeling flows involving wicking, often the liquid has to rise against gravity. One can easily include the effects of gravity through the following extension of Darcy's law:

$$\langle \vec{v} \rangle = -\frac{\mathbf{K}}{\mu} \cdot (\nabla \langle p \rangle^f - \rho \vec{g}) \tag{4.11}$$

One often finds it convenient to define a *modified* pressure P after including the effect of hydrostatic variation in pressure through the relation

$$P = p + \rho g h \tag{4.12}$$

with $g = |\vec{g}|$ while h is the elevation with respect to a certain datum. It is easy to see that a hydrostatic pressure distribution will result in a constant P. One can use this pressure definition given in Equation 4.12 with Equation 4.11 to develop the following form of Darcy's law after inclusion of the gravity effect:

$$\langle \vec{v} \rangle = -\frac{\mathbf{K}}{\mu} \cdot \nabla \langle P \rangle^f \tag{4.13}$$

Another approach to include the effect of gravity in porous media flows is through the use of a variable called piezometric head, Ø, which is defined as

$$\text{Ø} = \frac{p}{\rho g} + h \tag{4.14}$$

The use of this relation in Equation 4.11 results in the following form of Darcy's law:

$$\langle \vec{v} \rangle = -\frac{K_h}{\mu} \cdot \nabla \langle \varnothing \rangle^f \tag{4.15}$$

Here $\nabla \langle \varnothing \rangle^f$, gradient of the pore-averaged piezometric head, is called the hydraulic gradient. K_h, the hydraulic conductivity tensor, is defined as

$$K_h = \frac{K \rho g}{\mu} \tag{4.16}$$

In case the solid phase constituting the porous medium is moving and deforming, then the form of Darcy's law undergoes a change. Such a situation may occur during wicking if particles or fibers constituting the porous substrate absorb liquid and swell. In such cases, the flow resistance offered by the solid phase is proportional to the relative velocity between the liquid and solid phases, and the left-hand sides of the various forms of Darcy's law (Equations 4.9 through 4.11, and 4.13) can be replaced [65] by $\varepsilon(\langle \vec{v} \rangle^f - \langle \vec{v}_s \rangle^s)$ where $\langle \vec{v} \rangle^f$ is the intrinsic phase average of the fluid-phase velocity while $\langle \vec{v}_s \rangle^s$ is the intrinsic phase average of the solid-phase velocity. (Note that $\langle \vec{v} \rangle = \varepsilon \langle \vec{v} \rangle^f$ using the relations given in Equation 4.1.) In other words, an example of the most general form of Darcy's law derived from Equation 4.11 is

$$\varepsilon(\langle \vec{v} \rangle^f - \langle \vec{v}_s \rangle^s) = -\frac{K}{\mu} \cdot (\nabla \langle p \rangle^f - \rho \vec{g}) \tag{4.17}$$

Before closing this section, it is fruitful to note that the theoretical basis of Darcy's law has been provided by several researchers. (For example, using the volume-averaging method, Slattery [7], Whitaker [6], Gray and O'Neill [13], and Bear and Bachmat [10] derived Equation 4.17 from the Stokes flow occurring inside the pores. Other theoretical derivations of Darcy's law using other upscaling techniques, such as the homogenization method or the mixture theory, are also available in the literature.) Most of such derivations assume that a Newtonian fluid completely occupies the entire void space in a porous medium under the *isothermal* conditions so that the fluid properties do not change with the temperature. As all of the experiments that form the basis of the porous media models were conducted under isothermal conditions, a legitimate question is about the validity of such models when the fluid property severely changes with the temperature. This question has been partly answered by Narasimhan [14] and Hooman [15].

In the case of a non-Newtonian fluid flow, the shear stress–rate relation would be nonlinear and in order for the above equations to be valid, the fluid viscosity should be defined as a function of the velocity (with power law being a special case) [16].

4.3 Multiphase Flow through a Saturated Porous Medium

So far we have studied the flow of a single fluid through a porous matrix. An example is the flow of water in a geothermal reservoir. One can model a problem like this as a single-phase flow problem through a porous medium. However, the analysis becomes more complicated if there is more than one fluid occupying the pores; that is, simultaneous flow of more than one fluid through a porous medium. An example can be the concurrent presence of liquid water, vapor, and CO_2 in a geothermal reservoir [17–19] or water, oil, and a gas in a petroleum reservoir [20]. In general, multiphase fluid flow in porous media is of practical importance in reservoir engineering, fuel cells, heat pipes, hydrology and agriculture, food processing, and biomedical engineering. In view of the above, this section looks into the analysis of multiphase flow through a porous medium.

4.3.1 Continuum Approach

Similar to the case of single-phase flow in porous medium, one can apply a continuum approach to study the multiphase counterparts. Sophisticated imaging methods for observation of pore-scale geometry and fluid flow, as recently applied by Silin et al. [21], provide useful information for theoretical studies. Modeling the porous medium as a network of capillary channels connecting pores,* Nardi et al. [22] obtained relative permeabilities of a rock sample saturated with air, water and oil, and compared their findings with empirical data. Piri and Karpyn [23] used both approaches to study two-phase fluid flow in a rough-walled fracture. Lattice Boltzmann methods have also attracted attention in pore-scale modeling of multiphase flow in a porous medium [24–27]. A range of other numerical techniques have also been applied to solve the continuum models pertinent to multiphase flow through a porous medium with application in fuel cell systems. For instance, modeling liquid–gas transport in direct methanol fuel cells (DMFCs) are particularly interesting as some DFMCs use liquid methanol as the fuel in the anode, which leads to the formation of CO_2 bubbles in the catalyst layer [28–31].

---

* Such a simulation, also known as the network-model simulations, falls under the category of pore-scale simulations, which also include Lattice Boltzmann or Smooth Particle Hydrodynamics. Generally the former is computationally less expensive compared to the latter two techniques.

As pointed out by Dullein [32], depending on different saturation ratios, pore structure and surface tension, different flow conditions can be observed at the pore level, that is:

1. Some pores are occupied completely by one phase while the others are filled by the second phase.
2. Both phases flow concurrently through the same void volumes: the wetting phase covers pore walls while the nonwetting phase flows through pore centers.
3. The wetting phase is continuous while the nonwetting phase is discontinuous in the form of disjoint blobs.

Again, what makes these continuum models rather inefficient is the need for expensive computer simulation or sophisticated experimental measurements. Furthermore, these simulation techniques do not lend themselves to most of the industrial (especially large-scale) applications, and porous-continuum approaches tend to be more popular despite the lower resolutions they offer compared with continuum counterparts. In the remainder of this section we shift our attention to the development and application of such approaches to study multiphase flows in porous materials.

4.3.2 Porous-Continuum Approach

From the macroscopic point of view, the flow of immiscible fluids in porous media can be classified as steady and unsteady. The former refers to cases when neither of the two fluids is replacing the other one. For such cases, one can classify macroscopic flow regimes based on the general pattern and driving forces. According to the micro-model studies conducted using the network models by Lenormand et al. [33] and others [34], three flow regimes are expected:

1. Stable displacement, when a rather flat and simple flow front is observed moving in the direction of flow. In this regime it is very unlikely for islands or blobs of displaced phase to survive behind the replacing fluid front.
2. Capillary fingering, when narrow branches or "fingers" of replacing flow penetrate into replaced phase forming a tree-like flow pattern. Movement of fingers in different directions and formation of loops is likely in this regime.
3. Viscous fingering, when tree-like fingers are still present but no looping behavior is observed.

Two-dimensional parameters, the viscosity ratio, and the capillary number, were observed to control the microscopic flow regime. When the replacing phase is nonwetting, with high viscous forces in the replacing phase and relatively low capillary forces at the fluid–fluid interfaces, stable displacement is expected. On the other hand, high viscous forces in the replaced phase can lead to viscous fingering even when the capillary forces are appreciable. Capillary fingering is observed when the capillary forces are dominant, and the viscosity of the replaced phase is not significantly higher than that of the replacing phase.

Pioneering models for multiphase flow in porous media are based on generalization of the Darcy's equation proposed by Wyckoff and Botset [35] and Muskat [36] (see Bear [37] for a description), when the absolute permeability is corrected (lowered) by a relative permeability. This will be explained in the forthcoming discussion.

An interesting study was reported by Whitaker in 1986 [38,39], when a theoretical ground for the multiphase flow models was presented based on volume averaging the Navier–Stokes equations. In Whitaker's porous-continuum model, each phase has its permeability tensor, while a viscous drag tensor has been introduced by which momentum equations of the two fluid phases, phases w and nw representing the wetting and nonwetting fluids, respectively, are coupled. The momentum equation for the wetting phase can be expressed as

$$\langle \vec{v}_w \rangle = -\frac{\boldsymbol{K}_w}{\mu_w} \cdot \nabla \langle p_w \rangle^f - \bar{K} \cdot \langle \vec{v}_{nw} \rangle \qquad (4.18)$$

Here $\langle \vec{v}_w \rangle$ and $\langle \vec{v}_{nw} \rangle$ are the phase-averaged (Darcy) velocities for the wetting and nonwetting phases, respectively, while $\langle p_w \rangle^f$ is the intrinsic phase-averaged (pore-averaged) pressure of the wetting phase. Although the viscous drag tensor $\bar{K}$ can only be obtained by solving a very complex and coupled boundary-value (closure) problem at the micro-scale, Whitaker interpreted it to be the viscous drag coefficient between the two phases. On the basis of Whitaker's order-of-magnitude analysis, the contribution of velocity of the nonwetting phase in the momentum equation of the wetting phase is proportional to $\mu_{nw}/\mu_w (S_w/S_{nw})^2$.

Further experimental [40] and numerical [41] research supported the idea that fluid flow is better described by including the coupling terms between different phases. In a notable study, Gunstensen and Rothman [42] applied an immiscible Lattice Boltzmann method in a microscale geometry to investigate the coupling between flows of different phases and to examine the linear dependence of fluid flow-rate on the pressure gradient. They argued that when the linear relation between the flow rate and pressure gradient holds, the interaction between phases obeys the above mentioned theories. However, with weak forces being applied at low flow rates, the nonlinear forces, mainly due to the capillary effects, are significant and thus the conventional linear theories fail.

4.3.3 Two-Phase Immiscible Flow

Let us focus on a basic case of a two-phase flow of immiscible fluids with no mass transfer between the two fluid phases. There are, however, applications where mass can be transferred from one phase to another. A simple example is when the liquid water and vapor are present in a porous medium (such as during wicking)—evaporation of liquid water (or condensation of the vapor) can be mentioned as a reason for the transfer of mass from one phase to another.

Here, we consider two fluids in which one phase wets the porous medium more than the other phase, and is called the *wetting phase* and indicated by a subscript *w*. The other *non-wetting* phase is indicated by subscript *nw*. For instance, water is the wetting fluid relative to oil and gas in a petroleum reservoir.

In order to study two-phase flow, we need to introduce definitions such as *saturation, capillary pressure*, and *relative permeability*. The saturation of a fluid phase is defined as the fraction of the void space of a porous material occupied by this phase. Furthermore, as the two fluids jointly fill the voids one can write [20]

$$S_w + S_{nw} = 1 \tag{4.19}$$

where S_{nw} and S_w are the saturations of the nonwetting and wetting phases, respectively. The pressure difference between the *w* and *nw* phases is given by the capillary pressure (which is known to be a function of saturation [20])

$$\langle p_c \rangle = \langle p_{nw} \rangle^f - \langle p_w \rangle^f \tag{4.20}$$

This relation is based on the assumption that the two phases are in thermo-dynamic equilibrium.* Note that the phase pressures here are equal to the intrinsic phase-average pressures for the two fluid phases. (See Section 4.4.3 for more information on this important quantity.)

As there is no transfer of mass between phases in the immiscible flow, and mass is conserved within each phase, the continuity equation for each phase ($i = w, nw$) in a rigid, nondeforming porous medium can be written as

$$\varepsilon \frac{\partial(\rho_i S_i)}{\partial t} + \nabla \cdot (\rho_i \langle \vec{v}_i \rangle) = 0 \tag{4.21}$$

---

* Nonequilibrium effects can be important in applications such as drying of paper pulp that are marked with large gradients in fluid velocities and pressures. For such nonequilibrium conditions, the equation for the difference in the two fluid pressures can be of the form $\langle p_{nw} \rangle^f - \langle p_w \rangle^f = \langle p_c \rangle - \tau(\partial S_w / \partial t)$ where τ, a nonequilibrium capillarity coefficient, is a material property that may still be a function of saturation and fluid–fluid properties (e.g., interfacial area, contact line) [43].

where each phase is distinguished by its density ρ_i and its Darcy (or phase-averaged) velocity $\langle \vec{v}_i \rangle$.

Similarly, with $i = (w, nw)$, the Darcy velocity of each phase can be related to the pressure gradient as

$$-\nabla \langle p_i \rangle^f = \frac{\mu_i \langle \vec{v}_i \rangle}{K_i} \qquad (4.22)$$

Note that K_i is known as the *effective permeability* for phase i and is different from the absolute permeability of the porous medium. As simultaneous flow of two fluids leads to interference between the flow of the two phases, the effective permeability of each phase is lower than the absolute (overall) permeability (K) of the porous medium and is given by

$$K_i = k_{r,i} K \qquad (4.23)$$

where $k_{r,i}$ is called the *relative permeability* and indicates the tendency of phase i to wet the porous medium.

4.3.4 Unsaturated Flow

A special case of two-phase flow in porous media is when the non-wetting fluid (air) in a liquid–air system can be considered as stationary. Such a situation can arise during the spontaneous migration of water in soils in the capillary-fringe region above the water table inside the ground [37]. A similar situation can also arise during the wicking of liquids in certain porous media. Under such conditions, the saturation of the wetting phase, S_w, gradually drops to zero after starting with a value of unity in a fully saturated zone.

Under unsaturated flow, it is customary to consider the liquid-phase saturation, S_w, as the primary variable for describing the fluid distribution. The mass balance equation for the liquid phase is given as

$$\varepsilon \frac{\partial S_w}{\partial t} + \nabla \cdot \langle \vec{v}_w \rangle = 0 \qquad (4.24)$$

where the liquid-phase Darcy velocity, from Equations 4.14 and 4.15, is given as

$$\langle \vec{v}_w \rangle = -k_{r,w} \mathbf{K}_h \cdot \nabla \left(\frac{\langle p_w \rangle^f}{\rho g} + z \right) \qquad (4.25)$$

Recalling that K_h, the hydraulic conductivity tensor, is expressed in terms of K, the permeability tensor, as $K_h = (K\rho g/\mu)$ through Equation 4.16, the above given expression can be modified to

$$\langle \vec{v}_w \rangle = -k_{r,w} \frac{K}{\mu} \cdot \nabla \left(\langle p_w \rangle^f + \rho g z \right) \tag{4.26}$$

On substituting $\langle \vec{v}_w \rangle$ from Equation 4.26 into Equation 4.24, one can tentatively begin to develop an equation for S_w:

$$\varepsilon \frac{\partial S_w}{\partial t} = \nabla \cdot k_{r,w} (S_w) \frac{K}{\mu} \nabla \left(\langle p_w \rangle^f + \rho g z \right) \tag{4.27}$$

Note that in this equation, we have incorporated an important assumption: the porous medium under consideration is *isotropic* in nature, and hence the permeability, K, is a scalar quantity. This is a reasonable assumption since most porous media where the wicking can be modeled as the unsaturated flow, such as soil or paper, are isotropic in nature. (However, the case in which the medium is anisotropic is discussed in Chapter 6 of this book.) Using the relation $\nabla z = \vec{k}$, one can rearrange Equation 4.27 as

$$\varepsilon \frac{\partial S_w}{\partial t} = \nabla \cdot k_{r,w} (S_w) \frac{K}{\mu} \nabla \langle p_w \rangle^f + \frac{K}{\mu} \rho g \frac{\partial}{\partial z} \left[k_{r,w} (S_w) \right] \tag{4.28}$$

after assuming the liquid density and the medium permeability to be constants. The capillary pressure, given by Equation 4.20, which relates the pore-averaged pressures of the liquid and air, can be expressed for the present case as

$$\langle p_w \rangle^f = p_{\text{atm}} - \langle p_c \rangle (S_w) \tag{4.29}$$

where the air phase is assumed to be interconnected and has its pressure equal to the atmospheric pressure, p_{atm}. It is possible to render Equation 4.28 entirely in terms of the primary variable S_w if we use Equation 4.29 to obtain $\nabla \langle p_w \rangle^f$, and recognize that the relative permeability is generally assumed to be a function of the liquid saturation only:

$$\varepsilon \frac{\partial S_w}{\partial t} = \nabla \cdot k_{rw} (S_w) \frac{K}{\mu} \left(-\frac{d\langle p_c \rangle}{dS_w} \right) \nabla S_w + \frac{K\rho g}{\mu} \left(\frac{dk_{rw}}{dS_w} \right) \frac{\partial S_w}{\partial z} \tag{4.30}$$

For rigid, nonswelling porous media with a constant porosity, one can consolidate the second term on the left-hand side of this equation for moisture migration as

$$\varepsilon \frac{\partial S_w}{\partial t} = \nabla \cdot \bar{D}(S_w) \nabla S_w + \frac{K\rho g}{\mu}\left(\frac{dk_{r,w}}{dS_w}\right)\frac{\partial S_w}{\partial z} \tag{4.31}$$

where $\bar{D}(S_w)$, defined as the diffusivity coefficient or capillary diffusivity, is given as

$$\bar{D}(S_w) = k_{r,w}(S_w)\frac{K}{\mu}\left(-\frac{d\langle p_c\rangle}{dS_w}\right) \tag{4.32}$$

Equation 4.31 is the classic form of Richard's equation for tracking the movement of moisture in porous media [37]. Similar alternate forms of the Richard's equation are used in soil physics as well ([44], pp. 153–155).

If one ignores the effect of gravity (such as for wicking in a horizontally kept paper or napkin), then Equation 4.31 can be reduced to a rather elegant form of the heat equation as

$$\frac{\partial S_w}{\partial t} = \nabla \cdot D(S_w) \nabla S_w \tag{4.33}$$

where the saturation distribution is the solution of this parabolic equation. Under such circumstances, like heat, the saturation "diffuses" through a porous medium under the influence of capillary suction. Note that D is a modified form of the Diffusivity coefficient such that $D = \bar{D}/\varepsilon$. The dependence of D on saturation, that is, $D = D(S_w)$, transforms Equation 4.33 into a nonlinear "heat conduction" equation. Under such circumstances, a convenient functional form that may be sought for the diffusivity coefficient is $D(S_w) = D_o \exp[\beta(S_w - S_{w,o})]$ where D_o is the value of the coefficient at a reference saturation, $S_{w,o}$ ([45], p. 116). A numerical solution to Equation 4.33 for predicting the diffusive moisture migration under the influence of capillary pressure in two- or three-dimensions can be easily obtained. As outlined in Crank [45], one can also seek analytical or semi-analytical solutions to Equation 4.30 for various possible functional forms.

4.3.5 Some Applications of the Multiphase Flow Models

Here we review the application of multiphase flow modeling to a number of industrial applications including the fuel cell systems, the geothermal energy generation, and the coal industry.

The study of multiphase multispecies fluid flow is an important topic in fuel cell analysis since water is an inherent by-product of all fuel cells. In particular, there has been a renewed interest recently in modeling two-phase multispecies flow because of its immediate application in the gas diffusion layer (GDL) of proton exchange membrane (PEM) fuel cells, see Refs. [46–48] wherein the momentum conservation equation based on mass-averaged velocity is solved, as well as the mass conservation equations for different components with taking into account the molecular diffusion of all components and capillary effects. The relative velocities of phases also were considered using the macroscopic concept of relative permeability. Furthermore, water management is important in the performance of fuel cells; in many cases, dryness of the membrane may lead to the failure of fuel cell. On the other extreme, water flooding in the GDL of a cathode (for PEM fuel cells, in particular) can hinder gas transport and cause a severe drop in fuel cell performance [49–51]. It is shown that in a typical flow of moist air and liquid water in the GDL of PEM fuel cells, the capillary forces are more significant than the gravity and viscous counterparts [31]; therefore, it is believed that the capillary forces control the movement of water droplets formed within the porous domain. Li et al. [52] presented a comprehensive review of the problem of flooding in PEM fuel cells.

Conventional hydrothermal geothermal systems benefit from buoyancy-induced circulation of water in the reservoir. Depending on the reservoir pressure and depth, there could be localized boiling of water in the reservoir. This clearly forms a two-phase region inside a porous medium—the reservoir. More interestingly, the presence of other fluids such as CO_2, can add to the complexity of modeling as phase-transition can occur. Focusing on enhanced (or engineered) geothermal systems (EGS), the Queensland Geothermal Energy Centre of Excellence is currently investigating the possibility of a geothermosiphon with CO_2 as the working fluid (instead of water) [53–55]. The idea is to inject supercritical CO_2 ($T_{crit} = 31.1°C$) from the injection well and collect heated CO_2 from the other end (production well). The changes in the compressibility of CO_2 at the production well lowers the required pumping power as the fluid tends to move up the well with less or even no suction at the wellhead. Since some water already exists in some EGS reservoirs, the injection of CO_2 into those reservoirs makes it a multiphase flow through a porous medium (consisting of a network of natural fractures that are stimulated and thus connected), see Ref. [56] for more details.

Drying a porous medium has extensive industrial applications ranging from food processing to coloring of fabrics. One application is the modeling of a coal stockpile where initially moist coal is piled and exposed to ambient air. With time, the internal chemical reactions (oxidation) generate heat inside the stockpile, which induces a buoyant flow of air inside the pile and suction of the cold outside air into the pile. This fresh air adds more oxygen to the coal and the pile further heats up (wind can sometimes promote the oxidation process). As a result of this self-heating process, the internal moisture evaporates and the coal eventually reaches its thermal run-away temperature—this is when spontaneous combustion can occur with significant safety and economic consequences.

The list of applications of the multi-phase model to model flow and transport in porous media is not exhaustive by any means. Several important and traditional applications such as modeling oil-water flow during the secondary recovery of oil, modeling oil-water-steam flow during steam flooding under the tertiary recovery of oil, and modeling the flow of ground water, are well known and can be easily studied by the reader through numerous text books on these topics. Applications of unsaturated flow to model wicking in soils and paper-like media are also quite well-developed and will be covered to some extent in this book. Other novel applications such as those in biological and biomedical fields [57] continue to develop rapidly.

4.4 Measurement and Estimation of Various Flow Quantities

To use various laws of nature, it is important to measure various properties incorporated in these laws. For example, the permeability of a porous medium should be known if Darcy's law is to be used. This section is devoted to outlining different possible ways of measuring such flow-related properties of porous materials employed for wicking applications.

4.4.1 Permeability

The permeability of a porous medium can be described as the ease with which a fluid can be forced through a porous medium on the application of a certain pressure gradient. As we shall see below, we will use simple flow geometries involving single-phase flow with good control over flow and applied pressures. Most of such flow geometries involve creating in a porous sample either a simple one-dimensional (1-D) flow or a radial flow with liquid radiating away from an injection port. A simplified form of Darcy's law can be applied to such flow geometries in order to determine the permeability. However, the permeability used in Darcy's law can be either a scalar quantity, as in the case of an isotropic medium (Equation 4.9), or it can be a second-order tensor, as in the case of an anisotropic medium (Equation 4.10). The scalar form of permeability is much easier to determine compared with the permeability tensor, which has several independent components. The following sections describe various techniques available to measure the permeability in both of these forms.

4.4.1.1 Measurement of Permeability

Estimation of permeability remains an active area of research in the field of porous media. Although there are several techniques available for measuring permeability, here we list only those which can be used on relatively soft materials used in industrial wicking applications. First, we describe the techniques to obtain the permeability as a scalar quantity for the isotropic porous media.

4.4.1.2 Techniques for Determining Permeability as a Scalar for Isotropic Porous Materials

4.4.1.2.1 *Constant Head (1-D Flow) Permeameter*

Constant head permeameter is typically employed to measure the permeability of granular materials such as soil or sand [37], but it can be easily modified to measure the permeability of porous materials used in wicking applications such as paper, wicks, or fabrics. As shown in Figure 4.2a, a steady, 1-D flow of a test liquid (typically water) is established through a cylindrical sample of soil or sand. The piezometric head, defined as $\varnothing = p/\rho g + h$, is equal to the height of the water levels at the inlet and outlet that are maintained constant. Using the form of Darcy's law employing $\varnothing$, that is, Equation 4.15, we get

$$\frac{Q}{A} = K_h \frac{\Delta\varnothing}{L} \rightarrow K_h = \frac{QL}{A\Delta\varnothing} \rightarrow K = \frac{\mu QL}{\rho g A\Delta\varnothing} \tag{4.34}$$

Note that we equated Darcy's velocity to the net flow rate through the sample, Q, divided by the sample cross-section area, A, which is actually how the specific discharge (another name for Darcy's velocity) is estimated. We used

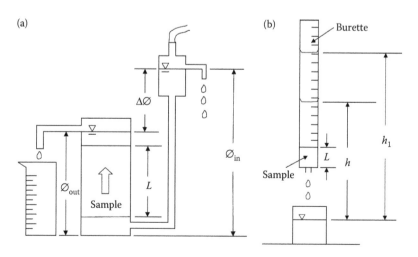

FIGURE 4.2
(a) A schematic of constant head permeameter. (b) A schematic of falling head permeameter.

the relation between hydraulic conductivity and permeability, Equation 4.16, to effect the final change in the above given relation. Note that the drop in the piezometric head over the sample length, $\Delta\varnothing$, is equal to the difference in the piezometric heads, which in turn is equal to the difference in the water-level elevations (Figure 4.2a) at the inlet and outlet.

Unlike the fixed-head permeameter where water level at the inlet and outlet are maintained at constant elevations, the falling-head permeameter deals with a water level at the inlet that is progressively decreasing with time due to the passage of water through the sample. As shown in Figure 4.2b, the continuously decreasing head at the inlet implies that inlet pressure driving the flow is decreasing; as a result, the Darcy velocity inside the porous sample is decreasing with time, as well. If the initial height of water column in the burette is h_1 at time $t = 0$, and the height reduces to h_2 at the final time of the experiment $t = t_{exp}$, then the permeability of the porous sample, K, can be estimated to be

$$K = \frac{\mu a L}{\rho g A t_{exp}} \ln \frac{h_1}{h_2}$$ (4.35)

where A is the cross-section area of the porous sample while a is the cross-section area of the burette. As described in Chapter 5 of this book, this method has been employed to estimate the permeability of polymer wicks.

4.4.1.2.2 *Constant Injection-Pressure (1-D Flow) Permeameter*

Here we describe a technique that has been popular within the composites processing community. Porous material in the form of a stack of layers of fabric is placed inside a flat rectangular mold, shown in Figure 4.3. A viscous test liquid in the form of motor oil or glycerine or diluted corn syrup is made to pass through the porous medium created by the compression of the stack between the upper and lower plates of the mold. Often, the mold plates are made from a transparent material such as polycarbonate or Plexiglas, and

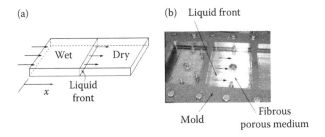

FIGURE 4.3
(a) A schematic of a constant injection-pressure permeameter based on 1-D flow through a flat rectangular mold. (b) A picture of the metallic mold with a transparent upper plate for visualizing the moving flow front.

consequently it is possible to see the progress of liquid front inside the fabric. The injection pressure during the forcing of liquid through the fabric stack is kept a constant through the use of a "pressure pot" where the test liquid in a container pressurized by a constant-pressure gas. Since the flow is occurring in a horizontal mold, the effect of gravity can be neglected during its analysis.

There are two possible ways of measuring permeability in such a mold. One is the *transient* 1-D flow method where the progress of liquid front is tracked with time as the fabric stack is being wetted by the injected liquid. The other is the *steady-state* 1-D flow method where passage of liquid through the fully-wetted fabric stack under constant conditions is studied.

Under the transient 1-D flow method, the front progress through the fabric stack is measured with a video camera. Note that the analytical solution for the front progress [58] is given as

$$x_f(t) = \sqrt{\frac{2Kp_{in}}{\epsilon\mu}}\sqrt{t} \tag{4.36}$$

where x_f is the x coordinate of the liquid front in the fabric stack (Figure 4.3) while p_{in} is the constant inlet pressure. While deriving this formula, perfect 100% saturation of fabric is assumed behind the moving flow-front while capillary suction at the front is neglected. One can use Equation 4.36 to fit a straight line through x_f versus $\sqrt{t}$ plot; slope of such a line will be equal to $\sqrt{2Kp_{in}/\epsilon\mu}$, which then can be used to estimate the permeability K if all other parameters are known. Note that the assumption of full saturation behind the moving is problematic in woven, stitched, or braided fabrics because of the bimodal pore-size distribution (also referred to as the dual-scale or two length-scales effect) which leads to partial saturation behind the front due to the delayed wetting of fiber tows [59,60]; however, it works quite well for a fiber mat made from a random arrangement of fibers or any other irregular porous media. In the steady-state 1-D flow method, the whole fabric stack is completely saturated with the test liquid while a steady flow is being maintained through it. As a result, this method does not suffer from the weakness of partial saturation during flow as in the previous method. One can use Darcy's law as given by Equation 4.7 to obtain the expression for the permeability as

$$K = \frac{\mu Q L}{p_{in} A} \tag{4.37}$$

where Q is the measured flow rate of the test liquid through the fabric stack, L is the x-direction length of the stack, while A is cross-section of the stack perpendicular to the flow direction.

4.4.1.2.3 *Constant Flow-Rate (1-D Flow) Permeameter*

One can modify the setup described in the previous section such that the test liquid is injected at a constant flow rate. This can be done by replacing the pressure pot with a constant flow-rate pump that can either be constructed from a gear pump or from a linear-displacement-type piston-cylinder pump actuated by an Instron machine [58].

As in the last section, we can have two variants of this test: the transient type or the steady-state type. The transient 1-D flow will have the inlet pressure change according to

$$p_{in}(t) = \frac{\mu}{\varepsilon K}\left(\frac{Q}{A}\right)^2 t \qquad (4.38)$$

where symbols have the same meaning as before except that Q is now the injected flow rate. As before, one can estimate the permeability K after fitting Equation 4.38 into the p_{in} versus t plot and estimating the slope. For the steady-state version of the test, Equation 4.35 can be used once again to estimate the permeability K.

4.4.1.2.4 *Constant Inlet-Pressure (Radial Flow) Permeameter*

One can modify the 1-D flow technique such that the fluid is injected from a hole punched into the fabric stack and the liquid radiates out from the hole instead of injecting the fluid from a side of the rectangular fabric-stack, as shown in Figure 4.3. If the fabric is an *isotropic* medium, the liquid radiates out in the form of a growing, circular liquid-front under a constant inlet-pressure and the radius of the front, r_f as a function of time, t, can be obtained from an intrinsic relation

$$\frac{r_f^2}{2}\ln\frac{r_f}{r_i} - \frac{r_f^2}{4} = \left(\frac{Kp_{in}}{\varepsilon\mu}\right)t - \frac{r_i^2}{4} \qquad (4.39)$$

where r_i is the radius of the hole. One can estimate the permeability K after fitting the analytically obtained curve from the above relation on the measured r_f versus t plot.

4.4.1.2.5 *Constant Flow-Rate (Radial Flow) Permeameter*

While conducting the injection at a constant rate, the inlet pressure is usually measured. Note that the inlet pressure for this case, obtained by combining Darcy's law and the continuity equation in the radial coordinates, can be predicted by the formula

$$p_{in} = \frac{Q}{2\pi b}\frac{\mu}{K}\ln\left(\frac{\sqrt{r_i^2 + (Q/\varepsilon\pi b)t}}{r_i}\right) \qquad (4.40)$$

Of the new symbols used here, b is the thickness of the fabric stack and r_i is the radius of the hole. One can then estimate the permeability K after fitting Equation 4.40 into the recorded p_{in} versus t plot.

Note that the radial flow tests are mostly performed in a transient mode where the measurements in terms of the front location or inlet pressure are made while the growing circular front has not touched one of the edges of the rectangular fabric stack. However, as described later, there is one radial rest that is conducted under steady-state conditions and is used to estimate the full permeability tensor for an anisotropic fabric; however, this test requires the fabric stack to be circular in shape, which differs from the usual rectangular shape of a typical fabric stack.

4.4.1.3 Techniques for Determining Permeability Tensor for Anisotropic Porous Materials

4.4.1.3.1 Steady-State 1-D Flow Technique

As before, the technique involves creating flow through a flat, rectangular stack of fabrics in the experimental setup shown in Figure 4.3. The flow of test liquid in a thin stack of fabrics can be treated as two-dimensional (2D) in-plane flow in an anisotropic medium. The form of Darcy's law without gravity, Equation 4.10, can be expanded in the matrix form as

$$\begin{Bmatrix} \langle u \rangle \\ \langle v \rangle \end{Bmatrix} = -\frac{1}{\mu} \begin{bmatrix} K_{xx} & K_{xy} \\ K_{yx} & K_{yy} \end{bmatrix} \begin{Bmatrix} \dfrac{\partial \langle p \rangle}{\partial x} \\ \dfrac{\partial \langle p \rangle}{\partial y} \end{Bmatrix} \tag{4.41}$$

where K_{xx}, K_{xy}, K_{yx}, K_{yy} are the four components of the in-plane permeability tensor for the fabric stack. Since the permeability tensor is symmetric for most porous media, $K_{xy} = K_{yx}$; hence, the number of independent permeability components to be determined through experiments is three.

Assuming the flow in the fabric stack to be unidirectional along the x-direction, the y-direction averaged-velocity component can be set to zero, that is, $\langle v \rangle = 0$. As a result, a relation between the x and y direction pressure gradients can be established as

$$\frac{\partial \langle p \rangle}{\partial y} = -\frac{K_{yx}}{K_{yy}} \frac{\partial \langle p \rangle}{\partial x} \tag{4.42}$$

On substituting Equation 4.42 into Equation 4.41, the expression for velocity $\langle u \rangle$ can be expressed as

$$\langle u \rangle = -\frac{1}{\mu} \left(K_{xx} - \frac{K_{xy}^2}{K_{yy}} \right) \frac{\partial \langle p \rangle}{\partial x} \tag{4.43}$$

On comparing Equation 4.43 with Equation 4.8, the effective permeability along the flow (x) direction is given as

$$K_{eff} = K_{xx} - \frac{K_{xy}^2}{K_{yy}} \tag{4.44}$$

If one of the principal axes of the permeability tensor (i.e., axis 1) is oriented at an angle θ to the flow or x direction (Figure 4.4), we can express all components of the permeability tensor in $x - y$ coordinates in terms of the principal permeability components (i.e., K_1, K_2) and θ as

$$\begin{aligned}
K_{xx} &= K_1 \cos^2\theta + K_2 \sin^2\theta \\
K_{yy} &= K_1 \sin^2\theta + K_2 \cos^2\theta \\
K_{xy} &= (K_2 - K_1)\sin\theta\cos\theta
\end{aligned} \tag{4.45}$$

where K_1 and K_2 are eigenvalues of the permeability tensor while directions 1 and 2 correspond to the respective eigenvectors. (Note that when direction 1 is aligned with the x axis, i.e., when $\theta = 0$, K_{xx} reduces to K_1, K_{yy} to K_2, while K_{xy} vanishes; as a result, the permeability K acquires a diagonal form.) Use of Equation 4.45 in Equation 4.44 allows us to express the effective permeability [61] as

$$K_{eff} = K_1\cos^2\theta\left[1 + \frac{K_2}{K_1}\tan^2\theta - \frac{((K_2/K_1)-1)^2}{(K_2/(K_1\tan^2\theta))+1}\right] \tag{4.46}$$

This expression indicates that the effective permeability along the flow direction is a function of three unknown quantities: K_1, K_2, and θ. In order to determine these three unknowns, we need three independent equations. Using Equations 4.43 and 4.44, we get

$$K_{eff} = \frac{\mu Q L}{P_{in} A} \tag{4.47}$$

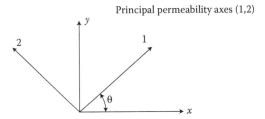

Principal permeability axes (1,2)

FIGURE 4.4
A schematic showing the orientation of the principal permeability axes (1,2) with respect to the lab axes (x,y).

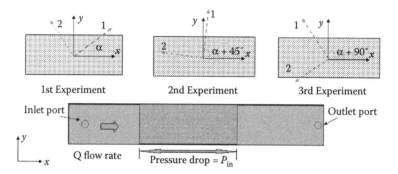

FIGURE 4.5
Three different steady 1-D flow experiments conducted after rotating the fabric to three different values of θ. These experiments can fully determine the in-plane permeability tensor. (Help
by Dr. Hua Tan in the preparation of this figure is gratefully acknowledged.)

for a steady 1-D flow experiment. As shown in Figure 4.5, we can have three
different values of K_{eff} corresponding to three different values of θ: α, α + 45°,
α + 90°, which are obtained by rotating the fabric by 45° after the first and
second experiments, and making the fabric stack from the rotated fabric.
Once K_1, K_2, and θ are known, one can easily find the three independent
components of the permeability tensor from Equation 4.45. (A variant of this
method exists where the solution of this problem becomes much easier [62].)

4.4.1.3.2 Transient Radial Flow Technique

Let us now consider the case when a transient radial experiment is being conducted. As mentioned before, the principal permeabilities K_1 and K_2 along
the principal directions 1 and 2 are different for an anisotropic medium. As
a result, when the test liquid is injected from a hole in the anisotropic fabric,
it will travel outwards in the form of an ellipse due to the differential flow
resistances along the principal directions (Figure 4.6a). If the mold plates are
transparent and a good contrast between the test liquid and fabric can be
established, then one can measure the aspect ratio as well as the orientation
of the ellipse. If $m = r_1/r_2$ is the aspect ratio with r_1 and r_2 being the major and
minor axes of the elliptical flow-front, then a "composite" permeability can
be estimated through the formula [63]

$$K_{comp} = \sqrt{K_1 K_2} = \frac{\mu Q}{2\pi h P_{in}} \left[\ln\left(\frac{r_1}{r_i}\right) + \ln\left(\frac{1 + \sqrt{1 + (r_i^2/r_1^2)\,(m^2 - 1)}}{m + 1} \right) \right] \quad (4.48)$$

where r_i is the radius of the injection port. Then the principal permeability
values can be derived through the relations $K_1 = mK_{comp}$ and $K_2 = K_{comp}/m$.
Note that the components of the in-plane permeability tensor corresponding

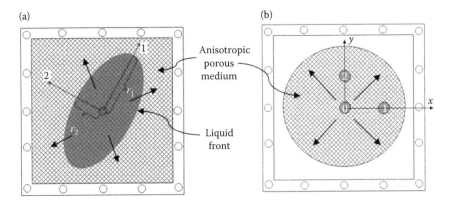

FIGURE 4.6
(a) A schematic of the *transient* radial filling experiment in an anisotropic porous medium: 1 and 2 correspond to the two in-plane principal directions. (b) A schematic of the *steady-state* radial filling experiment in an anisotropic porous medium: 0, 1, and 2 correspond to locations of the three different pressure transducers employed to measure the liquid pressure. The small circles are the nuts and bolts used for clamping the mold plates separated by a spacer. (Help by Dr. Hua Tan in the preparation of this figure is gratefully acknowledged.)

to a certain orientation of the principal directions vis-à-vis the lab frame can be estimated from Equation 4.45.

4.4.1.3.3 Steady-State Radial Flow Technique

We will now describe the radial-flow-based technique to estimate the permeability developed by Han et al. [63]. Figure 4.6b describes the circular geometry of the stack of fabrics used to measure the principal components of the permeability tensor found in anisotropic porous media. Note the gap surrounding the fabric stack is open to the atmospheric pressure. If $\alpha = K_2/K_1$ and $K_{eff} = \sqrt{K_1 K_2}$, then one needs to solve the following two coupled equations for the two unknowns, α and K_{eff}:

$$\log\left[\sqrt{\left(\frac{x_1}{x_0}\frac{1}{\sqrt{1-\alpha}}\right)^2 - 1} + \frac{x_1}{x_0}\frac{1}{\sqrt{1-\alpha}}\right] - \log\left(\frac{(1+\alpha)^{0.5}}{(1-\alpha)^{0.5}}\right) - \frac{2\pi h(p_0 - p_1)}{Q\mu}K_{eff} = 0$$

(4.49a)

$$\log\left[\sqrt{\left(\frac{y_2}{y_0}\frac{1}{\sqrt{1-\alpha}}\right)^2 - 1} + \frac{y_2}{y_0}\frac{1}{\sqrt{1-\alpha}}\right] - \log\left(\frac{(1+\alpha)^{0.5}}{(1-\alpha)^{0.5}}\right) - \frac{2\pi h(p_0 - p_2)}{Q\mu}K_{eff} = 0$$

(4.49b)

Once α and K_{eff} are known, one can estimate the principal permeability components through the relations $K_1 = K_{eff}/\sqrt{\alpha}$ and $K_2 = K_{eff}\sqrt{\alpha}$. Then as mentioned before, the components of the in-plane permeability tensor corresponding to a certain orientation of the principal directions vis-à-vis the lab frame can be estimated from Equation 4.45.

Further details on the permeability measurement techniques can be found in numerous publications on liquid composites molding process, some of which are [61,64–67].

4.4.1.4 Theoretical Models for Permeability

Apart from experimental measurement of the permeability, numerous attempts over the last century have been made to develop theoretical models for permeability based on Stokes flow through a geometry representative of a small portion of the considered porous medium. As described in several well-known texts on porous media such as Bear [37], Dullien [32], Scheidegger [68], and Greenkorn [69], numerous such permeability models have been developed for different types of porous media, including the consolidated (low porosity) media such as rocks, concrete, and so on, and the unconsolidated (higher porosity) media such as fibrous beds, glass-bead beds, lattice of fibers and spheres, and so on. As described by Dullien [32], the expressions for permeability can be placed into several categories, including phenomenological flow models for beds of particles and fibers, conduit or tube flow models, statistical models, pure empirical models, network models, deterministic models based on Stokes flow, and flow-around-submerged-objects models. Although a complete description of such models is beyond the scope of the present text, an attempt will be made to present a representative sample of such models without any derivation of their theoretical expressions.

Let us now list some empirical and theoretical formulas for estimating the permeability, which are well established in the literature on porous media. These formulas are generally of the form

$$K = D_f^2 \, \phi(\varepsilon_f) \tag{4.50}$$

where $\phi(\varepsilon_f)$ is a function of porosity. Several empirically obtained or theoretically derived formulas are listed in Table 4.1 which shows different suggested relations for $\phi(\varepsilon_f)$ for porous media consisting of packed *particles* or *fibers*.

Several of the listed theoretical models for the permeability of fibrous porous media assume the fibers to be arranged in a regular, parallel fashion. Since the collection of parallel, aligned fibers creates a transversely isotropic porous medium (i.e., the permeability perpendicular to the fiber axis is isotropic in that plane). For such porous media, the full permeability tensor can be represented by two independent principal components:

TABLE 4.1

Different Suggested Relations for $\phi(\varepsilon_f)$ for Porous Media Consisting of Packed Particles with Permeability Given as $K = D_f^2 \, \phi(\varepsilon_f)$ Such That D_f Is the Mean Particle or Fiber Diameter

Author (Comments)	$\phi(\varepsilon_f)$
Kozeny (1927), Carman (1937) [37]	$\dfrac{1}{180}\dfrac{\varepsilon_f^3}{(1-\varepsilon_f)^2}$
Duplessis and Masliyah (1991) [70] (consolidated rock-like media, $\varepsilon_f > 0.1$)	$\dfrac{\varepsilon_f\left[1-(1-\varepsilon_f)^{\frac{1}{3}}\right]\left[1-(1-\varepsilon_f)^{\frac{2}{3}}\right]}{63\,(1-\varepsilon_f)^{\frac{4}{3}}}$
Davies (1952) [32] (Empirical correlation, fibrous bed, $\varepsilon_f < 0.98$)	$\dfrac{1}{64(1-\varepsilon_f)^{1.5}[1+56(1-\varepsilon_f)^3]}$
Chen (1955) [32] (Empirical correlation, fibrous bed)	$0.129\dfrac{\varepsilon_f}{1-\varepsilon_f}\ln\dfrac{0.64}{(1-\varepsilon_f)^2}$
Berdichevski and Cai (1994) [71] (Analytical Model, Flow axial to parallel fibers)	$\dfrac{1}{32(1-\varepsilon_f)}\left[\ln\dfrac{1}{(1-\varepsilon_f)^2}-\varepsilon_f(2+\varepsilon_f)\right]$
Berdichevski and Cai (1994) [71] (Analytical Model, Flow transverse to parallel fibers)	$\dfrac{1}{32(1-\varepsilon_f)}\left[\ln\dfrac{1}{1-\varepsilon_f}-\dfrac{\varepsilon_f(2-\varepsilon_f)}{1+(1-\varepsilon_f)^2}\right]$
Rumpf and Gupte (1971) [32] ($K \cong 1$, spherical particles, $0.35 < \varepsilon_f < 0.7$)	$\dfrac{1}{5.6K}\varepsilon_f^{5.5}$
Bruschke and Advani (1993) [72] (Analytical Model, Flow transverse to parallel fibers in square arrangement, $\varepsilon_f < 0.4$)	$\dfrac{1}{3}\dfrac{(1-\eta)^2}{\eta^3}\left(\dfrac{3\eta\cdot\tan^{-1}\sqrt{(1+\eta)/(1-\eta)}}{\sqrt{1-\eta^2}}+\dfrac{\eta^2}{2}+1\right)^{-1}$ where $\eta = \dfrac{4}{\pi}(1-\varepsilon_f)$
Gebart (1992) [73] (Analytical model, Flow perpendicular to fibers, hexagonal arrangement of fibers)	$\dfrac{4}{9\pi\sqrt{6}}\left(\sqrt{\dfrac{1-\varepsilon_{f\min}}{1-\varepsilon_f}}-1\right)^{5/2}$ where $\varepsilon_{f\min} = 1-\dfrac{\pi}{2\sqrt{3}}$

the permeability along the fiber axis and the permeability across the fiber axis.

In general, the permeability obtained from the theoretical models is not very accurate when applied to a real porous medium. However, if we choose our theoretical models carefully, they are able to furnish the permeability values correct to within an order of magnitude. Therefore, it is important to conduct experiments in order to confirm the predictions of the theoretical permeability models.

4.4.2 Relative Permeability

As mentioned earlier, the simultaneous flow of more than one fluid through a porous medium during the multi-phase flow is modeled using the generalized Darcy's law where the permeability, K, of the porous medium is modified through multiplication by the dimensionless relative permeability, k_r, for each fluid.

4.4.2.1 Measurement of Relative Permeability

There are numerous methods available in the porous-media literature to estimate the relative permeability which can be characterized into several categories. According to the classification adopted by Honarpour et al. [74], the following categories are possible: (1) steady-state methods, (2) unsteady-state methods, (3) capillary pressure methods, (4) centrifuge methods, and (5) field data method. Of these five broad categories, the categories (1), (2), and (4) are the methods based on experimentation. In a more recent classification proposed by Christiansen et al. [75], the experimental methods for estimating relative permeability can be placed into two categories: (1) steady-state methods, (2) unsteady-state methods. The techniques described under these two types will be discussed in this section.

Within the category of *steady-state methods*, four subgroups can be identified [75]: the multiple-core methods (also known as the Penn State method), the high-rate methods, the stationary-liquid methods, and the uniform capillary-pressure methods (also known as the Hassler method). Of these four, the high-rate method is used the most and will be described in the following paragraph.

In the steady-state high-rate method, two fluids are forced to pass through a porous-medium sample at constant and sufficiently high flow rates. Typically, the sample is cylindrical in shape; a mixture of the two fluids enters from one end of the cylinder and exits from the other end after traversing the cylinder-length in a 1-D flow fashion. During the flow, care is taken to maintain a given saturation in the sample. After pressure drop and saturation in the sample reach a constant value, the relative permeabilities for the two (wetting and nonwetting) fluid phases can be computed as

$$k_{r,w} = -\frac{\mu_w L Q_w}{K A \Delta P}, \quad k_{r,nw} = -\frac{\mu_{nw} L Q_{nw}}{K A \Delta P} \tag{4.51}$$

where μ is the fluid viscosity, L is the sample length, Q is the flow rate of a fluid, K is the permeability of the sample (measured through a single-phase flow experiment described earlier), A is the cylinder (sample) cross-section area, and ΔP is the pressure drop (which is made equal for the two fluids). The relative permeabilities can easily be computed once the measured pressure drops for the imposed flow rate is entered into the above formulas.

(Note that fluid viscosities can be easily measured or are available in handbooks.) A schematic of the typical steady-state device, such as the Hassler's apparatus, is shown in Figure 4.7.

One advantage of the steady-state method, as reflected in the simple structure of Equation 4.51, is the easy analysis of the experimental data—typically, it takes less time to get the final relative-permeability values compared with the unsteady-state methods. The downside of this method is that in consolidated media such as rocks, it takes a long time to establish the steady-state flow conditions (unlike the shorter time needed to conduct the unsteady-flow experiments).

Within the category of *unsteady-state methods*, three subgroups can be identified [75]: high-rate methods, low-rate methods, and centrifuge methods. Of these, the high-rate methods are the oldest and have been used extensively by the researchers in the petroleum engineering and geosciences areas. The other two are relatively new and are gradually coming into prominence.

In the unsteady high-rate method, a displacing fluid is injected into a porous-medium sample fully saturated with another fluid. The injection is usually done at a high rate to minimize the problems associated with the capillary end effects. Throughout the experiment, the imposed pressure, the volume of displaced fluid produced, and the volume of displacing fluid injected are recorded and monitored. Differentiation of the pressure and volume data is required to obtain the relative permeability of the sample. Numerical programs based on the two-phase flow equations as applied to 1-D flow geometry are employed to analyze the data [75].

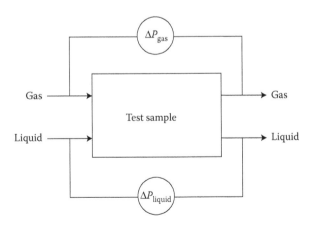

FIGURE 4.7
A schematic of a typical steady-state method employed to measure the relative permeability. A gas (non-wetting) and liquid (wetting) phases are mixed and introduced at the beginning of the test sample (also known as the "core"). After traversing the sample, they are separated at the end and the pressure drop for the two phases is measured as shown. Note that for the high-rate method, the two pressure drops are equated after neglecting any capillary pressure difference between the two phases.

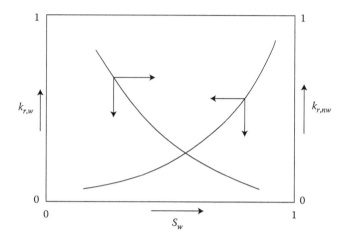

FIGURE 4.8
A typical plot of the relative permeability of wetting and nonwetting fluids as a function of saturation.

The general shape of the relative permeability versus saturation curves, which is typically of the shape shown in Figure 4.8, can be described by the equations

$$k_{r,w} = A(S_w)^n, \quad k_{r,nw} = B(1 - S_w)^m \tag{4.52}$$

where A, B, m, and n are constants. The above equation can be used to fit the experimental data to produce analytical expressions for relative permeability, which can be used in theoretical analysis as well as numerical simulations.

It is important to realize that most of the research conducted to measure relative permeability involve rock or soil samples [74,75]—virtually no research has been reported on measuring this quantity in softer materials used in typical wicking applications such as fabrics or paper. There could be several reasons for this gap in the literature. One reason could be that most devices developed for measuring relative permeability are designed to accept hard, rock-like samples with very low porosities and permeabilities. As a result, when a softer material with much higher porosity and permeability is used in these devices, the results are perhaps not acceptable, or the material's porous structure is damaged under the application of high pressure. Another reason could be the reliance on limited flow models, such as the Washburn equation which do not require the use of relative permeability, for modeling flows in softer materials (see Chapter 3).

Fortunately, several theoretical models have been developed that can be used to estimate this important quantity.

4.4.2.2 Theoretical Models for Relative Permeability

Honarpour et al. [74] describe numerous theoretical or mathematical models for relative permeability that have been proposed in the literature and can be classified into the categories of capillary, statistical, empirical, and network models. Of such models, the equations proposed by Corey are perhaps most useful, and thus very widely used. In the simplest form, Corey's equations, as proposed for oil and gas system (Equations 4.24 through 4.26 [74]), can be generalized for a nonwetting fluid and wetting fluid system as

$$k_{r,w} = \left(S_{w,e}\right)^4, \quad k_{r,nw} = \left(1 - S_{w,e}\right)^2 \left(1 - S_{w,e}^2\right) \tag{4.53}$$

where $S_{w,e} = (S_w - S_{w,r})/(1 - S_{w,r})$ while $S_{w,r}$ represents the lowest saturation of the nonwetting fluid at which the tortuosity of the nonwetting phase is infinite. The two relative permeabilities are related through

$$\frac{k_{r,w}}{\left(S_{w,e}\right)^2} + \frac{k_{r,nw}}{\left(1 - S_{w,e}\right)^2} = 1 \tag{4.54}$$

4.4.3 Capillary Pressure

During the multiphase flow in a porous medium, the simultaneous flow of two or more immiscible fluids occurs through the pore space of the medium, and two fluids can touch each other during the course of their motion. A discontinuity in pressure (Δp), also called the capillary pressure p_c, exists in the fluids across such interfaces due to their curvatures as

$$\Delta p = p_c = p_{nw} - p_w = \sigma\left(\frac{1}{r_1} + \frac{1}{r_2}\right) \tag{4.55}$$

where σ is the interfacial tension between the two fluids and r_1, r_2 are the two principal radii of curvatures [37]. In a capillary tube, using the principles of statics, the capillary pressure at the interface between the wetting and non-wetting fluid can be expressed as

$$p_c = p_{nw} - p_w = \frac{2\sigma}{r}\cos\theta \tag{4.56}$$

with r being the radius of the capillary tube and θ being the contact angle formed at the contact line corresponding to the meeting of the three (solid, wetting, and nonwetting) phases. Through the capillary models of porous media, where pores of different sizes are represented by capillary tubes of corresponding radii, it is easy to see that small pores corresponding to small tube radii have larger capillary "suction" pressures. As a result, the smaller pores have a larger tendency of retaining wetting liquids during a drainage experiment.

Inside a porous medium, since each point of contact between the two fluids will have a different interface with unique radii of curvatures, p_c will change from point-to-point within the porous medium. One can use the concept of REV to express the capillary pressure in an averaged form as

$$\langle p_c \rangle = \langle p_{nw} \rangle^f - \langle p_w \rangle^f \tag{4.57}$$

where $\langle p_{nw} \rangle^f$ and $\langle p_w \rangle^f$ are the phase-averaged pressures of the nonwetting and wetting fluids, respectively. As is clear from Equation 4.56, the capillary pressure is higher in smaller pores. Since a porous medium has a whole range of pore diameters, we can expect to see a whole range of capillary pressures inside the porous medium, which is displayed during the imbibition (i.e., suction of wetting liquid into a porous medium) or drainage (i.e., draining of a wetting liquid from a porous medium) processes. A typical plot of capillary pressure as a function of saturation is given in Figure 4.9. As we can see, the capillary pressure is high at low saturations. This implies that most of the big pores are empty and the wetting fluid is mainly residing inside small pores; as a result, radii of the wetting–nonwetting interface curvatures are small, and hence the capillary pressure (as given by Equation 4.56) is high. However,

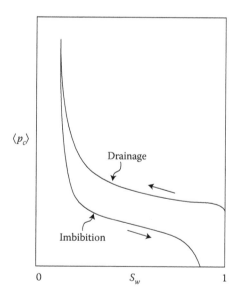

FIGURE 4.9
A typical plot of capillary pressure as a function of saturation of the wetting liquid. A difference in flow behavior during de-wetting and wetting of porous medium during drainage and imbibition leads to hysteresis in the plot.

as more of the wetting fluid flows into the porous medium and its saturation increases, the interface moves into larger pores and consequently the capillary pressure drops. The hysteresis shown in Figure 4.9 is typically attributed to the hysteresis in the contact angle,* as well as the ink-bottle effect[†] [37].

4.4.3.1 Measurement of Capillary Pressure

The several methods available for the experimental determination of the capillary pressure can be classified into two main categories: (1) displacement methods and (2) dynamic methods.

The displacement methods involve displacement of one fluid lodged within the pore space of a porous medium by another fluid [37]. The methods are based on the establishment of successive states of hydrostatic equilibrium within the sample and therefore can also be characterized as the "static" methods for determining the capillary pressure. These techniques are much more commonly used compared to the dynamic methods discussed later, and yield mostly drainage type curves. We will briefly describe three such methods: (a) porous diaphragm method, (b) mercury injection method, and (c) centrifuge method.

In the porous diaphragm method [37], a porous-medium sample soaked with the wetting fluid is placed tightly on a porous ceramic diaphragm within a chamber filled with the nonwetting fluid. The diaphragm allows only the wetting fluid to pass through. As a result, when the pressure applied to the nonwetting fluid increases, the nonwetting fluid moves into the sample, thereby pushing the wetting fluid through the diaphragm into a collecting setup below. This way the applied pressure (equal to the capillary pressure) can be correlated with the wetting-phase saturation determined from the volume of wetting fluid pushed out of the sample. A drawback of this method is that it takes a long time to establish equilibrium; thus, a typical capillary-pressure-versus-saturation plot takes several days to prepare.

In the mercury injection method [37], the mercury porosimeter is employed. First, a dry porous sample is placed in an evacuate chamber, which is later filled with liquid mercury. Since mercury is a nonwetting liquid for most solid surfaces, it does not imbibe into the porous sample spontaneously—a gas pressure has to be applied on the mercury to force it into the small pores of the sample. The volume of mercury disappearing into the sample yields the saturation while the applied pressure is equal to the capillary pressure. This method is quite fast—equilibrium is typically achieved in minutes, while the whole $<p_c>$ versus S_w curve takes only a few hours.

---

* As seen in a rain drop moving down an inclined roof, the advancing contact angle is larger than the receding contact angle. As a result, the capillary pressure given by Equation 4.56 is smaller for the advancing (imbibition) case compared to the receding (drainage) case.
[†] The interface has a tendency to get "stuck" in narrow throats of channels formed between particles during the drainage process. The smaller radii of the throats results in a larger capillary "suction" pressure that keeps the interface pinned to these throats during drainage.

The centrifuge method is becoming quite popular nowadays for determining the capillary pressure as a function of saturation. Here, a sample soaked with a wetting liquid is placed in rotary centrifuge. As the device rotates with a constant angular velocity, the centrifugal force acting (as the body force) on the wetting fluid forces the liquid out only to be collected by a receptacle. The volume of the fluid released at a given angular velocity yields the saturation while the centrifugal force decides the capillary pressure. One can generate the full $<p_c>$ versus S_w curve by traversing a whole range of angular velocity. Like the mercury porosimeter technique, this method is quite fast and the whole curve can be determined in a few hours.

The dynamics methods, which form the other set of techniques for determining the capillary pressure as a function of saturation, rely on the simultaneous flow of the wetting and nonwetting fluids through the porous sample. During such measurements, successive states of steady-state flows, corresponding to different saturations, are established. One of the methods uses the Hassler's apparatus, which has already been described as a device to measure relative permeability. Typically, a gas (nonwetting phase) and a liquid (wetting phase) are made to flow simultaneously, and the difference in the pressures of the two fluids allows one to determine the capillary pressure (Figure 4.7). The saturation, in turn, is determined by weighing the sample. Such dynamic methods are less common since they are more difficult to implement compared to the displacement methods.

4.5 Summary

In this chapter, the salient features of the physics required to model flow of liquids in a porous medium are presented. First, the saturated flow in a porous medium, with only a single fluid passing through the pore space of the porous medium, is described. The importance of representing the porous-media flows in the averaged form is highlighted with a brief mention of the volume-averaging method. For the pure viscous-flow regime, the macroscopic forms of the mass and momentum balance equations are presented where the latter is described using the Darcy's law. Here, the inclusion of gravity through the modified pressure and piezometric head is mentioned. The flow resistance offered by the porous medium is captured through the parameters of permeability and hydraulic conductivity. Later, the simultaneous flow of two fluids through the porous medium is described using the appropriate mass and momentum balance equations. Here the Darcy's law, as generalized for multiphase flow, after employing the parameter of relative permeability to condition the absolute permeability as a function of saturation is found to be appropriate. The capillary pressure, which relates the pressures of the two fluids and is indispensable to bringing closure to the mathematical problem of two-phase flow, is introduced. The unsaturated flow is considered next as the special case of the two-phase, air–liquid flow with air considered stationary and at ambient conditions throughout. Slight manipulation of the governing equations results in a heat equation such as Richards equation, which can predict the capillary-pressure-driven movement of moisture in a dry porous medium as the diffusion of liquid saturation in the porous medium. It was followed by a brief description of industrial applications of the two-phase flow physics, which include examples of modeling the processes including the water flooding problem in fuel cells, energy harvesting in geothermal systems, drying of porous substrates, ground-water flow, and secondary and tertiary recovery of oil.

In the final section, the techniques for measuring the parameters important for porous-media flows are described. To measure the permeability, the steady state as well as transient techniques, such as the constant-head, falling-head, and constant-pressure permeameters and the radial flow techniques, are presented. A brief description of the theoretical permeability models and a few details on handling additional complexities associated with anisotropic porous media are also provided. The property of relative permeability is considered next where the steady-state methods (the high-rate and Hassler methods) and the transient methods are described. A note on the empirical/theoretical models, such as the Corey's equations, follows. Finally, a synopsis of the methods to measure the capillary pressure using the techniques, such as the mercury porosimetry and the centrifuge method, is provided. It is expected that this short description of the physics employed to model the porous-media flows will provide an uninitiated reader enough motivation to launch into a deeper study and enable the development of effective models for predicting the wicking flows in porous media.

Nomenclature

Roman Letters

A	Cross-section area of the test sample or the stack of fiber mats (m²)
a	Cross-section area of the burette (m²)
$\bar{D}$	Diffusivity coefficient or capillary diffusivity (m²/s)
D	Modified diffusivity coefficient (m²/s)
g	Acceleration due to gravity (m/s²)
$\vec{g}$	Acceleration due to gravity (vector) (m/s²)
h	Elevation (m)
$\hat{i}, \hat{j}, \hat{k}$	Unit vectors along the x, y, and z coordinate axes, respectively
K	Permeability of an isotropic porous medium (m²)
$\mathbf{K}$	Permeability tensor for an anisotropic porous medium (m²)

$\bar{K}$	Viscous drag tensor (Equation 4.18)
K_1, K_2	Principal components of the in-plane permeability tensor (m², Equation 4.45)
K_{comp}	Composite permeability (m², Equation 4.48)
K_{eff}	Effective permeability (m²)
K_h	Hydraulic conductivity of an isotropic porous medium (m²)
$\mathbf{K}_h$	Hydraulic conductivity tensor for an anisotropic porous medium (m/s)
K_i	Effective permeability for phase $i = (w, nw)$ in a two-phase system (m², Equation 4.23)
k_r	Relative permeability
$K_{xx}, K_{xy},$ K_{yx}, K_{yy}	Four components of the in-plane permeability tensor (m², Equation 4.41)
L	Length of the porous sample or fabric stack (m)
m	A constant, Equation 4.52; aspect ratio of ellipse, Equation 4.48
n	A constant in Equation 4.52
P	Modified pressure (Pa)
p	Hydrodynamic fluid pressure (Pa)
p_{in}	Inlet pressure (Pa)
Q	Flow rate
q	A general flow-related quantity (Equation 4.1)
r_1	Major axis of the elliptical flow front (m, Equation 4.48)
r_2	Minor axis of the elliptical flow front (m, Equation 4.48)
r_f	Radius of the front (m)
r_i	Radius of the injection hole (m)
REV	Representative elementary volume
S	Saturation of the fluid phase
t	Time (s)
u, v, w	Components of velocity along the x, y, and z axes, respectively (m/s)
V	Volume of REV (m³)
$\vec{v}$	Point-wise fluid velocity vector (m/s)
V_f	Volume of fluid in REV (m³)

Greek Letters

ϵ	Porosity
$\emptyset$	Piezometric head (m)
ρ	Fluid density (kg/m³)
μ	Fluid viscosity (Pa s)

Subscripts

e	Effective (saturation)
exp	Experimental

i	Index ($i = nw, w$)
nw	Of the nonwetting phase
o	Of reference conditions
r	Residual (pertaining to the saturation)
s	Of the solid phase
w	Of the wetting phase

Other Symbols

$\langle \ \rangle$	Volume-averaged or phase-averaged quantity
$\langle \ \rangle^f$	Pore-averaged or intrinsic phase-averaged quantity of fluid phase
$\langle \ \rangle^s$	Pore-averaged or intrinsic phase-averaged quantity of solid phase (Equation 4.17)

References

1. Merrikh, A.A. and J.L. Lage, From continuum to porous continuum: The visual resolution impact on modeling natural convection in heterogeneous media, in *Transport Phenomena in Porous Media*, D.B. Ingham and I. Pop, Editors. 2005, Elsevier Science: Oxford.
2. Atkin, R.J. and R. E. Craine. 1976. Continuum theory of mixtures: Applications. *Journal of the Institute of Mathematics and its Applications*, 17: 153–207.
3. Bowen, R., 1976. Theory of mixtures, in *Continuum Physics*, New York: Academic Press.
4. Sanchez-Palencia, E., 1980. Non homogeneous media and vibration theory, in *Lecture Notes in Physics*, Berlin: Springer-Verlag.
5. Bakhvalov, N. and G. Panasenko, 1989. *Homogenization: Averaging Processes in Porous Media*, Dordrecht: Kluwer Academic Publishers.
6. Whitaker, S., 1996. *The Method of Volume Averaging*, Dordrecht: Kluwer Academic Publishers.
7. Slattery, J., 1969. Single-phase flow through porous media. *Am. Inst. Chem. Eng.*, 15(6): 866–872.
8. Gray, W.G. and P.C.Y. Lee, 1977. On the theorems for local volume averaging of multiphase systems. *International Journal of Multiphase Flow*, 3: 333–340.
9. Gray, W.G., et al., 1993. *Mathematical Tools for Changing Spatial Scales in the Analysis of Physical Systems*, Boca Raton, FL: CRC Press, Inc.
10. Bear, J. and Bachmat, Y., 1990. *Introduction to Modeling of Transport Phenomena in Porous Media*, Dordrecht: Kluwer Academic Publishers.
11. Darcy, H.P.G., 1856. *Les Fontaines Publiques de la Ville de Dijon*, Paris: Victor Dalmont.
12. Lage, J.L., The fundamental theory of flow through permeable media from Darcy to turbulence, in *Transport Phenomena in Porous Media*, D.B. Ingham and I. Pop, Editors. 1998. Oxford: Elsevier Science.

13. Gray, W.G. and K. O'Neill, 1976. On the general equations for flow in porous media and their reductions to Darcy's Law. *Water Resources Research*, 12(2): 148–154.

14. Narasimhan, A., 2002. *Unraveling, Modeling and Validating the Temperature Dependent Viscosity Effects in Flow through Porous Media*, Dallas, Texas: Southern Methodist University.

15. Hooman, K., 2009. *Modeling Variable Viscosity Forced and Free Convection in Porous Media*, Brisbane, Queensland: The University of Queensland.

16. Shenoy, A.V., 1994. Non-Newtonian fluid heat transfer in porous media. *Advances in Heat Transfer*, 24: 101–190.

17. Straus, J.M. and G. Schubert, 1979. Effect of CO_2 on the buoyancy of geothermal fluids. *Geophysical Research Letters*, 6(1): 5–8.

18. Schubert, G. and J.M. Straus, 1981. Thermodynamic properties for the convection of steam–water–CO_2 mixtures. *American Journal of Science*, 281(3): 318–334.

19. Donaldson, I.G., 1982. Heat and mass circulation in geothermal systems. *Annual Review of Earth and Planetary Sciences*, 10: 377–395.

20. Chen, Z., G. Huan, and Y. Ma, 2006. *Computational Methods for Multiphase Flows in Porous Media*, Philadelphia: Society for Industrial and Applied Mathematics.

21. Silin, D., et al., Microtomography and pore-scale modeling of two-phase fluid distribution. *Transport in Porous Media*, 86(2): 525–545.

22. Nardi, C., et al., 2009. Pore-scale modeling of three-phase flow: Comparative study with experimental reservoir data, in *International Symposium of the Society of Core Analysts*, The Netherlands: Noordwijk.

23. Piri, M. and Z.T. Karpyn, 2007. Prediction of fluid occupancy in fractures using network modeling and x-ray microtomography. II: Results. *Physical Review E*, 76(1): 016316-1–016316-11.

24. Ramstad, T., P.E. Oren, and S. Bakke, Simulation of two-phase flow in reservoir rocks using a Lattice Boltzmann method. *SPE Journal*, 15(4): 923–933.

25. Kang, Q.J., et al., 2005. Numerical modeling of pore-scale phenomena during CO_2 sequestration in oceanic sediments. *Fuel Processing Technology*, 86(14–15): 1647–1665.

26. Paunov, V.N., et al., Lattice-Boltzmann simulation of ideal and nonideal immiscible two-phase flow in porous media, in *Computational Methods in Water Resources Xi, Vol 1: Computational Methods in Subsurface Flow and Transport Problems*, A.A. Aldama, et al., Editors. 1996, pp. 457–464, England: Computational Mechanics Publications Ltd.

27. Yiotis, A.G., et al., 2007. A lattice Boltzmann study of viscous coupling effects in immiscible two-phase flow in porous media. *Colloids and Surfaces a-Physicochemical and Engineering Aspects*, 300(1–2): 35–49.

28. Mukherjee, P.P., Q.J. Kang, and C.Y. Wang, 2011. Pore-scale modeling of two-phase transport in polymer electrolyte fuel cells-progress and perspective. *Energy & Environmental Science*, 4(2): 346–369.

29. Chen, R. and T.S. Zhao, 2005. Mathematical modeling of a passive-feed DMFC with heat transfer effect. *Journal of Power Sources*, 152(1): 122–130.

30. Bahrami, H. and A. Faghri, 2010. Transport phenomena in a semi-passive direct methanol fuel cell. *International Journal of Heat and Mass Transfer*, 53: 2563–2578.

31. Yang, W.W., T.S. Zhao, and C. Xu, 2007. Three-dimensional two-phase mass transport model for direct methanol fuel cells. *Electrochimica Acta*, 53(2): 853–862.

32. Dullien, F.A.L., 1992. *Porous Media : Fluid Transport and Pore Structure*, San Diego, CA: Sydney Academic Press.
33. Lenormand, R., E. Touboul, and C. Zarcone, 1988. Numerical models and experiments on immiscible displacements in porous media. *Journal of Fluid Mechanics*, 189: 165–187.
34. Dias, M.M. and A.C. Payatakes, 1986. Network models for two-phase flow in porous media Part 1. Immiscible microdisplacement of non-wetting fluids. *Journal of Fluid Mechanics*, 164: 305–336.
35. Wyckoff, R.D. and H.G. Botset, 1936. The flow of gas–liquid mixtures through unconsolidated sands. *Physics-A Journal of General and Applied Physics*, 7(1): 325–345.
36. Muskat, M., 1937. *The Flow of Homogeneous Fluids through Porous Media*, New York: McGraw-Hill.
37. Bear, J., 1972. *Dynamics of Fluids in Porous Media*, New York: Dover.
38. Whitaker, S., 1986. Flow in porous media. 2. The governing equations for immiscible 2-phase flow. *Transport in Porous Media*, 1(2): 105–125.
39. Whitaker, S., 1986. Flow in porous media. 1. A theoretical derivation of Darcy's Law. *Transport in Porous Media*, 1(1): 3–25.
40. Kalaydjian, F., 1990. Origin and quantification of coupling between relative permeabilities for 2-phase flows in porous media. *Transport in Porous Media*, 5(3): 215–229.
41. Rothman, D.H., 1990. Macroscopic Laws for immiscible 2-phase flow in porous media—Results from numerical experiments. *Journal of Geophysical Research-Solid Earth and Planets*, 95(B6): 8663–8674.
42. Gunstensen, A.K. and D.H. Rothman, 1993. Lattice-Boltzmann studies of immiscible 2-phase flow through porous media. *Journal of Geophysical Research-Solid Earth*, 98(B4): 6431–6441.
43. Joekar-Niasar, V. and S.M. Hassanizadeh, 2011. Effects of fluids properties on non-equilibrium capillary effects: Dynamic pore-network modeling. *International Journal of Multiphase Flow*, 37: 198–214.
44. Pinder, G.F. and W.G. Gray, 2008. *Essentials of Multiphase Flow and Transport in Porous Media*, Hoboken, NJ: Wiley.
45. Crank, J., 1975. *The Mathematics of Diffusion*, 2nd ed., Oxford, UK: Clarendon Press.
46. Wang, Z.H., C.Y. Wang, and K.S. Chen, 2001. Two-phase flow and transport in the air cathode of proton exchange membrane fuel cells. *Journal of Power Sources*, 94(1): 40–50.
47. Natarajan, D. and T. Van Nguyen, 2001. A two-dimensional, two-phase, multicomponent, transient model for the cathode of a proton exchange membrane fuel cell using conventional gas distributors. *Journal of the Electrochemical Society*, 148(12): A1324–A1335.
48. Mazumder, S. and J.V. Cole, 2003. Rigorous 3-d mathematical modeling of PEM fuel cells—II. Model predictions with liquid water transport. *Journal of the Electrochemical Society*, 150(11): A1510–A1517.
49. Fuller, T.F. and J. Newman, 1993. Water and thermal management in solid-polymer-electrolyte fuel cells. *Journal of the Electrochemical Society*, 140(5): 1218–1225.
50. Gurau, V., H. Liu, and S. Kakac, 1998. Two-dimensional model for proton exchange membrane fuel cells. *AIChE Journal*, 44: 2410–2422.

51. Granovskii, M., I. Dincer, and M.A. Rosen, 2006. Environmental and economic aspects of hydrogen production and utilization in fuel cell vehicles. *Journal of Power Sources*, 157(1): 411–421.

52. Li, H., et al., 2008. A review of water flooding issues in the proton exchange membrane fuel cell. *Journal of Power Sources*, 178(1): 103–117.

53. Fard, M.H., K. Hooman, and H.T. Chua, 2010. Numerical simulation of a super-critical CO_2 geothermosiphon. *International Communications in Heat and Mass Transfer*, 37(10): 1447–1451.

54. Atrens, A.D., H. Gurgenci, and V. Rudolph, Electricity generation using a carbon-dioxide thermosiphon. *Geothermics*, 39(2): 161–169.

55. Atrens, A.D., H. Gurgenci, and V. Rudolph, 2009. CO_2 Thermosiphon for competitive geothermal power generation. *Energy & Fuels*, 23(1): 553–557.

56. Tester, J.W., et al., 2007. Impact of enhanced geothermal systems on US energy supply in the twenty-first century. *Philosophical Transactions of the Royal Society a-Mathematical Physical and Engineering Sciences*, 365(1853): 1057–1094.

57. Vafai, K., 2010. *Porous Media, Applications in Biological Systems and Biotechnology*, Boca Raton, FL: CRC Press.

58. Advani, S. and E.M. Sozer, 2010. *Process Modeling in Composites Manufacturing*. 2nd ed., Boca Raton, FL: CRC Press, Inc.

59. Tan, H., T. Roy, and K.M. Pillai, 2007. Variations in unsaturated flow with flow direction in resin transfer molding: An experimental investigation. *Composites Part A-Applied Science and Manufacturing*, 38(8): 1872–1892.

60. Tan, H. and K.M. Pillai, 2010. Fast liquid composite molding simulation of unsaturated flow in dual-scale fiber mats using the imbibition characteristics of a fabric-based unit cell. *Polymer Composites*, 31(10): 1790–1807.

61. Parnas, R.S., et al., Permeability characterization. Part 1: A proposed standard reference fabric for permeability. *Polymer Composites*, 16(6): 429–445.

62. Weitzenbock, J.R., R.A. Shenoi, and P.A. Wilson, 1998. Measurement of principal permeability with the channel flow experiment. *Polymer Composites*, 20(2): 321–335.

63. Han, K.K., C.W. Lee, and B.P. Rice, 2000. Measurements of the permeability of fiber preforms and applications. *Composites Science and Technology*, 60(12–13): 2435–2441.

64. Lee, L.J., Liquid composite molding, in *Advanced Composites Manufacturing*, T.G. Gutowski, Editor, 1997. New York: John Wiley and Sons.

65. Parnas, R.S., 2000. *Liquid Composite Molding*, Munich, Germany: Hanser Publishers.

66. Parnas, R.S. and A.J. Salem, 1993. A comparison of the unidirectional and radial In-Plane flow of fluids through Woven composite reinforcements. *Polymer Composites*, 14(5): 383–394.

67. Chan, A.W. and S.-T. Hwang, 1991. Anisotropic in-plane permeability of fabric media. *Polymer Engineering and Science*, 31(16): 1233–1239.

68. Scheidegger, A.E., 1974. *The Physics of Flow through Porous Media*, University of Toronto Press.

69. Greenkorn, R.A., 1983. *Flow Phenomena in Porous Media*, New York: Marcel Dekker.

70. Duplessis, J. and J. Masliyah, 1991. Flow through isotropic granular porous media. *Transport in Porous Media*, 6: 207–221.

71. Berdichevsky, A.L. and Z. Cai, 1993. *Polymer Composites*, 14: 132–143.

72. Bruschke, M.V. and Advani, S.G., 1993. Flow of generalized Newtonian fluids across a periodic array of cylinders. *Journal of Rheology*, 37(3): 479–498.
73. Gebart, B.R., 1992. Permeability of unidirectional reinforcements for RTM. *Journal of Composite Materials*, 26(8): 1100–1133.
74. Honarpour, M., L. Koederitz, and A.H. Harvey, 1986. *Relative Permeability of Petroleum Reservoirs*, Boca Raton, FL: CRC Press, Inc.
75. Christiansen, R.L., J.S. Kalbus, and S.M. Howarth, 1997. *Evaluation of Methods for Measuring Relative Permeability of Anhydrite from the Salado Formation: Sensitivity Analysis and Data Reduction*, SANDIA Labs, Report No. SAND94–1346 (Distribution Category UC-721).

5

Single-Phase Flow (Sharp-Interface) Models for Wicking

Reza Masoodi and Krishna M. Pillai

CONTENTS

In porous media studies, wicking is the spontaneous movement of a wetting liquid into a porous medium under the influence of capillary pressure. In this chapter, a relatively new approach of using the single-phase flow assumption behind a clearly defined liquid front in a porous medium is introduced to model the wicking process. Such an approach employs Darcy's law in conjunction with the continuity equation to model liquid flow behind the liquid front. In the first part of the chapter, governing equations along with pressure, velocity, and suction boundary conditions for modeling wicking in a rigid porous medium are presented. It is accompanied by details on experimental measurement of various properties used in the model. In the following part, the sharp-front model is expanded to include the effects of matrix swelling due to liquid absorption. In the last part, numerical simulation of liquid imbibition using the sharp-front model after employing the finite element/control volume method is introduced to model wicking in more complex wick-geometries, as well as to incorporate nonlinearities arising due to swelling of the solid matrix. Examples of analytical and numerical predictions compared with experimental results are added at the end of each section. The comparisons show that the sharp-front flow modeling approach is an accurate and robust method of predicting wicking that can be applied to many practical engineering applications.

5.1 Introduction

Wicking, which is the spontaneous suction of a wetting liquid into a porous medium, is also known as imbibition (i.e., displacement of a nonwetting phase (air) by a wetting phase (liquid)) in porous media literature. The driving force behind the wicking motion is a negative capillary (suction) pressure created at liquid–air interfaces. As the surfaces of solid particles in a porous medium become wet, the surface energies of particles change; for wetting liquids, this change in surface energies appears as a negative capillary pressure that draws liquid–air interfaces deeper into porous substrates and leads to the phenomenon of wicking.

In some porous materials, there is a clear, sharp boundary between the wet and dry regions. It means there is a clear liquid front across which the saturation* jumps sharply from 0% ahead of it to close to 100% after it, that is, the porous material is completely dry before the front reaches it, while becoming

---

* Saturation is the ratio of liquid volume in the pores to the total volume of the pores.

almost completely wet as soon as the front passes it. As an example, the clear liquid-front assumption is good in the case of wicking in thin, paper-like materials. When one end of a napkin is put in touch with a wetting liquid, a clear liquid front moves into the napkin (see Figure 5.1a). A similar example is seen during composites processing, in which the injection of a monomer thermosetting resin is characterized by the presence of a sharp front (see Figure 5.1b).

The existence of a sharp front during the imbibition (wicking) process may come as a surprise to some because of the common misconception that imbibition, due to its nature as a two-phase flow of a wetting fluid phase (such as water) displacing a nonwetting fluid phase (i.e., air), has to be marked by a gradation in saturation and, therefore, has to be governed by the traditional two-phase flow model in porous media. However, such a notion has been disproved in the remarkable study conducted by Lenormand et al. (1988) on the immiscible displacement of one fluid by the other conducted

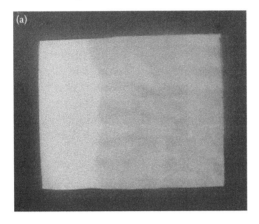

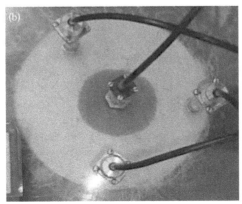

FIGURE 5.1
(a) An example of sharp front seen during the wicking of water in a paper napkin. (b) Another example of sharp front observed during the center-point injection of viscous resin in a mold packed with glass fibers.

with a two-dimensional network. It was observed that if the viscosity of the injected fluid is much higher than the viscosity of the displaced fluid (i.e., the mobility ratio is high) and if the velocity of the fluids is relatively high (i.e., the capillary number is relatively high), such an *immiscible displacement* is characterized by the presence of a sharp interface* between the injected and displaced fluids. In such cases, residual displaced-fluid pockets or "blobs" behind the front are only as large as a few pores. In other words, saturation behind the moving interface is close to 100%. The study also noted that principal forces during the displacement process occur due to the viscosity of the injected fluid, and the pressure drop in the displaced fluid can be considered negligible. Similar conclusions were reached in an independent network simulation conducted by Dias and Payatakes (1986). The advanced pore–network simulation of Chapter 9, conducted using the two-phase approach after including most pore-level phenomena arising at the liquid–air interface, also points to the existence of sharp front during wicking. But the study warns about the presence of tiny trapped bubbles behind the liquid front, as well as tiny fingers near the front. However, from a macroscopic viewpoint, ignoring these miniscule pore-level phenomena and treating the liquid front as sharp, and considering the region behind the front as fully saturated, is perhaps advisable—this will allow one to develop quick and convenient solutions to the wicking phenomena seen in industrial and natural porous media.

Under the sharp-front assumption, one can model wicking in porous media as a single-phase flow; such a flow model is also known as the "plug flow" or "piston flow" models in the literature. The sharp-front model makes flow modeling easy, robust, and reliable, especially in complex wick geometries. Extensive literature exist in the area of numerical modeling of such sharp-interface flows observed during the processing of polymer matrix composites (Tucker and Dessenberger 1994; Tan and Pillai 2010), as well as metal matrix composites (Mortensen et al. 1989; Tong and Khan 1996). Many of the modeling approaches described in this chapter are inspired by these similar modeling approaches employed by the composites processing community.

To model wicking as a single-phase flow, another approach is to use the traditional wicking models based on Lucas–Washburn (L–W) equation (Lucas 1918; Washburn 1921). As described in Chapter 3, the porous medium is assumed to be a set of parallel capillary tubes in this approach. The flow in each tube, due to a difference in pressure at the pipe entrance and the suction pressure at the liquid front, is modeled using the Hagen–Poiseuille equation (Sutera 1993). This model gives a linear relation between the absorbed liquid mass and the square root of time when the effect of gravity is neglected. Szekely et al. (1971) modified the L–W equation and included the inertial and

---

* Other flow regimes characterized by capillary and viscous fingers, which disrupt the stable displacement (sharp interface) regime, were also observed in the study. However, such fingering regimes were characterized by either low mobility ratios (i.e., the injected fluid has a lower viscosity compared with the displaced fluid) or low capillary numbers (i.e., the injection velocity is quite small).

gravity effects. In general, the L–W equation has been tested through experiments by several researchers and has been shown to be valid for most liquid-wicking porous materials (Kissa 1996; Chatterjee and Gupta 2002; Masoodi et al. 2008).

Although the L–W equation has been used widely for predicting wicking into porous substrates such as textiles (Chatterjee and Gupta 2002), it has some limitations. One of its main drawbacks is that it can predict only one-dimensional wicking flows, as it is based on Hagen–Poiseuille flow through a bundle of aligned, parallel capillary tubes. Therefore, the L–W equation *cannot* predict the 2D and 3D wicking flows occurring in wicks of complicated shapes. Moreover, since complicated and interconnected flow paths of liquids in a real porous medium are replaced by hypothetical straight and nonconnected flow paths through the parallel cylindrical tubes in the L–W equation, some type of fitting parameter in the form of hydraulic or capillary radius is invariably used to match model predictions with experimental results.

The other approach for handling wicking with sharp fronts is to model the single-phase wicking flow using Darcy's law, which gives a simple, linear relation between the average liquid velocity and the pressure drop in porous media. As shown in Chapter 4, the flow physics corresponding with Darcy's law is quite well established in the porous media literature, as it is regularly used to model flow in porous substances (Bear 1972; Scheidegger 1974; Dullien 1992). The wicking problem is treated as a quasi-steady boundary value problem, in which the pressure distribution in the fully saturated region behind the moving sharp front is solved through the Laplace equation, obtained after combining Darcy's law with the equation of continuity (Bear 1972). The capillary (suction) pressure is imposed at the sharp-front boundary as a pressure boundary condition. Later, the same pressure distribution is used to update the boundary using Darcy's law. Masoodi et al. (2007) used such a Darcy's law-based model for predicting wicking rate in polymer wicks created by sintering polymer beads. The discussion presented in Chapter 6 of this book mentions that the sharp-front models have enjoyed significant success while modeling wicking in radial coordinates.

In general, using the single-phase Darcy's law approach for modeling the wicking flow enables one to use the extensive published research available in the area of porous media flow and transport. Another advantage of this approach is the extension of the wicking flow modeling into 3D wicks: Masoodi et al. (2011) used POREFLOW©, a Fortran program based on the finite element/control volume (FE/CV) method (POREFLOW 2011), to model wicking flows in 2D or 3D porous media. Recently, Masoodi and Pillai adapted Darcy's law-based, single-phase flow approach for modeling wicking flows in swelling porous materials (Masoodi and Pillai 2010; Masoodi et al. 2012a). They modified the continuity equation to include the effects of swelling in solid matrix due to liquid absorption and considered the time-varying porosity and permeability functions in their wicking flow models.

In this chapter, details of the sharp-front flow model in rigid porous media are presented, followed by their experimental verification. The wicking in swelling porous materials is explored in the subsequent sections. Next, the numerical simulations of wicking in three-dimensional, rigid porous media and of wicking in swelling porous media are explained, followed by a comparison of the simulation results with some experimental data. Excellent validations of the numerical and analytical results demonstrate that the sharp-front flow modeling approach using Darcy's law is an accurate method of modeling flows in wicks that can be applied to many practical engineering applications.

5.2 Wicking in Rigid Porous Materials

5.2.1 Governing Equations

As mentioned in Chapter 4, governing equations for the single-phase flow of a Newtonian liquid in an isotropic and rigid porous medium under isothermal conditions are Darcy's law and continuity equation, respectively:

$$\langle \overrightarrow{V_f} \rangle = -\frac{K}{\mu} \nabla \langle P_f \rangle^f \tag{5.1}$$

$$\nabla \cdot \langle \overrightarrow{V_f} \rangle = 0 \tag{5.2}$$

Here, $\langle \overrightarrow{V_f} \rangle$ and $\langle P_f \rangle^f$ are the volume-averaged liquid velocity and pore-averaged modified pressure,* respectively, and K is the permeability of the porous medium (Bear 1972; Tucker and Dessenberger 1994; Whitaker 1998). The modified liquid pressure P_f, which includes the effect of hydrostatic variation in pressure, is defined as

$$P_f = p_f + \rho g h \tag{5.3}$$

where h is the height of a point within a saturated (or fully wetted) porous medium (Masoodi et al. 2007). If the permeability and viscosity are consid-

---

* Recall (from Chapter 4) that the volume average (or the phase average) and the pore average (or the intrinsic phase average) for any flow-related quantity q_f in a porous medium are defined, respectively, as $\langle q_f \rangle = 1/Vol \int_{Vol_f} q_f \, dV$ and $\langle q_f \rangle^f = 1/Vol_f \int_{Vol_f} q_f \, dV$ where q_f is integrated over an averaging volume (Vol) called the representative elementary volume or REV.

ered to be constants, then a combination of Equation 5.1 with Equation 5.2 leads to the Laplace equation for the modified pressure as

$$\nabla^2 \left\langle P_f \right\rangle^f = 0 \qquad (5.4)$$

The strategy would be to solve Equation 5.4 quasi-statically in the wet region of a wick using appropriate boundary conditions. Then, use the pore-averaged pressure distribution to predict the velocity of the sharp interface at any given time using Equation 5.1.

5.2.2 Example of an Application: Wicking in a Polymer Wick

In a recent study, Masoodi et al. (2007) studied wicking in porous wicks made from sintered polymer beads. Using a microbalance, the cylindrical wick was hung vertically over a pool of the wicking liquid such that the liquid could be imbibed up after a contact between the wick bottom and the liquid has been established. In approximately two-thirds of the wicks, a clear liquid front could be seen traveling up the wick. For such wicks, change in wick weight with time as measured by the microbalance was translated into change in the liquid-front height as a function of time using the sharp liquid-front approximation with 100% saturation behind the liquid front. Wicks made from three different polymers (polycarbonate (PC), polyethylene (PE), and polypropylene (PP)) were used with three different wicking liquids (hexadecane (HDEC), decane (DEC), and dodecane (DDEC)) in the experimental study.

In this section, we will present a simple model based on the governing equations described in Section 5.2.1 for predicting flow-front height as a function of time. A schematic of the flow domain, as well as wick geometry, is given in Figure 5.2.

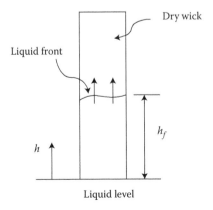

FIGURE 5.2
A schematic of the one-dimensional flow occurring in a cylindrical wick.

If we temporarily drop the brackets signifying the pore average in pressures as well as the subscript f, then using Equation 5.4 for the 1D flow shown in Figure 5.2, the pressure distribution in the wetted region of the vertical wick is given by

$$\frac{d^2P}{dh^2} = 0 \tag{5.5}$$

where h is the vertical coordinate with respect to the liquid level. Integration of Equation 5.5 yields a general solution of the form $P(h) = Ah + B$, where the constants A and B are estimated using the boundary conditions

$$p = p_{atm} \quad \text{at } h = 0 \tag{5.6a}$$

and

$$p = p_{atm} - p_s \quad \text{at } h = h_f \tag{5.6b}$$

which are in terms of the pore-averaged hydrodynamic pressure p and the atmospheric pressure p_{atm}. Note that p_s is the suction pressure generated at the liquid front due to capillary action after assuming the presence of a sharp interface at $h = h_f$. Equations 5.6a and 5.6b can be presented in terms of the modified pressure P as

$$P = p_{atm} \quad \text{at } h = 0 \tag{5.7a}$$

$$P = (p_{atm} - p_s) + \rho g h_f \quad \text{at } h = h_f \tag{5.7b}$$

Using these boundary conditions with the general solution of Equation 5.5 leads to the following expression for pressure distribution in the wetted region of the wick, that is, in $0 \le h \le h_f(t)$:

$$P(h) = p_{atm} + \rho g h - p_s \frac{h}{h_f} \tag{5.8}$$

Note that the front height h_f changes with time and needs to be determined. An expression for h_f can be developed by relating front speed dh_f/dt to the Darcy velocity (obtained from Equation 5.1) at front h_f as

$$\frac{dh_f}{dt} = \frac{V(h = h_f)}{\varepsilon} = -\frac{K}{\varepsilon \mu} \frac{dP}{dh} \tag{5.9}$$

Here, ε is the porosity of the porous medium, which is defined as the ratio of pore volume to the total REV volume in the porous substance. The use of Equation 5.8 changes Equation 5.9 to

$$\frac{dh_f}{dt} = \frac{K}{\varepsilon\mu}\left(\frac{p_s}{h_f} - \rho g\right)$$ (5.10)

On integrating while using the separation-of-variables technique and employing the initial value $h_f(t = 0) = 0$, we get

$$p_s \ln\left|\frac{p_s}{p_s - \rho g h_f}\right| - \rho g h_f = \frac{\rho^2 g^2 K}{\varepsilon\mu}t$$ (5.11)

One may neglect the effect of gravity on wicking in smaller samples. After ignoring the gravity term, integration of Equation 5.10 results in

$$h_f = \sqrt{\frac{2Kp_s}{\varepsilon\mu}t}$$ (5.12)

the form of which is similar to the L–W equation (see Chapter 3) for wicking in the absence of gravity.

Note that although we developed closed-form solutions to predict the wicking-front height, h_f, as a function of time (and thus developed the ability to predict the mass of liquid wicked into a wick as a function of time), the solutions are incomplete since the capillary suction pressure, p_s, is still unknown for porous wicks with their nontubular, convoluted microstructures. In the following section, we briefly describe a method for predicting p_s for different types of porous media.

5.2.3 Capillary Suction Pressure

The difference in the surface energies of the dry and wet solid matrices in a porous medium leads to the creation of a capillary pressure at the liquid front, which is responsible for the spontaneous sucking of wetting liquids into a porous medium. The Young–Laplace equation, as described in Chapter 1, is a well-known relation for capillary pressure that describes the capillary difference across the interface between two stationary immiscible fluids (Berg 1993; Finn 1999). The Young–Laplace equation was originally developed for estimating capillary pressure in the capillary tubes (Finn 1999); however, when applied to a general porous substrate, the capillary radius refers to the radius of the imaginary capillary tubes constituting the porous medium.

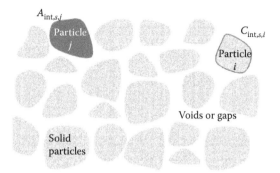

FIGURE 5.3
A schematic of particles and void areas in a cross section of the porous medium.

Moving away from the limitations imposed by the imagery of a porous medium as a bundle of capillary tubes, we will now discuss a general formula for the capillary suction pressure that was proposed by Masoodi and Pillai (2012) based on the cross-section area and perimeter of particles as traversed by the flat liquid–air interface during its motion through a porous substrate. Assume the porous medium is composed of solid particles of different sizes such that a hypothetical sharp liquid–air interface slicing the medium creates a cross-section, as shown in Figure 5.3. If $C_{int,s,i}$ is the cross-sectional perimeter of solid particle i at the interface, then $C_{int,s}$, the total cross-sectional perimeter of solid particles at the interface, can be expressed as $C_{int,s} = \Sigma C_{int,s,i}$, where the summation is over all the particles intersecting the liquid–air interface. It can be shown that the cross-sectional area of solid particles, $A_{int,s} = \Sigma A_{int,s,i}$, and the cross-sectional area of the void space, $A_{int,v} = \Sigma A_{int,v,j}$, are related through the expression

$$A_{int,v} = \frac{\varepsilon}{1-\varepsilon} A_{int,s} \qquad (5.13)$$

where ε is area porosity.[*]

Masoodi and Pillai (2012) equated the work done by capillary pressure equal to the interfacial energy released during the wetting of the porous medium, shown in Figure 5.3, by the sharp liquid–air interface traversing through the porous medium. As a result, the final relation for the capillary pressure is obtained as

$$p_s = \frac{2\gamma_l \cos(\theta)}{R_e} \qquad (5.14)$$

---

[*] The "area" porosity averaged over a length, perpendicular to the area, is identical to the volume-based porosity (Bear 1972).

where R_e is the equivalent capillary radius found through the expression

$$R_e = 2\frac{A_{int,s}}{C_{int,s}}\frac{\varepsilon}{1-\varepsilon} \tag{5.15}$$

Equation 5.14 is similar to the Young–Laplace expression for capillary pressure, while Equation 5.15 relates the local equivalent capillary radius to the microstructural features of the considered porous medium. Using Equation 5.13, one can relate the equivalent capillary radius to the geometry of void spaces through

$$R_e = 2\frac{A_{int,v}}{C_{int,v}} \tag{5.16}$$

Note that Equation 5.16 resembles the definition of the hydraulic radius. These derived formulas for suction pressure and equivalent capillary radius can be used wherever the traditional Young–Laplace equation for capillary pressure is applicable. Note that Equation 5.14 in conjunction with Equation 5.15 or Equation 5.16 is most appropriate for estimating the capillary suction pressure in any porous medium where 2D micrograph can be prepared from a thin slice of the medium.

However, Equation 5.14 in conjunction with Equation 5.15 can also be used to derive simple expressions for capillary suction pressure as applicable to wicking flows in different porous media made of simple microstructures. See Table 5.1 for a description of such expressions.

5.2.4 Measuring Wicking Parameters

In order to compare the prediction of the wicking models given by Equations 5.11 and 5.12 with experiments, properties of the liquids and wicks used were measured *independently* using the methods described in the present section (Masoodi et al. 2007). Most prominent of these parameters include porosity (ε), capillary radius (R_c), saturated permeability (K), particle size (R_s), liquid surface tension (γ_l), and dynamic contact angle (θ).

5.2.4.1 Porosity (ε)

The gravimetric test method can be used to find the mean porosity of a porous wick. In the adaptation of this method employed by Masoodi et al. (2007), the mass of the dry cylindrical wick is measured first. Then the wick is placed vertically such that its lower end is in contact with a liquid reservoir. As a result, the wicking liquid starts to move into the wick and the height of the wet region behind the liquid front increases with time. After a while, the liquid front reaches a steady-state location (L_{ss}), in which the force due to

TABLE 5.1

Experimentally Validated Expressions for the Capillary Suction Pressure, Developed at the Sharp Interface on the Head of a Wicking Flow, for Different Porous Media Microstructures

---

1. Flow along a bank of parallel, uniradial capillary tubes

$$A_{int,v} = n\pi R_c^2, C_{int,v} = n(2\pi R_c) \quad \text{such that } p_s = 2\frac{\gamma_l\cos\theta}{R_c}$$

2. Flow along a bank of parallel, uniradial fibers

$$A_{int,s} = n\pi R_{fb}^2, C_{int,s} = 2n\pi R_{fb} \quad \text{such that } p_s = 2\frac{(1-\varepsilon)}{\varepsilon}\frac{\gamma_l\cos\theta}{R_{fb}}$$

3. Flow across a bank of parallel, uniradial fibers

$$A_{int,s} = 2nl_{fb}R_{fb}, C_{int,s} = 2n(l_{fb} + 2R_{fb})[l_{fb} \gg 2R_{fb}] \quad \text{such that } p_s = \frac{(1-\varepsilon)}{\varepsilon}\frac{\gamma_l\cos\theta}{R_{fb}}$$

4. Flow in a porous medium made of multiradial spherical particles

$$A_{int,s} = \frac{\frac{4}{3}\pi\int_0^\infty r^3\varphi(r)dr}{L}, C_{int,s} = \frac{4\pi\int_0^\infty r^2\varphi(r)dr}{L} \quad \text{such that } p_s = 3\frac{(1-\varepsilon)}{\varepsilon}\frac{\int_0^\infty r^2\varphi(r)dr}{\int_0^\infty r^3\varphi(r)dr}\gamma_l\cos\theta$$

5. Flow in a porous medium made of uniradial spherical particles

$$A_{int,s} = \frac{n\frac{4}{3}\pi R_p^3}{L}, C_{int,s} = \frac{n4\pi R_p^2}{L} \quad \text{such that } p_s = 3\frac{(1-\varepsilon)}{\varepsilon}\frac{\gamma_l\cos\theta}{R_p}$$

Source: Masoodi, R. and Pillai, K.M. 2012. *Journal of Porous Media* 15(8): 775–783.
Note: n, number of capillary tubes or fibers intersecting the liquid–air interface; R_c, capillary radius; R_{fb}, radius of uniradial fibers; R_p, radius of uniradial particles; r, fiber radius; l_{fb}, length of uniradial fibers; φ, probability density function; L, length of the imaginary interfacial volume, created by extruding the interfacial area of Figure 5.3 along the length L.

capillary pressure balances the weight of the liquid column. Now, the difference between the current mass of the wick (wet mass) and the previously recorded mass (dry mass) is the liquid mass inside the pores under full saturation, m_{sat}. If A_{cs} is the cross-sectional area of the wick, one can use the expression

$$\varepsilon = \frac{m_{sat}}{\rho A_{cs}L_{ss}} \tag{5.17}$$

to estimate the wick porosity.

5.2.4.2 Capillary Radius (R_c)

For the suction pressure used in Equations 5.11 and 5.12, Masoodi et al. (2007) employed the capillary radius-based model described through Equation 1 of Table 5.1. The capillary radius is estimated from balancing the upward-pulling suction force in hypothetical capillary tubes by the downward-acting force of gravity in a porous wick. The test is exactly similar to the gravimetric experiment described previously for measuring porosity. After reaching the steady-state height, the capillary pressure balances the gravity "pressure," and the capillary radius can be estimated (Masoodi et al. 2008) using the formula

$$R_c = \frac{2\gamma \cos(\theta)}{\rho g L_{ss}} \tag{5.18}$$

5.2.4.3 Saturated Permeability (K)

Permeability is a measure of the ease with which a liquid can flow through a porous medium, and it is an important parameter in Darcy's law, Equation 5.1. It is one of the most important wicking parameters that should be measured accurately. As mentioned in Chapter 4, the fixed head and falling head permeameters are simple and effective devices for measuring the saturated permeability of porous materials such as polymer or ceramic wicks. Masoodi et al. (2007) created a falling head permeameter out of a syringe and a measuring cylinder to measure the saturation permeability in the polymer wicks (Figure 5.4). Assuming a steady Darcy flow in the wick during the

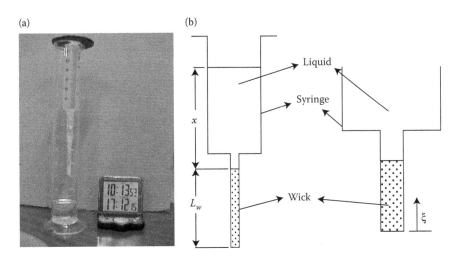

FIGURE 5.4
A photo and schematic of the falling head permeameter used for measuring the permeability of the polymer wicks. (From Masoodi, R., K.M. Pillai, and P. Varanasi. 2007. *Journal of AIChE* 53: 2769–2782. With permission.)

experiment, the following expression can be easily derived (Masoodi et al. 2007) for the setup shown in Figure 5.4:

$$\ln\left(\frac{x_0}{x}\right) = \frac{K\rho g R_w^2}{\mu L_w R_{sr}^2} t \qquad (5.19)$$

Here, R_w and R_{sr} refer to the radii of the wick and syringe, respectively. The slope of $\ln(x_0/x)$ versus time t plot yields the value of K once other parameters in the slope are substituted.*

5.2.4.4 Particle or Fiber Diameters (R_p or R_{fb})

For the suction pressure used in Equations 5.11 and 5.12, Masoodi et al. (2007) employed the spherical particle-based model described through Equation 4 of Table 5.1, in which micrography was used to measure the particle size distribution in a porous wick sintered out of polymer beads. Another method yielding this information is mercury porosimetry, which is not appropriate for soft, flexible materials since high pressure imposed on the porous matrix may lead to the deformation of the materials (Giesche 2006). Some researchers have also used nondestructive test methods, such as x-ray tomography, to characterize a porous substance (Al-Raoush and Willson 2005; Al-Raoush and Alshibli 2006).

5.2.4.5 Liquid Surface Tension (γ_l)

Surface tension is a property of a liquid surface that allows it to resist stretching. There are several methods that can be used to measure the surface tension of liquids, such as duNouy rod, DuNouy ring, Wilhelmy plate, Pendant drop, Spinning drop, and Bubble pressure (Berg 1993). Masoodi et al. (2007) used the Wilhelmy plate method to measure the surface tension of liquids in their experiments on wicking.

5.2.4.6 Dynamic Contact Angle (θ) in a Porous Medium

A tensiometer or a dynamic contact angle analyzer (DCA) is traditionally employed to measure the contact angle for any solid–liquid system on a flat surface using the Wilhelmy plate method. However, DCA can also be adapted to measure the average contact angle within the complicated pore geometry of a porous wick. Consider a porous wick with a constant cross-sectional

---

* If the porous substrate where the wicking is taking place is anisotropic, then the test will yield one-directional permeability. To decipher the full permeability tensor, additional directional permeabilities need to be determined (see Chapter 4), which can be obtained by rotating the substrate by known angles.

area of A_{cs}, which is suspended by a DCA over the wicking liquid. After combining Equations 5.12 and 5.14, the following expressions are obtained for liquid-front height and absorbed liquid-mass as functions of time:

$$L_{lf} = \left(\frac{4K\gamma\cos\theta}{\epsilon\mu R_e} \right)^{1/2} \sqrt{t} \tag{5.20}$$

$$m = \epsilon\rho A_{cs}L_{lf} \tag{5.21}$$

Eliminating L_{lf} from these two equations leads to

$$m^2 = \frac{\lambda\rho^2\gamma\cos\theta}{\mu}t \tag{5.22}$$

where λ, the material constant, is equal to

$$\lambda = \frac{4\epsilon A_{cs}^2 K}{R_e} \tag{5.23}$$

The measurement of contact angle is a two-step process. In the first step, a test liquid of *known* contact angle should be used.* The mass of liquid wicked as a function of time for the liquid should then be measured. Later, the viscosity, surface tension, and density of the test liquid are measured. Then, a plot of absorbed mass squared (m^2) versus time has to be created. In such a plot, the slope of the fitted line is identical to the coefficient of time in the right-hand side of Equation 5.22. As the liquid properties and contact angle are known, the material constant λ can then be estimated. In the second step, the test is to be repeated with the liquid selected for wicking and a fitted line should be developed for the m^2 versus t plot. This time, all parameters in the slope of the fitted line (i.e., $\lambda\rho^2\gamma\cos\theta/\mu$) are known except for θ; thus, the average contact angle is determined.

This method is valid when λ remains unchanged for both the steps. According to Equation 5.23, the material constant λ is a geometry-based term that is a function of the macrogeometric parameters (A_{cs}) as well as the microgeometric parameters (ϵ, R_e, and K). Since one needs to use two different wicks for the two steps, the wicks drawn from a batch should display almost identical geometrical parameters—such a consistency is achieved easily in industrially produced wicks. Note that the effect of gravity is

---

* A practical method is to use a liquid with its surface energy being much less than that of solid matrix; in such a situation, the contact angle will be very small and can be considered to be zero. For example, one can use hexane to measure the contact angle of polymer wicks (Masoodi et al. 2007).

neglected in the analysis; therefore, this method is not valid if the location of the measured liquid front reaches the maximum possible wicking height (L_{ss}), that is, $L_{lf} \ll L_{ss}$. In addition, Equation 5.22 can also be obtained from the L–W equation; in that case, the expression for the material constant λ will be different from the one presented in Equation 5.23 (Masoodi et al. 2008).

5.2.5 Validating the Sharp-Front Wicking Flow Model

Figure 5.5 compares the predictions of Equation 5.11 (in conjunction with Equation 4 of Table 5.1) with the early-time experimental data for the polyethylene wick—it is clear the sharp-front model predicts the experimentally measured trend remarkably well. Similar trends of very good predictions for the polycarbonate and polypropylene wicks are reported elsewhere (Masoodi et al. 2007). In these comparisons, it was also observed that the sharp-front models are not only accurate but also behave much better than the capillary model or the L–W equation. An important fact to remember is that all the parameters used in the sharp-front model were independently measured—*none of them was used as a fitting parameter*. This makes the sharp-front model an accurate and rigorously verifiable method to predict liquid flows in wicks.

5.3 Wicking in Swelling Porous Materials

In order to use the sharp-front flow model in swelling porous media, some modifications are necessary since the pore size, porosity, and permeability change as a result of swelling of the solid matrix due to liquid absorption. In fact, these parameters are functions of both time and space in a swelling porous medium. This section describes a new approach to modify the governing equations in rigid porous media for modeling wicking in swelling porous media.

5.3.1 Governing Equations

The governing equations for single-phase flow through a porous medium under isothermal conditions are Darcy's law and the continuity equation. These equations have traditionally been tested and theoretically derived for rigid porous media with the constituent particles (or fibers) remaining unchanged and nonabsorbing during the course of the flow. However, it is generally acknowledged, at least theoretically, that the form of these equations will change if one considers a porous medium that absorbs liquid and swells during the course of the flow (Gray and O'Neill 1976; Preziosi and Farina 2002). First, the modifications in the continuity equation will be discussed.

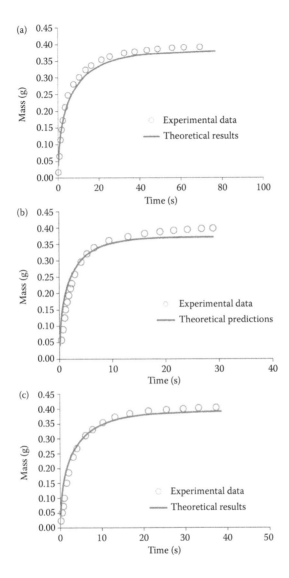

FIGURE 5.5
Mass absorption history in a polyethylene wick using three different alkanes: (a) Hexadecane, (b) decane, and (c) dodecane.

5.3.1.1 Mass Balance (Continuity) Equation

The point-wise continuity equation for single-phase flow through pores of a porous medium can be expressed as

$$\frac{\partial \rho_f}{\partial t} + \nabla \cdot (\rho_f \vec{V}_f) = 0 \tag{5.24}$$

where ρ_f and $\vec{V}_f$ are the density and velocity of the fluid phase, respectively (Tucker and Dessenberger 1994). The volume averaging of Equation 5.24 in a term-by-term fashion within a representative elementary volume (REV) leads to*

$$\left\langle \frac{\partial \rho_f}{\partial t} \right\rangle + \left\langle \nabla.(\rho_f \vec{V}_f) \right\rangle = 0 \tag{5.25}$$

For any fluid-phase quantity q_f in a porous medium, the two averaging theorems that are used in the volume averaging method can be expressed (Gray and Lee 1977) as

$$\left\langle \nabla q_f \right\rangle = \nabla \left\langle q_f \right\rangle + \frac{1}{Vol} \int_{A_{fs}} q_f \vec{n}_{fs}\, \mathrm{d}A \tag{5.26}$$

$$\left\langle \frac{\partial q_f}{\partial t} \right\rangle = \frac{\partial \left\langle q_f \right\rangle}{\partial t} - \frac{1}{Vol} \int_{A_{fs}} q_f \vec{V}_{fs}.\vec{n}_{fs}\, \mathrm{d}A \tag{5.27}$$

where A_{fs} is the fluid–solid interface area, $\vec{n}_{fs}$ is the unit normal vector on the fluid–solid interface pointing to the solid phase, and $\vec{V}_{fs}$ is the velocity of fluid–solid interface (see Figure 5.6). The first and second terms on the right-hand side of Equation 5.25 can be expanded using the average theorems, Equations 5.26 and 5.27, as

$$\frac{\partial \left\langle \rho_f \right\rangle}{\partial t} - \frac{1}{Vol} \int_{A_{fs}} \rho_f \vec{V}_{fs}.\vec{n}_{fs}\, \mathrm{d}A + \nabla.\left\langle \rho_f \vec{V}_f \right\rangle + \frac{1}{Vol} \int_{A_{fs}} \rho_f \vec{V}_f.\vec{n}_{fs}\, \mathrm{d}A = 0 \tag{5.28}$$

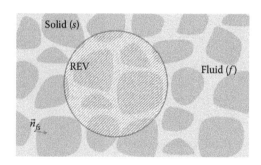

FIGURE 5.6
A schematic showing a representative elementary volume (REV) containing solid particles and fluid-filled pores in a porous medium.

---

* See Chapter 4 for a brief introduction to the volume averaging method.

After rearranging Equation 5.28, the continuity equation is expressed in the form

$$\frac{\partial \langle \rho_f \rangle}{\partial t} + \nabla \cdot \langle \rho_f \vec{V}_f \rangle + \frac{1}{Vol} \int_{A_{fs}} \rho_f (\vec{V}_f - \vec{V}_{fs}) . \vec{n}_{fs} \, dA = 0 \qquad (5.29)$$

where $\vec{V}_f - \vec{V}_{fs}$ is the relative velocity of the fluid with respect to the fluid–solid interface. In other words, the surface integral, which is evaluated on the surface of fluid–solid interface, is the rate of absorption (or release) of fluid by the solid matrix of the porous medium. Equation 5.29 is the general form of the continuity equation for single-phase flows in a deforming and swelling porous media. For an *incompressible* liquid, the density can be canceled out from Equation 5.29 and the continuity equation reduces to

$$\frac{\partial \langle 1 \rangle}{\partial t} + \nabla \cdot \langle \vec{V}_f \rangle + S = 0 \qquad (5.30)$$

where

$$S = \frac{1}{Vol} \int_{A_{fs}} (\vec{V}_f - \vec{V}_{fs}) \vec{n}_{fs} \, dA \qquad (5.31)$$

The *sink* term S^* is defined to be the ratio of volumetric rate of liquid absorption within an REV to the total volume of the REV. Using the definition of phase average given in Chapter 4 $(q_f = 1/V \int_{V_f} q_f \, dV)$, it is easy to show that $\langle 1 \rangle = \varepsilon$. As a result, the final form of the continuity equation for flow of an incompressible liquid in a swelling, liquid-absorbing porous medium reduces to

$$\nabla \cdot \langle \vec{V}_f \rangle = -S - \frac{\partial \varepsilon}{\partial t} \qquad (5.32)$$

Since S is the rate of liquid absorption by the solid matrix and dv_s/dt is directly related to the rate of increase of solid-matrix volume (v_s is defined as the ratio of solid volume to the total volume in an REV) Masoodi and Pillai (2010) proposed that S be proportional to dv_s/dt. Therefore, S can be expressed as

$$S = b \frac{dv_s}{dt} \qquad (5.33)$$

---

* It can also be a source term if the liquid effuses out of the solid matrix.

where $b = 0$ implies "no absorption of liquid by solid particles" and $b = 1$ implies "the rate of increase of solid-matrix volume is equal to the volumetric rate of liquid absorption by solid matrix." The constant of proportionality b, named *absorption coefficient*, must fall in the range $0 \leq b \leq 1$.* Note that the solid and fluid volume fractions in a porous medium are related through the expression

$$v_s + \varepsilon_f = 1 \qquad (5.34)$$

Substituting Equation 5.33 in Equation 5.32, and using Equation 5.34 to eliminate v_s, leads to the following relation for the continuity equation:

$$\nabla \cdot \left\langle \vec{V}_f \right\rangle = (b - 1)\frac{\partial \varepsilon_f}{\partial t} \qquad (5.35)$$

Masoodi and Pillai (2010) showed that $b = 1$ (or b being very close to 1) is valid for paper-like materials made of cellulose and superabsorbent fibers.† If b is *assumed* to be unity, the right side of Equation 5.35 vanishes and it reduces to Equation 5.2, the "regular" continuity equation for rigid, non-swelling porous media. Therefore, one can use Equation 5.2 as the continuity equation for swelling, liquid-absorbing porous media, as well.

5.3.1.2 Momentum Balance Equation

Unlike the derivation of the macroscopic mass balance equation, in which the application of the volume averaging method, after including the particle-surface level effects, easily yields eminently usable forms of the macroscopic continuity equation, the derivation of the macroscopic momentum balance equation is quite complex and beyond the scope of this chapter. However, we will summarize a few important developments in this area in order to propose an appropriate momentum balance equation for single-phase flows in a liquid-absorbing, swelling porous medium.

Gray and O'Neill (1976) applied the volume averaging method to obtain macroscopic momentum balance equations in a porous medium with a slowly deforming solid matrix. More recently, Preziosi and Farina (2002) applied the mixture theory to develop the same equations for growing porous media that absorb liquid, as well. For our case of the slow wicking flows where the inertial terms can be neglected, and where the mass exchange between

---

* One can allow b to be greater than 1 if the increase in solid volume is caused by some effect other than the migration of the liquid.

† $b = 1$ also works well for modeling forced flows of liquids through natural fiber-mats that swell on coming in contact with liquids (Masoodi and Pillai, 2011).

the liquid and solid phases is not that extreme, both these approaches suggest that macroscopic momentum balance equation yields Darcy's law of the form

$$\left\langle \vec{V}_f \right\rangle^f - \left\langle \vec{V}_s \right\rangle^s = -\frac{\mathbf{K}}{\varepsilon\mu} \cdot \left(\nabla \left\langle p_f \right\rangle^f - \rho\vec{g} \right)$$
(5.36)

where $\left\langle \vec{V}_f \right\rangle^f$ and $\left\langle \vec{V}_s \right\rangle^s$ are the intrinsic phase-averaged velocities of the fluid and solid phases, while $\left\langle p_f \right\rangle^f$ is the intrinsic phase-averaged fluid pressure. The other symbols retain their meanings as described previously. If one assumes that particle swelling is a local phenomena that causes only local changes in porosity and does not induce any translatory motion in the solid matrix, then one can impose the reasonable condition that $\left\langle \vec{V}_s \right\rangle^s = 0$. On noting that the Darcy or phase-averaged velocity is given by the relation $\left\langle \vec{V}_f \right\rangle = \varepsilon\left\langle \vec{V}_f \right\rangle^f$ as well as the fact that wicking media considered in this chapter are isotropic, Equation 5.36 reduces to Equation 5.1. This means that the classical form of Darcy's law as presented in Equation 5.1 is still applicable to the nonrigid, swelling porous materials. However, it is subject to the condition that the local permeability is changing gradually as a function of "wetting" time, that is, time since the onset of wetting of the local point— this means that permeability is a function of both time and space in the wetted region behind the liquid front. (Using the volume averaging method, Whitaker (1986) set up the theoretical determination of the permeability tensor through a closure problem, in which the position of the fluid–solid interface within an REV must be determined. However, we will be adopting a less complicated and more practical approach, in which the conventional permeability models as functions of the local time-varying porosity will be employed to predict the permeability as a function of time and space within the wetted region.)

5.3.2 Changes in Permeability and Porosity Due to the Swelling of Solid Matrix

When the liquid front passes a location in a porous medium, the local solid matrix made of particles or fibers gets wet and starts swelling, leading to a continuous reduction of local porosity as a function of the "wetting" time (i.e., the time the matrix has been wet). If t_{wet} is the time that a local segment of the porous wick has been made wet by the invading liquid front, then the following general formula holds for the local porosity:

$$\varepsilon = \begin{cases} \varepsilon_0 & t < t_{wet} \\ fn(t - t_{wet}) & t \geq t_{wet} \end{cases}$$
(5.37)

Here, fn is an unknown function while ε_0 is the initial local porosity, equal to the local porosity of the dry porous medium. Masoodi et al. (2012b) derived an expression for the local porosity changes in woven fibrous materials as

$$\varepsilon = \begin{cases} \varepsilon_0 & t < t_{wet} \\ 1 - (1 - \varepsilon_0)\left(\dfrac{D}{D_0}\right)^2 & t \geq t_{wet} \end{cases} \qquad (5.38)$$

where D is the changing fiber diameter that can be measured as a function of the wetting time using a microscope, while D_0 is the initial fiber diameter. This equation was found to be valid for up to 10% change in the initial fiber diameter (Masoodi et al. 2012b). Micrography, magnetic resonance imaging (MRI) (Leisen and Beckham 2008), and x-ray microtomography (Al-Raoush and Alshibli 2006) may be some of the techniques available for measuring local porosities; however, little published literature is available on accurately quantifying the real-time changes in porosity.

Swelling of the solid matrix in a porous medium also leads to a change in the local permeability due to a reduction in the local porosity. As was done for the porosity, one can write the general expression for local permeability changes in a swelling porous medium as

$$K = \begin{cases} K_0 & t < t_{wet} \\ fn(t - t_{wet}) & t \geq t_{wet} \end{cases} \qquad (5.39)$$

where fn is an unknown function of wetting time and K_0 is the local permeability of the dry porous medium. The accuracy of any theoretical or numerical wicking model highly depends on the accuracy of the estimated function for permeability. Masoodi et al. (2012b) used the empirical Kozeny–Carman formula for permeability and derived the following expression for local permeability changes in a swelling porous medium made of natural fibers.

$$K = \begin{cases} K_0 & t < t_{wet} \\ K_0 \left(\dfrac{1 - (1 - \varepsilon_0)(D/D_0)^2}{\varepsilon_0}\right)^3 \left(\dfrac{D_0}{D}\right)^2 & t \geq t_{wet} \end{cases} \qquad (5.40)$$

Masoodi et al. (2012b) observed that direct measurement of permeability changes is the most accurate method to predict local permeability changes during forced flows. They observed that a deterministic expression, such as Equation 5.40, is valid until 10% swelling of the individual fibers.

5.3.3 Application Example: Wicking in Paper-Like Swelling Porous Media

Schuchardt and Berg (1990) studied the swelling in paper-like porous media experimentally and modified the L–W equation for modeling wicking in such swelling materials. Their model was built on the assumption that capillary radii in a porous medium decrease linearly with time as a result of swelling. The predictions of the modified L–W equation compared better with the experimental data than the conventional L–W model.

Masoodi and Pillai (2010) used Darcy's law, Equation 5.1, in conjunction with the continuity equation, Equation 5.35, for the 1D flow, shown in Figure 5.7, to predict wicking rates in paper-like swelling porous media after employing the steps similar to those shown in Equations 5.5 through 5.10. Taking the parameter b to be unity, an analytical expression

$$L_{lf} = \sqrt{\frac{2p_s}{\varepsilon_{f0}\mu} \int_0^t K(t')\, dt'} \tag{5.41}$$

for the liquid-front location was developed after imposing the assumptions that (1) gravity is neglected and (2) the permeability K in the wetted region is homogeneous and only changes with time, that is, $K = K(t)$.

Masoodi and Pillai (2010) used different empirical and theoretical formulas for permeability in terms of the size of solid particles/fibers and the porosity (some of the formulas are described in Chapter 4) in Equation 5.41 to predict the wicking and compare it with the experimental data presented by Schuchardt and Berg (1990). Note the change in porosity was estimated from the reported data on the change in hydraulic radius with time.

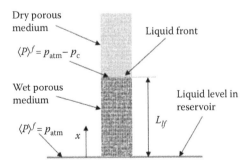

FIGURE 5.7
A schematic showing the wicking (liquid-front) height in a swelling paper strip after employing the sharp-front assumption. (From Masoodi, R. and K.M. Pillai. 2010. *AIChE Journal* 56(9): 2257–2267. With permission.)

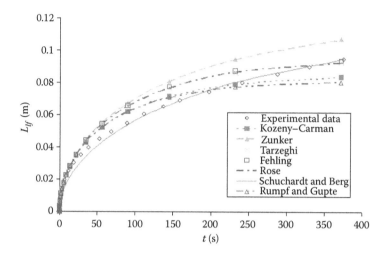

FIGURE 5.8

Prediction of wicking height as a function of time for the wicking of water in a paper strip made from 13%FC CMC (a superabsorbent) mixed with cellulose. The analytical predictions employ Equation 5.41, the Darcy's law-based model using different formulas for the permeability of *particulate* porous media. The predicted values are compared with the modified Washburn equation predictions and the experimental data of Schuchardt and Berg (1990). (From Masoodi, R. and K.M. Pillai. 2010. *AIChE Journal* 56(9): 2257–2267. With permission.)

The sharp-front model based on Darcy's law and the traditional form of continuity equation ($b = 1$ case) for different models of permeability is compared with the experimental results in Figure 5.8. It is clear that Schuchardt and Bergs' model based on the modified Washburn equation matches with the experimental data better that any other model.* (This is not surprising, since the model gets its values of its hydraulic radius parameters from fitting the experimental data.) However, the sharp-front model with different models of permeability exhibits a decent comparison with the experimental results. The formulas by Terzaghi, Fehling, and Rose best compare with the experimental data, although the predictions from other formulas are also reasonable. It is remarkable that the model based on an analytical solution predicts wicking with a fair degree of accuracy, despite suffering two limiting assumptions: (1) exclusion of gravity and (2) homogeneous permeability in the wet zone behind the liquid front. However, this sets the stage for a numerical solution of the sharp-front model described in the next section, in which the gravity and nonhomogeneous permeability distribution in the wet zone can be easily included.

---

* The sharp-front model gives the exact same predictions as the modified Washburn equation if a permeability model based on the capillary-tube approximation adopted by Schuchardt and Berg is employed (Masoodi and Pillai 2010).

5.4 Numerical Simulation of Wicking Flow

In many practical wicking applications, the governing equations for the sharp-front model for the wicking of liquid into porous substrates, Equations 5.1 and 5.2 [or (5.35) with $b = 1$], have to be solved numerically. The main reasons for using numerical simulations are (a) complexities in the geometry of a 3D wick where the simple 1D wicking flow equations given by the L–W equation, as well as the sharp-front analytical solutions (Equations 5.11, 5.12, and 5.40) are inaccurate; (b) nonlinearities in the form of a time- and space-dependent permeability of wet region behind the moving liquid front in swelling, liquid-absorbing porous wick; and (c) easy inclusion of the gravity effects. One may use computational fluid dynamics (CFD) software to solve the two governing equations in wet area behind the moving flow front and thus model the wicking process.

5.4.1 Algorithm for Wicking Simulation

As described in the introduction to this chapter, the single-phase Darcy flow behind a moving liquid front can be solved numerically using the quasi-steady approach. The finite element/control volume (FE/CV) algorithm is implemented to solve for the wicking flow with the flow domain (wick) discretized into an FE mesh, and with CVs defined around the FE nodes (Figure 5.9).

The transient liquid flow in a porous medium involving a moving-boundary is divided into multiple time steps. After assuming a quasi-steady condition during each time step, the elliptic pressure equation $\nabla \cdot [\mathbf{K} \cdot \nabla \langle P_f \rangle^f] = 0$,

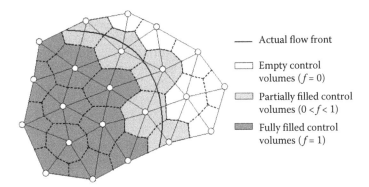

— Actual flow front

☐ Empty control volumes ($f = 0$)

▨ Partially filled control volumes ($0 < f < 1$)

▨ Fully filled control volumes ($f = 1$)

FIGURE 5.9
A typical finite element (FE) mesh with control volumes (CV) placed around FE nodes as an example of implementing the FE/CV method to model the motion of liquids in wicks. (From Masoodi, R., K.M. Pillai, and H. Tan. 2011. *AIChE Journal* 57(5): 1132–1143. With permission.)

obtained by combining Equations 5.1 and 5.2, is first solved for the modified pressure in the "saturated" FE nodes that are marked with CVs having a fill factor* of unity. The pressure field in the wet area is then used to estimate Darcy velocities at CV faces. The border CVs with a fill factor between 0 and 1 are then filled using this velocity, and the smallest fill-time determines the next time step. Note that filling of one of the border CVs advances the liquid front (Figure 5.9). Further details on this FE/CV-based wicking simulation can be obtained from Masoodi et al. (2011). Note the only difference between a wicking simulation in a rigid porous medium and that in a liquid-absorbing, swelling porous medium is that the local permeability in the latter is a function of time.[†]

5.4.2 POREFLOW: A Numerical Simulation for Wicking Flows

POREFLOW is a CFD code focused primarily on solving flow infiltration/wetting-type problems encountered in porous media (POREFLOW 2011). The FE/CV algorithm is implemented in the code to simulate the single-phase flow behind a moving boundary. The applied algorithm is very efficient and robust for solving the moving-boundary problems in complex domain geometries. The geometry may be 2D or 3D and the mesh may be structured or unstructured, giving maximum flexibility to the user. Note that the FE/CV algorithm has been employed in the past to model the wetting of fibrous porous media in liquid molding processes to manufacture polymer composites (Tan and Pillai 2010); however, using this simulation technique to model the wicking flow is new (Masoodi et al. 2011).

Currently, it is not easy to coax any commercial CFD software to model the *moving-front* wicking flow in swelling porous media using the *nonlinear* form of Darcy's law for single-phase flow; however, POREFLOW has the capability to include the effects of swelling in porous media by using the variable, time-dependent local permeability and porosity. POREFLOW remains the only numerical simulation with the experimentally validated capability to predict the liquid flow through swelling porous media, such as paper stripes and natural fibers (Masoodi et al. 2012a, 2012b).

5.4.3 Application Examples

5.4.3.1 Wicking in a 3D Polymer Wick

In this application, POREFLOW was used to solve the governing equation for wicking in *altered* rigid wicks made from sintering polymer beads. Note

---

* The fill factor, f, is defined as the pore volume in a CV filled with liquid divided by the total pore volume in the CV.
† It means the permeability of each element is a function of "wetting" time (i.e., the time elapsed since the passage of liquid front through an FE node).

that for a simple wick of cylindrical shape, the 1D flow-based analytical solutions (Equations 5.11, 5.12, and 5.41) are sufficient for modeling wicking along the axial direction. In an altered wick, there are two different cross sections (Figure 5.10a), which compels one to use 3D flow simulation, especially in the neck regions. In Figure 5.10, comparisons of the numerical predictions with the experimental data show an excellent match is obtained, while

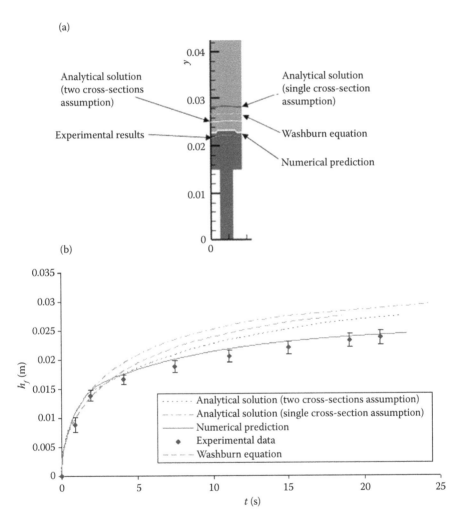

FIGURE 5.10
Wicking in an altered wick: a comparison of the numerical prediction with the analytical solutions corresponding with the one and two different cross sections for the wick, and the experimental result. (a) A comparison of the flow-front locations in the altered wick at time $t = 20$ s. (b) A comparison of the flow-front height h_f versus time t plots. (From Masoodi, R., K.M. Pillai, and H. Tan. 2011. *AIChE Journal* 57(5): 1132–1143. With permission.)

analytical solutions for the single and double* cross section cases along with the Washburn equation fail to predict the 3D wicking with sufficient accuracy (Masoodi et al. 2011).

5.4.3.2 Wicking in Paper-Like Swelling Porous Media

Another application of the sharp-front wicking flow model is the flow simulation in a porous paper-like substrate made of a network of cellulose fibers and powdered superabsorbents (Masoodi et al. 2012a). Once again, POREFLOW is employed to solve the governing equations with accompanying nonlinearities arising due to matrix swelling, and to conduct a 1D wicking flow simulation. Such a simulation allows one to overcome the limitations of the less-accurate analytical solution (Equation 5.41) described in Section 5.3.3. The accuracy of estimated local permeability is critical in calculating the wicking flow in porous media. Masoodi et al. (2012a) suggested a new approach for estimating the local permeability in a swelling paper-like porous material as a function of the wetting time. Figure 5.11 shows the estimated permeability for four different materials used by Wiryana and Berg (1991). Such permeability functions were later used to assign permeability to each element behind the moving liquid front in the FE/CV simulation.

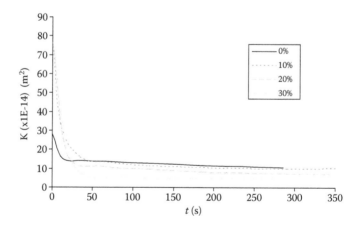

FIGURE 5.11
Estimated local permeabilities as a function of the wetting time for four different paper-like substrates used by Wiryana and Berg (1991). (The percentages correspond with the % of FC CMC (a superabsorbent) mixed with cellulose in the substrates.) (From Masoodi, R., H. Tan, and K.M. Pillai. 2012a. *AIChE Journal*, 58(8): 2536–2544. With permission.)

---

* In the double cross-sectional analytical solution, the wicking flow is assumed to be 1D in both the narrow as well as the wide sections, and the two flows are linked through the continuity of modified pressure and flow rate (Masoodi et al. 2011).

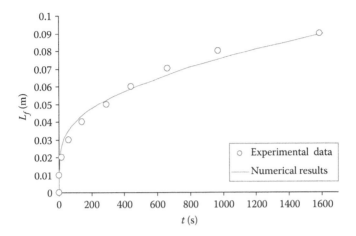

FIGURE 5.12
Comparing the numerical simulation with the experimental data for the 30% FC CMC paper-like substrate. (From Masoodi, R., H. Tan, and K.M. Pillai. 2012a. *AIChE Journal*, 58(8): 2536–2544. With permission.)

The flow simulation is then used to predict wicking in the swelling porous medium. Figure 5.12 shows a typical simulation result compared with the experimental data—we see better agreement with the experiments compared with the rather inadequate match shown in Figure 5.8 by the analytical solution. (See Masoodi et al. 2012a for such comparisons for the other remaining substrates.) It is clear that incorporation of gravity, as well as the local variations in permeability through the numerical simulation, leads to a dramatic improvement in the accuracy of the sharp-front model in predicting wicking in swelling porous materials.

5.5 Summary and Conclusions

This chapter describes a relatively new approach of using the single-phase flow behind a clearly defined liquid front to model wicking in porous substrates. Such an approach employs Darcy's law in conjunction with the continuity equation to model the liquid flow in rigid porous media. Darcy's law, along with a modified continuity equation after including the liquid-absorption and swelling effects through local permeability variations, is employed for modeling wicking in swelling porous media. A new expression for the capillary suction pressure, based on the energy balance principle applied at the liquid front, is also described.

A comparison of the experiments indicates that a high degree of accuracy is displayed by the sharp-front models for both rigid and swelling porous substrates. The wicking-flow governing equations are not easily solved analytically in 3D complex geometries or for swelling porous materials. POREFLOW, a numerical simulation based on FE/CV algorithm, is employed to simulate wicking (with or without swelling of solid matrix) in 2D and 3D geometries of porous materials. The numerical results compare well with experimental observations for the cases of an altered 3D wick and swelling paper strips, thereby establishing the accuracy of the suggested single-phase flow approach. An important advantage of the sharp-front approach is that all the involved model parameters can be measured or estimated independently while still obtaining a good match with experimental results, thereby obviating the need for any fitting parameter.

Nomenclature

Roman Letters

A	Area (m²)
b	Absorption coefficient (1/s)
C	Perimeter (m, Figure 5.3)
D	Fiber diameter (m, Equation 5.38)
f	Fill factor (dimensionless)
fn	A general function (Equations 5.37 and 5.39)
g	Acceleration due to gravity (m/s²)
h	Coordinate for elevation in the wick (m)
h_f	Height of the flow front in the wick (m)
K	Permeability (m²)
$\mathbf{K}$	Permeability tensor (m², Equation 5.36)
L	Length of wick (m, Figure 5.4) or elevation of liquid front (m)
m	Liquid mass (kg)
n	Number of capillary tubes or fibers passing through the liquid–air interface (Table 5.1)
$\bar{n}$	Unit normal vector (Figure 5.6)
P	Modified pressure (Pa, Equation 5.3)
p	Hydrodynamic pressure (Pa)
R	Radius (m)
r	Radius of individual spherical beads or fibers (m, Table 5.1)
S	Sink term in the modified continuity equation (1/s)
t	Time (s)
V	Upwards velocity (m/s, Equation 5.9)

v	Solid volume fraction in a porous medium (Equations 5.33 and 5.34)
$\bar{V}$	Velocity vector (m/s)
Vol	Volume of REV (representative elementary volume)
x	Depth of water in the syringe (Figure 5.4)

Greek Letters

θ	Contact angle (degree)
γ	Surface tension (N/m²)
μ	Viscosity of liquid (kg/m.s)
ε	Porosity (dimensionless)
∇	Gradient operator
φ	Probability density function (Table 5.1)
ρ	Density (kg/m³)
λ	Material constant (Equation 5.23)

Subscripts

0	Initial value
atm	Atmosphere
c	Capillary
cs	Cross section
ds	Dry surface
e	Effective capillary (Equations 5.15 and 5.16)
f	Fluid
fb	Corresponding to fibers (Table 5.1)
fs	Fluid–solid interface
i	Corresponding to the solid particle or void space *i* (Figure 5.3)
int	Interface surface of dry and wet matrix (Figure 5.3)
j	Corresponding to the solid particle or void space *j* (Figure 5.3)
l	Liquid
lf	Liquid front
p	Corresponding to particles (Table 5.1)
s	Solid
sat	Saturation
sr	Syringe (Figure 5.4)
sp	Spherical particle
ss	Steady state
v	Void space (Figure 5.3)
w	Wick
wet	Wet
ws	Wet surface

Other Symbols

$\langle\rangle$ Volume-averaged or phase-averaged quantity

$\langle\rangle^f$ Pore-averaged or intrinsic phase-averaged quantity of fluid phase

$\langle\rangle^s$ Pore-averaged or intrinsic phase-averaged quantity of solid phase (Equation 5.36)

References

Al-Raoush, R. and K.A. Alshibli. 2006. Distribution of local void ratio in porous media systems from 3D x-ray microtomography images. *Physica A* 361: 441–456.

Al-Raoush, R. and C.S. Willson. 2005. Extraction of physically-representative pore network from unconsolidated porous media systems using synchrotron microtomography. *Journal of Hydrology* 23(3): 274–299.

Bear, J. 1972. *Dynamics of Fluids in Porous Media*. New York: Elsevier Science.

Berg, J.C. 1993. *Wettability*. New York: Dekker.

Chatterjee, P.K. and B.S. Gupta. 2002. *Absorbent Technology*. Amsterdam: Elsevier.

Dullien, F.A.L. 1992. *Porous Media: Fluid Transport and Pore Structure*. San Diego: Academic Press.

Dias, M.M. and A.C. Payatakes. 1986. Network models for two-phase flow in porous media Part 1. Immiscible microdisplacement of non-wetting fluids. *Journal of Fluid Mechanics* 164: 305–336.

Finn, R. 1999. Capillary surface interfaces. *Notices of the AMS* 46(7): 770–781.

Giesche, H. 2006. Mercury porosimetry: A general (practical) overview. *Particle & Particle Systems Characterization* 23: 1–11.

Gray, W.G. and K. O'Neill. 1976. On the general equations for flow in porous media and their reductions to Darcy's law. *Water Resources Research* 12(2): 148–154.

Gray, W.G. and P.C. Lee. 1977. On the theorems for local volume averaging of multiphase systems. *International Journal of Multiphase Flow* 3: 333–340.

Kissa, E. 1966. Wetting and wicking. *Textile Research Journal* 66: 660–668.

Leisen, J. and H.W. Beckham. 2008. Void structure in textiles by nuclear magnetic resonance, Part I. Imaging of imbibed fluids and image analysis by calculation of fluid density auto-incorrelation functions, *Journal of Textile Institute* 99: 243–251.

Lenormand, R., E. Touboul, and C. Zarcone. 1988. Numerical models and experiments on immiscible displacements in porous media. *Journal of Fluid Mechanics* 189: 165–187.

Lucas R. 1918. Rate of capillary ascension of liquids. *Kollid Z* 23: 15–22.

Masoodi, R. and K.M. Pillai. 2011. Modeling the processing of natural fiber composites made using liquid composite molding. In: *Handbook of Bioplastics and Biocomposites Engineering Applications*, ed. S. Pilla, 43–74, New Jersey: Wiley-Scrivener.

Masoodi, R. and Pillai, K.M. 2012. A general formula for capillary suction pressure in porous media. *Journal of Porous Media* 15(8): 775–783.

Masoodi, R. and K.M. Pillai. 2010. Darcy's law-based model for wicking in paper-like swelling porous media. *AIChE Journal* 56(9): 2257–2267.

Masoodi, R., K.M. Pillai, and P. Varanasi. 2007. Darcy's law based models for liquid absorption in polymer wicks. *Journal of AIChE* 53: 2769–2782.

Masoodi, R., K.M. Pillai, and P. Varanasi. 2008. Role of hydraulic and capillary radii in improving the effectiveness of capillary model in wicking. *ASME Summer Conference*. Jacksonville, FL, USA, August 10–14.

Masoodi, R., K.M. Pillai, and H. Tan. 2011. Darcy's law based numerical simulation for modeling 3-D liquid absorption into porous wicks. *AIChE Journal* 57(5): 1132–1143.

Masoodi, R., H. Tan, and K.M. Pillai. 2012a. Numerical simulation of liquid absorption in paper-like swelling porous media. *AIChE Journal* 58(8): 2536–2544.

Masoodi, R., K.M. Pillai, N. Grahl, and H. Tan. 2012b. Numerical simulation of LCM mold-filling during the manufacture of natural fiber composites. *Journal of Reinforced Plastics and Composites* 31(6): 363–378.

Mortensen, A., L.J. Masur, J.A. Cornie, and M.C. Flemings. 1989. Infiltration of fibrous preforms by a pure metal: Part I. theory. *Metallurgical Transactions A*. 20A: 2535–2547.

Preziosi, L. and A. Farina. 2002. On Darcy's law for growing porous media. *International Journal of Non-Linear Mechanics* 37: 485–491.

POREFLOW. 2011. A software to model liquid infiltration in industrial porous media. http://www4dev.uwm.edu/porous/.

Scheidegger, A.E. 1974. *The Physics of Flow through Porous Media*. Toronto: University of Toronto.

Schuchardt, D.R. and J.C. Berg. 1990. Liquid transport in composite cellulose-super-absorbent fiber network. *Wood and Fiber Science* 23(3): 342–357.

Sutera, S.P. 1993. The history of Poiseuille's law. *Annual Review of Fluid Mechanics* 25: 1–19.

Szekely J., A.W. Neumann, and Y.K. Chuang. 1971. The rate of capillary penetration and the applicability of the Washburn equation. *Journal of Colloid Interface Science* 35: 273–278.

Tan, H. and K.M. Pillai. 2010. Processing composites for blast protection. In: *Blast Protection of Civil Infrastructures and Vehicles Using Composites*. ed. N. Uddin Cambridge: Woodhead.

Tong, X. and J.A. Khan. 1996. Infiltration and solidification/remelting of a pure metal in a two-dimensional porous preform. *Journal of Heat Transfer* 118: 173–180.

Tucker, C.I. and R.B. Dessenberger. 1994. Governing equation for flow and heat transfer in stationary fiber beds. In: *Flow and Rheology in Polymer Composites Manufacturing*, ed. S.G. Advani, Chapter 8. Amsterdam: Elsevier.

Washburn, E.V. 1921. The dynamics of capillary flow. *Physical Review* 17:273–283.

Whitaker, S. 1986. Flow in porous media III: Deformable media. *Transport in Porous Media* 1: 127–154.

Whitaker, S. 1998. *The Method of Volume Averaging*. Dordrecht: Kluwer Academic Publishers.

Wiryana, S. and J.C. Berg. 1991. The transport of water in wet-formed networks of cellulose fibers and powdered superabsorbent. *Wood and Fiber Science* 23(3): 456–464.

6

Modeling Fluid Absorption in Anisotropic Fibrous Porous Media

Hooman Vahedi Tafreshi and Thomas M. Bucher

CONTENTS

In this chapter, we present a novel methodology for modeling fluid imbibition in fibrous porous media. The modeling methodology is based on obtaining mathematical expressions for the variation of a medium's capillary pressure and relative permeability with saturation, via 3-D numerical simulations conducted on scales comparable to the dimensions of the constituting fibers. This information is then used in Richards' equation for two-phase flow in porous media to predict the rate of fluid absorption in a fibrous material as a function of time and space. A peculiar feature of the simulation method presented here is the possibility of isolating different microstructural parameters of the medium (fiber diameter, fiber orientation, etc.) and studying their influence on the rate of fluid absorption.

6.1 Virtual Three-Dimensional Fibrous Media

The fibrous structures subject to numerical characterization by the methods described in this chapter can be obtained either directly through imaging techniques applied to actual fibrous samples, or computationally using algorithms that produce virtual fibrous structures with the desired properties. Once generated, these structures can be used for numerical analysis in a number of applications related to fluid, particle, and/or heat transport in fibrous media.

6.1.1 Imaged Fibrous Structures

A number of different methods for imaging have been developed and utilized in the literature, such as x-ray microtomography and magnetic resonance imaging (MRI), both of which are nondestructive techniques based on the principles of these imaging methods as they are used in other applications (Hyvaluoma et al., 2006; Thoemen et al., 2008; Koivu et al., 2009). In x-ray microtomography, radiation is passed through a fabric sample at varying angles. The local absorption coefficients for the x-rays within the sample allow "shadowgrams" representing 2-D cross sections from which a 3-D map of the sample can be constructed (Hyvaluoma et al., 2006). For the MRI method, the imaging technique requires a material in the sample readily receptive to the influence of the applied magnetic field. In medical applications, the human body's large water composition facilitates this. Fibers, on the other hand, provide a weak signal output on their own (Hoferer et al., 2007), so a sample must therefore be submerged in water to allow imaging. Void spaces in the water signal (the solid fibers) is then quantified and visualized. From this, a 3-D virtual medium can be constructed for analysis.

A destructive method used for constructing the microstructures of porous samples is digital volumetric imaging (DVI), which is a serial-sectioning technique (Jaganathan et al., 2008a–d, 2009a,b). In DVI, a given sample is first impregnated and cured with a polymeric resin to preserve its microstructure. It is then placed in a machine that repeatedly slices the fabric and captures an image of the cross section thus revealed. These cross-section images are then used to reconstruct a 3-D image that can be utilized for numerical simulations.

Once a 3-D image is obtained, whether via DVI, MRI, or x-ray microtomography, additional postprocessing is required. First, the given images must be "thresholded." That is, a cutoff value is assigned to the initially grayscale images, which rigidly divides regions of the given image into that of either black or white (solid or pore space, respectively). From there, a meshing scheme must be applied to the geometry, the resolution of which must be a balance between simulation box size and accuracy, and computational memory and time (Jaganathan et al., 2008b).

The primary advantage associated with producing virtual fibrous structures via imaging is that the simulation domain is actually a sample taken from an existing material. Fiber shape, size distribution, spacing, and orientation are those of the sample itself. There are, however, several disadvantages to this method. First, the application of the threshold value by which solid and pore space are separated from one another depends heavily on the operator (or the setting of a software program) in the absence of an established standard. Varying the chosen threshold value changes all the parameters and calculations associated with that structure. Another limitation is in regard to the resolution of the image. For example, in the work of Jaganathan et al. (2008b), the chosen resolution was 1.77 μm. If the diameter of the fibers is much smaller than that, the resolution for accurate computational results cannot be achieved, which rules out applying this method to nano-fibers entirely. Finally, while this method can characterize the structure and behavior of an actual sample with reasonable accuracy, impregnating a sample with the resin, sectioning it, and reconstructing and thresholding the images are a considerably time-consuming process to perform repeatedly for the number of samples under consideration.

6.1.2 Modeled Fibrous Structures

The alternative to imaging is to develop computer-generated fibrous structures. This method eliminates the errors associated with image processing, but being based on series of simplifying assumptions, may introduce new sources of error. In this approach, each fiber is essentially the output of a parametric equation for a line that is set to run through a domain of set dimensions. The placement of the fibers is the result of one of several available randomness algorithms (e.g., μ-randomness, S-randomness, or I-randomness) (Pourdeyhimi et al., 1996). The shape, size, and orientation of the generated fibers are dependent on the constraints placed on the structure's desired parameters.

Figure 6.1 shows examples of different configurations for computer-generated fibrous structures. A 3-D isotropic medium (Figure 6.1a) in which placement and orientation of the fibers are purely random can be considered as a representative of air-laid nonwoven mats. For a typical anisotropic layered nonwoven mat as shown in Figure 6.1b (e.g., electro-spun nanofiber mats, melt-blown or spun-bonded sheets), fibers have

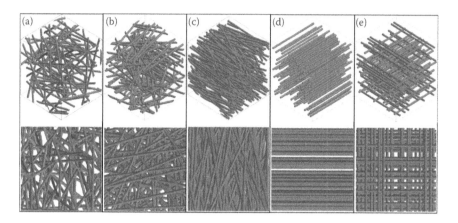

FIGURE 6.1
Examples of modeled fibrous geometries with different fiber orientations (isometric and over-head views in the top and bottom, respectively): (a) 3-D isotropic or purely random fiber orientation, (b) layered fabric with random in-plane orientation, (c) layered with highly oriented "machine-directional" fibers, (d) medium with unidirectional fibers, and (e) highly oriented medium with fibers in a given layer orthogonal to those in the adjacent layers.

very little orientation in the thickness direction, but remain somewhat isotropic in their orientation within a layer. However, machine direction-ality (Figure 6.1c), in which fiber orientation favors the direction in which the fabric was fed through its manufacturing process (e.g., carded mats or hydroentangled fabrics), leading also to in-plane anisotropy, is also typical. Figure 6.1d represents the extreme case of unidirectional-oriented fibers, like in the case of fiber bundles and tows, which also provide an important numerical benchmark for characterizing the influence of fiber orientation on absorption (e.g., Ashari et al., 2010a) or permeability (Tahir and Tafreshi, 2009).

Figure 6.1e shows an example of modeled structures in which fibers are laid on top of one another in an orthogonal fashion. Such structures can be produced by cross-lapping orientated mats (e.g., carded fiberwebs) or AC-electrospinning (Sarkar et al., 2007) among many others. The fibers in this geometry are arranged in layers orthogonal to one another.

While meshing a computer-generated structure remains a concern that must be addressed, one does not have the physical restrictions imposed by the imaging hardware used. This also means that for modeled geometries, there is no restriction on how small the fibers can be. One must of course still be aware of the validity of the continuum assumption, which can be compromised when modeling very small fibers (see Hosseini and Tafreshi, 2010). Along with resolution, the domain size itself is also important. Clague and Phillips (1997) proposed a minimum size of $14\sqrt{k}$ for reducing the statistical errors associated with the in-plane size of the computational domain. Nonetheless, regardless of the means by which simulation domains were obtained, repetition is important for improving the statistical confidence in results.

Computer-generated fibrous structures often do not have a constraint forbidding fibers from overlapping and crossing through each other—except for Figure 6.1d and 6.1e, in which orientation is tightly controlled (not random). This is because previous studies have shown that such a constraint, while allowing for a more realistic representation of fiber-to-fiber interplay, has no significant influence in the fluid transport in a fibrous medium. In fact, allowing the fibers to interpenetrate provides a means for controlling the solid volume fraction (SVF) of the structure when modeling the fibers with straight cylinders (see Wang et al., 2006; Tahir and Tafreshi, 2009 for more information).

6.2 Modeling Capillary Pressure in 3-D Fibrous Media

Capillary pressure refers to the difference in pressure across the interface of two immiscible fluids in a porous medium. Its relationship to saturation (ratio of the void space occupied by a given fluid to the total available void volume) is ultimately dependent on the void geometry within the medium (Dullien, 1992). In a traditional capillary pressure experiment, a porous sample of some known thickness and microstructure is fully saturated with a wetting-phase (WP) fluid from a reservoir connected to one side of the sample, and situated with a reservoir of non-wetting-phase (NWP) fluid contacting the other side. A differential pressure (capillary pressure) is then applied across the sample from the NWP side, and the NWP begins to penetrate into the sample, displacing the WP. Upon reaching a steady-state condition, the volume of wetting fluid displaced under the pressure is measured, and based on the medium's porosity, a saturation value corresponding to that pressure is determined.

The intrusion of WP into the medium at a given pressure is dependent on the size distribution of the interfiber spacing therein, with pressure p_c related to pore radius r_{cap} via the Young–Laplace equation (Dullien, 1992):

$$p_c = \frac{2\sigma \cos\theta}{r_{cap}} \tag{6.1}$$

where σ and θ are the surface tension and WP contact angle, respectively. The NWP at a given pressure can only penetrate into a medium as deeply as the corresponding capillary size will allow, with more constrictive pores acting as a bottleneck, preventing further penetration (Jaganathan et al., 2008a).

6.2.1 Full-Morphology Approach

The full morphology (FM) simulation method is a quasi-static geometric approach for characterizing the relationship between saturation and capillary pressure within a medium, which operates based on the Young–Laplace equation (6.1). It was first developed by Hazlett (1995), and later used by Hilpert and Miller (2001) for soil science applications, and has been used extensively for fibrous media (e.g., Becker et al., 2008; Jaganathan et al., 2009a,b; Ashari and Tafreshi, 2009b; Bucher et al., 2012). In the FM algorithm, the void spaces in a given fibrous structure (be it imaged or generated) are fitted with spheres with a radius corresponding to a predetermined capillary pressure value (Equation 6.1). These spheres represent the NWP, and are all overlaid and interconnected, forming a virtual continuum that originates from one side of the structure designated as the nonwetting reservoir. By maintaining connectivity to the NWP source and having their placement limited by void size (more constrictive voids acting as a bottleneck against further NWP intrusion), the insertion of different sized spheres corresponding to different pressures represents the drainage process in the medium. For each incremental pressure, WP or NWP saturation can be measured based on the volume of the structure's void space that is occupied by the spheres. The FM algorithm is implemented in the GeoDict code developed by Math2Market GmbH, Germany (www.geodict.com). Figure 6.2 illustrates this process for a sample structure at four different applied capillary pressures, with the top face acting as the nonwetting reservoir.

As can be seen in Figure 6.2, the size distribution of the interfiber spaces in a typical porous medium are such that there exists a pressure threshold beyond which there is little subsequent resistance against further NWP penetration. In terms of the FM methodology, this means that the spheres have become sufficiently small as to be able to reach far into the domain upon an incremental pressure rise beyond this threshold. Filling the remaining space in the medium requires the continuous raising of pressure, allowing the spheres to fill the small, unoccupied pockets that are left unaffected (pressure must go to infinity to fill them all). A typical result for the relationship between capillary pressure and saturation can be seen in Figure 6.3. When dealing with absorption, the interfacial pressure difference between WP and NWP is taken from the wetting side, so pressure is negative. The FM data in Figure 6.3 are fitted with existing empirical correlations from the experimental works of Landeryou et al. (2005), Haverkamp et al. (1977), and Van Genuchten (1980). As can be seen, despite the fact that the FM method is purely based on geometrical calculations—lacking rigorous physics—it shows reasonable agreement with empirical correlations. This is especially important considering that the alternative methods for producing such information—for example, the volume of fluid method—are prohibitively slow in 3-D. A few studies aimed at evaluating the accuracy

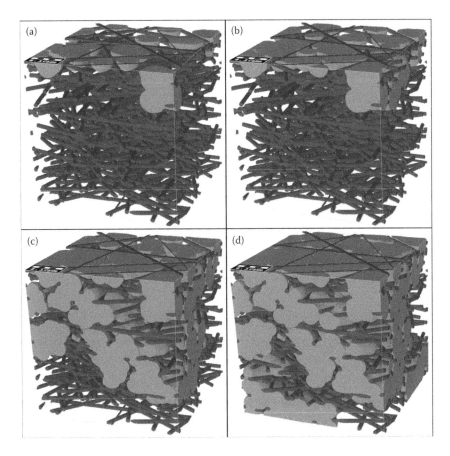

FIGURE 6.2
A visualization of different stages of drainage for a fibrous structure with SVF of 10%, fiber diameter of 10 µm, thickness of 480 µm, and random in-plane fiber orientation, using FM analysis. Corresponding pressures are: (a) 2949.15 Pa, (b) 3411.75 Pa, (c) 3866.65 Pa, and (d) 4046.50 Pa. The nonwetting fluid represented in the gray region is made up of spheres fitted into the domain.

of the FM method in comparison with a method based on balance of forces across a meniscus are given in the work of Bucher et al. (2012) and Emami et al. (2011).

The capillary pressure–saturation relationship can also be obtained via the vertical height rise (VHR) experiment. In this method, a long strip of fabric is hung vertically with its tip in contact with a fluid reservoir, such that capillarity draws the liquid up the column. After allowing time for the system to stabilize, the fabric is frozen and cut into strips. The weight of each strip is recorded, and from this, saturation as a function of pressure through $p_c = \rho g h$ is established. Ashari et al. (2010b) presented a comparison between the data they

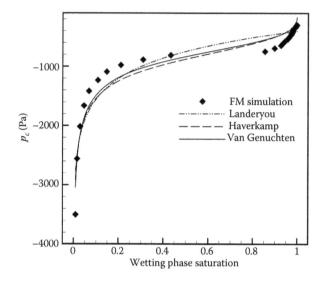

FIGURE 6.3
Example FM analysis results fitted against established empirical correlations from the litera-ture compared curves are from the works of Landeryou et al. (2005), Haverkamp et al. (1977), and Van Genuchten (1980). (From Ashari, A and Tafreshi, HV. 2009a. *Chemical Engineering Science*, 64, 2067.)

obtained from VHR and their FM simulations (see Figure 6.4). The VHR test is a very useful method for estimating the capillary pressure–saturation rela-tionship when working with a medium comprised of porous (swelling) fibers. This is because, with swelling fibers, the morphology of the medium changes upon fluid absorption, thereby compromising the utility of the FM simulation method. The correlations used in Figure 6.4 and described further in the next subsection can also be used to establish an effective value for the contact angle θ to be inserted into Equation 6.1 when it is not known ahead of time.

6.2.2 Effects of Microstructural Parameters on Capillary Pressure

As was stated above, capillary pressure as a function of saturation in a fibrous medium is ultimately dependent on the amount and distribution of empty space therein. The effects that variations in the particulars of a medium's microstructure have on $p_c(S)$ are manifest through their effect on interfiber spacing. Figure 6.5 shows capillary pressure plotted versus saturation for various layered fibrous media using the FM method. The trend observed in these plots is typical of the pressure–saturation relationship for porous media in general. When the medium is dry, strong capillarity draws in liquid from a given reservoir. This capillary effect is strongest in the smaller voids

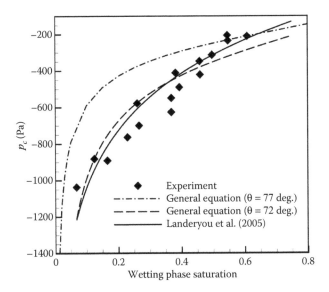

FIGURE 6.4
Results of vertical height rise experiment performed on a sample comprised of nonswelling fibers, fitted against empirical correlations from the literature. The general equation used (Equation 6.3a) is a modification to the empirical correlation of Haverkamp et al. (1977), which will be discussed in Section 6.2.2 (Ashari and Tafreshi, 2009b). (From Ashari, A, Bucher, TM, and Tafreshi, HV. 2010b. *Computational Materials Science*, 50, 378–390.)

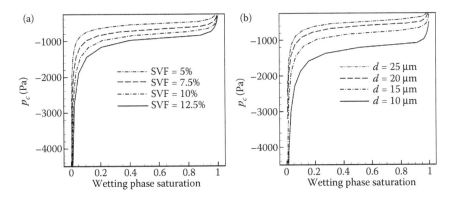

FIGURE 6.5
Capillary pressure versus saturation using the FM method for modeled virtual fibrous media with dimensions of $1500 \times 1500 \times 1000$ μm: (a) varying SVF while fiber diameter is held at 15 μm, (b) varying fiber diameter while SVF is held at 10%. The surface tension and wetting-phase contact angle used in Equation 6.1 are equal to 0.7275 N/m and 80 degrees, respectively, for both plots.

within the medium, as surface tension has a greater influence. The wetting fluid then works its way into larger spaces. The capillary pressure relates to pore space through Equation 6.1, with smaller overall void spaces resulting in a higher magnitude for pressure. The effect of a medium's microstructure can be seen in Figure 6.5a, in which SVF is varied for a given fiber diameter. As can be seen, a higher SVF results in a more negative capillary pressure for a given saturation (stronger driving force). The same effect can be seen in Figure 6.5b, in which fiber diameter is varied while keeping the SVF constant. A larger fiber size consolidates the solid mass in the medium into fewer fibers, thereby enlarging the void spaces and weakening the capillary drawing force.

Empirical correlations between capillary pressure and saturation S for a given porous medium have been developed in the literature in several studies. Three such correlations are those of Landeryou et al. (2005), Haverkamp et al. (1977), and Van Ganuchten (1980), and are expressed here as

$$p_c^{Lan} = C_1 \ln(S) + b_1 \tag{6.2a}$$

$$p_c^{Hav} = C_2(S^{-1} - 1)^{1/b_2} \tag{6.2b}$$

$$p_c^{Gen} = C_3(S^{b_3/(1-b_3)} - 1)^{1/b_3} \tag{6.2c}$$

where b and C are obtained via curve fitting. Ashari and Tafreshi (2009b) modified Equations 6.2b and 6.2c in order to develop general expressions for $p_c(S)$ given as

$$p_c^{Hav} = C_2 \frac{\sigma \cos \theta}{\sigma_w \cos \theta_w}(S^{-1} - 1)^{1/b_2} \tag{6.3a}$$

$$p_c^{Gen} = C_3 \frac{\sigma \cos \theta}{\sigma_w \cos \theta_w}(S^{b_3/(1-b_3)} - 1)^{1/b_3} \tag{6.3b}$$

where σ_w and θ_w are water–air surface tension and water contact angle, respectively. In this work, the authors also performed a large number of numerical simulations in order to relate the coefficients b and C to SVF (range of 5–12.5%) and fiber diameter (range of 10–25 µm). With these expressions, it is possible to circumvent the necessity for computationally expensive simulations to obtain $p_c(S)$ for fibrous media comprised of non-swelling fibers.

6.3　Modeling Permeability in 3-D Fibrous Media

6.3.1　FM–Stokes Simulation Method

Permeability in a fibrous medium can be predicted by solving the continuity and Navier–Stokes equations in the pore space between the fibers:

$$\nabla \cdot \bar{u} = 0 \tag{6.4}$$

$$\rho \left(\frac{\partial}{\partial t} + \bar{u} \cdot \nabla \right) \bar{u} = -\nabla p + \mu \nabla^2 \bar{u} \tag{6.5}$$

where u is the point-wise velocity, and ρ and μ are density and viscosity, respectively. The pressure gradient ∇p in this case is the local or point-wise pressure gradient. As $Re \ll 1$ in the case of absorption, the inertia term in Equation 6.5 may be neglected; hence the momentum equation simplifies to the Stokes equation, given as

$$\nabla p = \mu \nabla^2 \bar{u} \tag{6.6}$$

Viscous flow through a fibrous medium can also be characterized using Darcy's law (Mao and Russell, 2003; Ashari et al., 2010a,b), which states

$$\langle \bar{u} \rangle = -\frac{k(S)}{\mu} \nabla \langle p \rangle^f \tag{6.7}$$

where $\langle \bar{u} \rangle$ and $\nabla \langle p \rangle^f$ are the volume-averaged WP velocity and pressure, respectively, over a representative elementary volume, and $k(S)$ is the medium's permeability, a second-order tensor in units of m^2 which relates pressure gradient and fluid velocity.

The permeability through a medium for a given saturation can be ascertained by using Equations 6.6 and 6.7. The FM–Stokes method, as the name implies, uses the FM method described in the last section, and establishes a given quasi-static saturation for the analyzed medium, and then solves the Stokes and Darcy equations for flow in the x-, y-, and z-directions in order to develop a relation between permeability and saturation. By solving for flow in all three directions with accompanying pressure gradients in all three directions, one obtains the nine permeability elements in the tensor. The method assumes that flow between the wetting and nonwetting phases can be decoupled, with one phase not becoming entrained in the other. This allows the system to be treated as a pair of single-phase flow problems, with each level of saturation essentially behaving as a separate medium (Ashari et al., 2010a). As WP permeability is of interest, the method treats the NWP as a solid, with a no-slip boundary condition at the WP–NWP interface.

The permeability term $k(S)$ can be further resolved into single-phase and relative components, shown as

$$k(S) = k_{ij}^s \cdot k^r(S) \tag{6.8}$$

$k^r(S)$ is relative permeability, a value between 0 and 1 that represents permeability normalized against that of complete saturation. A second-order tensor, permeability for any level of saturation requires nine values to be fully expressed. Several studies in the literature showed that when the flow directions are the same as the principle directions of the medium, this tensor is almost symmetric, with the off-diagonal terms being negligibly smaller than the diagonal elements. Thus, $k_{xx}(S)$, $k_{yy}(S)$, and $k_{zz}(S)$ become the only terms of interest (Jaganathan et al., 2008d; Ashari et al., 2010a).

6.3.2 Analytic Expressions for Saturated and Relative Permeability

Several analytic expressions for single-phase permeability k_{ij}^s as a function of SVF and fiber diameter can be found in the literature that can be used to circumvent computationally expensive numerical simulations in certain cases. One such correlation is the empirical function developed by Davies (1973), which applies specifically to through-plane (z-direction) permeability in layered fibrous media with random in-plane fiber orientation. The correlation is given as

$$k_{zz}^s = \frac{r^2}{16\varepsilon^{3/2}(1 + 56\varepsilon^3)} \tag{6.9}$$

where r is equal to the fiber radius, and ε is equal to SVF. The Davies correlation has been used widely as a basis for comparison in a number of more recent numerical studies, and has shown close agreement for media with fiber diameters greater than a few micrometers (Tahir and Tafreshi, 2009; Hosseini and Tafreshi, 2010).

Spielman and Goren (1968) developed a series of expressions that relate k_{ij}^s directly to SVF and fiber radius for three of the cases shown in Figure 6.1. The first is a 3-D isotropic medium in which fiber orientation is purely random in all three dimensions (Figure 6.1a), in which case $k_{xx}^s = k_{yy}^s = k_{zz}^s = k_{iso}^s$, where x, y, and z refer to the machine direction, cross direction, and thickness direction (through-plane), respectively. Saturated permeability for such a medium in any direction is given by

$$\frac{1}{4\varepsilon} = \frac{1}{3} + \frac{5}{6}\frac{\sqrt{k_{iso}^s}}{r}\frac{K_1\left(\dfrac{r}{\sqrt{k_{iso}^s}}\right)}{K_0\left(\dfrac{r}{\sqrt{k_{iso}^s}}\right)} \tag{6.10}$$

where K_0 and K_1 are zero- and first-order modified Bessel functions of the second kind. Note that Equation 6.10 (and Equations 6.11 and 6.12 to follow) must be solved numerically. Layered fibrous media, the fibers of which have effectively no through-plane orientation, but random in-plane orientation (Figure 6.1b), are another case to which Spielman and Goren's relations apply. The relation for this configuration is given by

$$\frac{1}{4\varepsilon} = \frac{1}{4} + \frac{3}{4}\frac{\sqrt{k_{xx}^s}}{r}\frac{K_1\left(\dfrac{r}{\sqrt{k_{xx}^s}}\right)}{K_0\left(\dfrac{r}{\sqrt{k_{xx}^s}}\right)} \quad \text{(In-plane)} \tag{6.11a}$$

$$\frac{1}{4\varepsilon} = \frac{1}{2} + \frac{\sqrt{k_{zz}^s}}{r}\frac{K_1\left(\dfrac{r}{\sqrt{k_{zz}^s}}\right)}{K_0\left(\dfrac{r}{\sqrt{k_{zz}^s}}\right)} \quad \text{(Through-plane)} \tag{6.11b}$$

where $k_{xx}^s = k_{yy}^s$, since such a geometry is isotropic in the in-plane directions.

The third case to which the Spielman and Goren relations can be applied is that of a fibrous structure with disordered unidirectional fiber orientation (Figure 6.1d). The correlation for the unidirectional geometry is

$$\frac{1}{4\varepsilon} = \frac{1}{2}\frac{\sqrt{k_{xx}^s}}{r}\frac{K_1\left(\dfrac{r}{\sqrt{k_{xx}^s}}\right)}{K_0\left(\dfrac{r}{\sqrt{k_{xx}^s}}\right)} \tag{6.12}$$

where the x-direction represents the direction along the fibers.

Spielman and Goren (1968) did not develop a correlation for permeability in the directions perpendicular to fiber orientation ($k_{yy}^s = k_{zz}^s$ in this case). Tahir and Tafreshi (2009) and Fotovati et al. (2010) performed series of flow simulations in a number of virtual fibrous structures to conclude that the through-plane permeability of a layered fibrous medium is not affected by the in-plane orientation of its fibers, which is consistent with the lack of a specific correlation for permeability perpendicular to the fibers in the case of unidirectional fibers.

Ultimately, permeability is inversely related to SVF, as fluid penetration into a medium is a more tortuous process when more solid volume obstructs flow. Likewise, permeability being directly related to fiber diameter is intuitive. For a constant SVF, a larger fiber diameter consolidates the given solid mass into fewer fibers, thereby reducing the solid surface area in contact with the fluid, leading to less friction.

Relative permeability $k^r(S)$ has also been studied in the past few decades. A widely used power-law relation between relative permeability and saturation is that of Brooks and Corey (1964), which is given as

$$k^r = S^n \tag{6.13}$$

with $n = (2 + 3\lambda)/\lambda$. The pore-size distribution index λ tends to infinity for media containing a single pore size ($n = 3$). This index tends to smaller values for nonuniform media with a wider pore-size distribution. Most porous media have an n coefficient of 4. The above correlation was originally developed for soil applications, but was later used for fibrous media by Landeryou et al. (2005) and Mao (2009). In their work, they define a correlation for total permeability $k(S)$ in a given direction as

$$k(S) = k^s S^3 \tag{6.14}$$

where the exponent 3 is essentially in agreement with the work of Brooks and Corey (1964). Ashari and Tafreshi (2009b) and Ashari et al. (2010a) performed a number of numerical permeability analyses on modeled layered fibrous media with varying degrees of in-plane anisotropy using the FM–Stokes method. They found, via curve fitting, that in all cases tested, the n exponent in Equation 6.13 was in the neighborhood of 3 or 4. Ashari and Tafreshi (2009b) also observed through a series of FM–Stokes analyses that relative permeability is not influenced appreciably by SVF in a range of parameters typical of fibrous media.

Landeryou et al. (2005) also suggested that there exists a "percolation threshold" at a saturation of about $S = 0.4$ below which total permeability drops rapidly, resulting in a sharper, more distinct fluid front in an experiment with fibrous media. Why this occurs can simply be due to the wetting fluid breaking up and losing its continuity at such low saturations, although the numeric value of this threshold can be very different in fibrous sheets with different microstructures. While the Brooks and Corey relation does not directly capture this, the formula has been modified in some works when applied directly to numerical simulation by setting the n exponent to a larger value at saturations below 0.4 (Ashari et al., 2010b).

6.4 Bimodal Fibrous Media

6.4.1 Modeled Bimodal Fibrous Geometries

Many fibrous media are actually bimodal, that is, they consist of fibers of two different sizes. This adds a new level of complexity to modeling the absorption properties of a medium, as two additional microstructure parameters must be taken into account: the size difference between the respective fiber diameters and the fraction of the fiber population represented by each fiber type.

The process by which fibrous fabrics are produced allows for a population of different fiber sizes to be more easily accounted for in terms of mass percentage of one over the other, as the amount of material used in fabrication can easily be measured. When generating bimodal fibrous geometries (depending on the nature of the algorithm used), it is sometimes more convenient to express population representation of a given fiber size in terms of a number-based fraction instead of a mass-based fraction. Assuming all fibers to be of the same length, the relationship between the mass and number-of-fibers interpretation for a given fibrous structure (m and n, respectively) can be derived through geometry as

$$m_c = \frac{n_c d_c^2}{n_c d_c^2 + n_f d_f^2} \tag{6.15a}$$

$$m_f = \frac{n_f d_f^2}{n_c d_c^2 + n_f d_f^2} \tag{6.15b}$$

where d is fiber diameter, and the subscripts c and f refer to coarse or fine fibers, respectively. These equations can of course be rearranged if the number fraction must be determined from a given mass fraction.

Figure 6.6a is a fibrous geometry representing a layered bimodal medium with random in-plane fiber orientation—an example of a fibrous mat that can be produced by spun-bonding, melt-bonding, or DC electrospinning, among others. The disorder in the system allows the fibers to be generated and positioned randomly, so a mass-based approach to fiber population is preferred. Figure 6.6b represents a layered bimodal medium with orthogonal fiber orientation. Such fiber mats can be produced, for instance, via biased-AC electrospinning (Sarkar et al., 2007). In this case, the high degree of order achievable in the manufacturing process requires the placement of the fibers to be much less random. Consequently, a number-based fiber population becomes ideal for generating such structures.

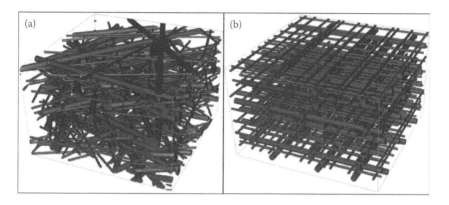

FIGURE 6.6

Sample modeled bimodal fibrous geometries: (a) layered fabric with isotropic in-plane fiber orientation, generated using mass-based fiber population; (b) layered fabric with highly ordered orthogonally oriented fiber layers, generated using number-based fiber population. Both have an SVF of 8.5%, a coarse fiber diameter of 15 μm, a fine fiber diameter of 5 μm, and number-based coarse-fiber fraction of 0.2.

6.4.2 Effect of Bimodality on Capillary Pressure and Permeability

As was the case with media of a single fiber size discussed earlier, the relationship between saturation and capillary pressure p_c is ultimately dependent on the amount and size distribution of the empty space therein, with smaller voids more easily availing themselves to the effect of the fluid surface tension force. In the work of Bucher et al. (2012), the authors performed an extensive parameter study on a host of generated bimodal fibrous geometries to isolate the relative contributions of each relevant parameter (see Figure 6.7). Figure 6.7a and 6.7b shows a behavior similar to that discussed in Figure 6.5. Note, however, that for Figure 6.7b, while fiber diameter is being varied, the size ratio between the two fiber types is retained. A higher SVF results in smaller pores, which in turn strengthens the capillary effect. Likewise, a smaller fiber diameter for a given SVF results in more fibers crossing through the geometry, thereby reducing the average spacing between fibers, and so strengthening the capillary effect.

Figure 6.7c illustrates what happens when the size difference between the two fiber types is increased. A similar trend can be seen as there was in Figure 6.7b, that being: for a given SVF and fine fiber diameter, a larger coarse fiber diameter consolidates the given solid volume into fewer fibers, thereby opening the available pore space and weakening the capillary effect in the medium (capillary pressure becomes less negative). Figure 6.7d depicts the effect of coarse-fiber number fraction, which is essentially caused by the same mechanism as was the case for Figure 6.7b and 6.7c. Of particular interest in Figure 6.7d is the sudden reduction in capillarity within the given media even when only 10% of the fiber

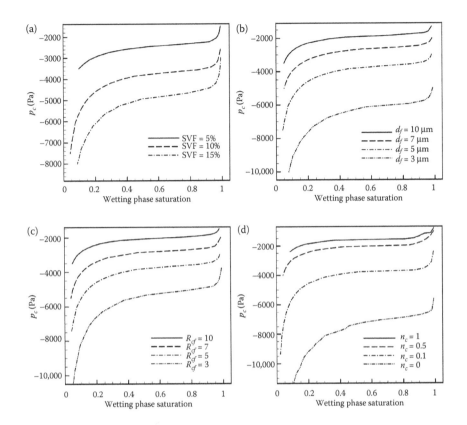

FIGURE 6.7
Capillary pressure–saturation curves for layered bimodal fibrous geometries with random in-plane fiber orientation, to illustrate the effect of various microstructural parameters on the relation: (a) SVF is varied, while fine fiber diameter, coarse-to-fine fiber-diameter ratio, and coarse-fiber number fraction are held at 5 μm, 5, and 0.1, respectively; (b) fiber diameter is varied, while SVF, coarse-to-fine fiber-diameter ratio, and coarse-fiber number fraction are held at 10%, 5, and 0.1, respectively; (c) coarse-to-fine fiber-diameter ratio is varied, while SVF, fine fiber diameter, and coarse-fiber number fraction are held at 10%, 5 μm, and 0.1, respectively, and (d) coarse-fiber number fraction is varied, while SVF, fine fiber diameter, and coarse-to-fine fiber-diameter ratio are held at 10%, 5 μm, and 5, respectively. Surface tension and wetting-phase contact angle are equal to 0.7275 N/m and 60 degrees for all curves.

population is comprised of coarse fibers when SVF and fiber diameters are held constant. One may easily infer that large fibers and bimodality should be avoided wherever possible in the application of absorption into fibrous media, but in many cases, a population of coarser fibers is necessary to impart mechanical strength and durability to the medium in its designed use, while fine fibers limit the resulting compromise in absorptivity. Moreover, addition of coarser fibers increases the permeability of the medium (rise in cross-sectional area), which actually improves the fluid transport therein.

Concerning permeability calculations, Clague and Phillips (1997) developed an analytical expression to establish an equivalent unimodal fiber radius r_{eq} for bimodal fibrous media, which can be simplified to

$$r_{eq} = \sqrt{n_c r_c^2 + n_f r_f^2} \tag{6.16}$$

Mattern and Deen (2008) later referred to their expression as the "volume-weighted resistivity" equation. Brown and Thorpe (2001) as well as Jaganathan et al. (2008c) compared results of their 2-D simulations of bimodal fibrous media with those of some existing unimodal equivalent diameter. With the exponent β being 1, 2, or 3, corresponding to approaches dubbed "number-weighted," "area-weighted," and "volume-weighted" average diameters, respectively, one can obtain a unimodal fiber diameter given as

$$r_{eq}^{(\beta)} = \frac{n_c r_c^\beta + n_f r_f^\beta}{n_c r_c^{\beta-1} + n_f r_f^{\beta-1}} \tag{6.17}$$

Tafreshi et al. (2009) later performed a series of 3-D simulations to develop a new unimodal equivalent fiber diameter for predicting permeability of bimodal fibrous media. Their equivalent fiber diameter, "cube-root of weighted mean cube," is given as

$$r_{eq} = \sqrt[3]{n_c r_c^3 + n_f r_f^3} \tag{6.18}$$

Tafreshi et al. (2009) found that the volume-weighted resistivity, area-weighted average (β = 2), and cube-root expressions are all acceptable for estimating an equivalent fiber radius. With this equivalent radius, the bimodal medium can be treated as unimodal in the context of permeability. Later, Fotovati et al. (2010) reported reasonable agreement among these equivalent diameter definitions when used for predicting the collection efficiency of aerosol filters. For capillary pressure prediction, however, no correlation yet exists in the literature for simplifying a given bimodal medium of known fiber parameters to an equivalent unimodal structure for that context.

6.5 Richards' Equation of Two-Phase Flows in Porous Media

6.5.1 Richards' Equation versus the Lucas–Washburn Equation

The Lucas–Washburn relation has been used to model fluid absorption in porous media (Lucas, 1918; Washburn, 1921; Hollies et al., 1957; Miller and Jansen, 1982; Hodgson and Berg, 1987; Zhmud et al., 2000; Landeryou et al., 2005). Three crucial assumptions the approach makes are: flow in one direction only, separated regions of fully wet and fully dry, and pore space modeled as an assembly of small parallel tubes of equal radius. A general derivation for the Lucas–Washburn equation for applications such as paper and granular porous media, for which the derivation begins with Newton's second law (Hyvaluoma et al., 2006) is given here as

$$\rho\left(x\frac{d^2x}{dt^2} + \frac{1}{2}\left(\frac{dx}{dt}\right)^2 \right) = \frac{2\sigma\cos\theta}{r_{cap}^{eq}} - \frac{8\mu x}{r_{cap}^{eq\,2}}\frac{dx}{dt} \qquad (6.19)$$

where r_{cap}^{eq} is taken to apply for all the pores in the medium (an assumption of the derivation). While this expression ignores gravity, Hyvaluoma et al. (2006) explain that spread in the direction of gravity for their applications is small enough to neglect, making spread in the system a balance between capillary and viscous forces. Neglecting the inertial term and integrating yields a correlation for height as a function of time, shown as (Hyvaluoma et al., 2006)

$$x(t) = \left(\frac{r_{cap}^{eq}\sigma\cos\theta}{2\mu}\right)^{1/2} t^{1/2} \qquad (6.20)$$

Despite the assumptions made in its derivation, the Lucas–Washburn relation has been shown to work quite well for applications in which intrusion in only one dimension is of interest, and in cases where a sharp fluid front between wetting and nonwetting fluids is typical, that is, partial saturation can be ignored (e.g., paper).

The Richards equation characterizes two-phase flow in porous media as a diffusive, continuum-based model, in which partial saturation and multidimensional analysis can occur (Richards, 1931). It begins with the 3-D unsteady continuity equation, given as

$$\phi\frac{\partial S}{\partial t} + \frac{\partial\langle u\rangle}{\partial x} + \frac{\partial\langle v\rangle}{\partial y} + \frac{\partial\langle w\rangle}{\partial z} = 0 \qquad (6.21)$$

where ϕ is equal to porosity. The volume-averaged velocity terms in Equation 6.21 are substituted using Darcy's law, shown here with gravity included as

$$\langle\bar{u}\rangle = -\frac{k}{\mu}\left(\nabla\langle p\rangle^f - \rho\bar{g}\sin\alpha\right) \qquad (6.22)$$

Assuming that gravitational acceleration is in the negative z-direction, Equation 6.21 becomes (Richards, 1931)

$$\phi \frac{\partial S}{\partial t} - \frac{1}{\mu} \left(\frac{\partial}{\partial x} \left(k_{xx}(S) \frac{\partial p_c}{\partial x} \right) + \frac{\partial}{\partial y} \left(k_{yy}(S) \frac{\partial p_c}{\partial y} \right) \right.$$

$$\left. + \frac{\partial}{\partial z} \left(k_{zz}(S) \frac{\partial p_c}{\partial z} + \rho g k_{zz}(S) \right) \right) = 0 \tag{6.23}$$

Equation 6.23 assumes that the flow directions are the principal directions for the permeability. Also, note that total permeability—the product of single-phase and relative permeability—is in the equation. As capillary pressure is a function of saturation, the chain rule can be applied. Expanding the permeability and pressure terms, the equation becomes

$$\phi \frac{\partial S}{\partial t} - \frac{1}{\mu} \left(\frac{\partial}{\partial x} \left(k_{xx}^s k^r(S) \frac{\partial p_c}{\partial S} \frac{\partial S}{\partial x} \right) + \frac{\partial}{\partial y} \left(k_{yy}^s k^r(S) \frac{\partial p_c}{\partial S} \frac{\partial S}{\partial y} \right) \right.$$

$$\left. + \frac{\partial}{\partial z} \left(k_{zz}^s k^r(S) \frac{\partial p_c}{\partial S} \frac{\partial S}{\partial z} + \rho g k_{zz}^s k^r(S) \right) \right) = 0 \tag{6.24}$$

It is in this form that saturation emerges as the variable being solved for as function of time and space. From here, the equation can be re-expressed to group terms more conveniently as

$$\frac{\partial S}{\partial t} + \frac{\partial}{\partial x} \left(D_{xx}(S) \frac{\partial S}{\partial x} \right) + \frac{\partial}{\partial y} \left(D_{yy}(S) \frac{\partial S}{\partial y} \right) + \frac{\partial}{\partial z} \left(D_{zz}(S) \frac{\partial S}{\partial z} \right) + D_g(S) \frac{\partial S}{\partial z} = 0$$

$$\tag{6.25}$$

where

$$D_{ij}(S) = -\frac{k_{ij}^s}{\phi \mu} \frac{\partial p_c}{\partial S} k^r(S) \tag{6.26a}$$

and

$$D_g(S) = -\frac{\rho g k_{zz}^s}{\phi \mu} \frac{\partial k^r}{\partial S} \tag{6.26b}$$

Jaganathan et al. (2009b) solved the Richards equation in one dimension for the case of the inclined fibrous sheets examined previously by Landeryou et al. (2005), and were able to predict the region of partial saturation. They also validated the partial saturation experimentally by cutting and weighing strips corresponding to different heights along a thin fibrous sheet in order to determine the saturation at these different heights.

The terms in the Richards equation that are themselves functions of saturation (capillary pressure and permeability) can be substituted with the appropriate expressions discussed in the previous sections of this chapter. Thus, when the equation is closed using the results of FM and FM–Stokes simulations and empirical correlations, a powerful modeling technique emerges, possessing the ability not only to model saturation distribution in multidimensional media, but also to capture the influence of individual microstructural parameters of the media on overall absorption. Figure 6.8 is an example showing the results of solving the Richards equation for a thin fibrous sheet (Ashari et al., 2010a). The domain represents a quarter of a fibrous sheet with a small circle held at $S = 1$, while the fluid diffuses into the rest of the sample according to the Richards equation. The capillary pressure term is the result of FM simulations fitted against the empirical correlation of Landeryou et al. (2005) (Equation 6.2a). Permeability $k(S)$ was determined using the FM–Stokes method, normalizing intermediate permeability values and curve fitting using the correlation of Brooks and Corey (1964) (Equation 6.14). As can be seen, a region of partial saturation exists in which a gradient from wet to dry (dark to light) is observed.

There are some cases in which the Richards equation can directly be compared to the Lucas–Washburn approach, despite multidimensionality

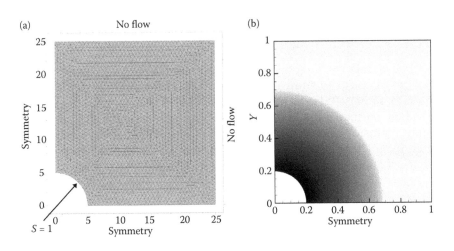

FIGURE 6.8
Two-dimensional simulation of a thin fibrous sheet (SVF of 10% and fiber diameter of 15 μm) with isotropic in-plane fiber orientation: (a) the computational domain mesh and boundary conditions, (b) contour plot showing the domain at $t = 1.3$ s into absorption, saturation from 1 to 0 moves from dark to light.

and partial saturation. One such case is the thin fibrous sheet with random in-plane fiber orientation. Since in-plane spread is isotropic and thickness is negligible, the x- and y-directions can be resolved to a polar-coordinate system direction R. Marmur (1988) developed an analytical expression based on Newton's second law for radial capillary spread that uses the same principles and assumptions as the Lucas–Washburn equation. This relation, characterizing radial spread from an infinite fluid source, is given here as

$$\left(\frac{R}{R_0}\right)^2\left(\ln\frac{R}{R_0}-\frac{1}{2}\right)+\frac{1}{2}=\frac{\sigma r_{cap}^{eq}\cos\theta}{6\mu R_0^2}t \tag{6.27}$$

where R_0 is equal to the radius of the fluid source. Equation 6.27 has been used extensively in the literature, and has been validated experimentally in several instances. Borhan and Rungta (1993) performed a series of radial capillary spread experiments with an ensemble of several fluids (e.g., water, ethylene glycol, and glycerol) and substrates (e.g., filter paper and glass beads). Equation 6.27 was a basis of comparison for their results. Danino and Marmur (1994) also tested the relation—as well as cases of unidirectional horizontal penetration, the spread of drops as opposed to a continuous source, and radial spread from a finite liquid source—using paper. Hyvaluoma et al. (2006) also used the correlation, rederived from Darcy's law (Equation 6.22), and used it for validation of 2-D lattice–Boltzmann simulation of liquid penetration into paper. In all cases of simulation and experiment, good agreement with Equation 6.27 was observed.

Ashari et al. (2010a) solved the 2-D Richards equation for the case of a thin fibrous sheet with random in-plane fiber orientation, a case that lends itself to comparison against Equation 6.27. For this comparison, Ashari et al. (2010a) used data on the total volume absorbed into the sheet with respect to time to calculate an equivalent radius the liquid would occupy if a clear wet–dry boundary did exist in the domain. They then compared this equivalent radius, normalized with the radius of the fully saturated liquid source, directly against the predictions of Marmur's correlation (1988) for the same system parameters. Figure 6.9 shows the comparison of the two methods. Note that Equation 6.27 contains the effective capillary radius term r_{cap}^{eq}. For this, Ashari et al. (2010a) fitted the equation with an effective capillary radius of 20 μm, and observed good agreement.

6.5.2 Effect of Media's Microstructure on Absorption

As was stated earlier, simulation of absorption in fibrous media using the Richards equation allows for capturing the effects of individual microstructure properties of the associated medium on its overall absorption

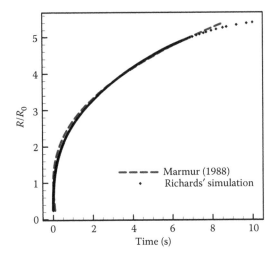

FIGURE 6.9
Simulation of the two-dimensional Richards equation fitted against the prediction of the analytical expression of Marmur (1988). The simulation domain is modeled as a thin fibrous sheet
(ignoring the z-direction) with an SVF of 10%, a fiber diameter of 15 μm, and random in-plane
fiber orientation. Radius R calculated by consolidating known absorbed volume into a region
of full saturation. Equation 6.27 is fitted with an effective pore radius of 20 μm.

properties. When the FM and FM–Stokes simulation methods are utilized on
the microscale, the effect of microstructural parameters such as porosity and
fiber diameter/orientation as a function of saturation are directly captured.
Inserting these correlations into the Richards equation at the macroscale
allows for visualization and computation of the amount and nature of fluid
absorption into a given fibrous medium. Figure 6.10 illustrates the effect of
fiber orientation alone on the overall spread characteristics of a thin fibrous
sheet (Ashari et al., 2010a). The media represented by the sheets in Figures
6.10a, 6.10b, and 6.10c are layered with random in-plane fibers, layered with
machine-directional fibers favoring the x-direction, and totally unidirectional fibers in the x-direction, respectively.

As can be seen in Figure 6.10, in-plane fluid spread is observed to favor
the direction of the fibers, if they favor a direction. Ashari et al. (2010a) also
observed that sheets whose fibers favor one in-plane direction over another
absorb more liquid over time. Conceptually, these two effects occur because
more highly ordered fibers better facilitate channels for liquid menisci to
advance, with fewer fibers intercrossing, as well as more consistent interfiber spacing along those channels. Conversely, fibers crossing each other at
large angles, or fibers more predominantly perpendicular to a given direction of interest for fluid spread, act as obstacles that impede the spread in
that direction relative to another, and are not as conducive to the capillary
effect. Numerically, these effects are manifest through the impact of fiber

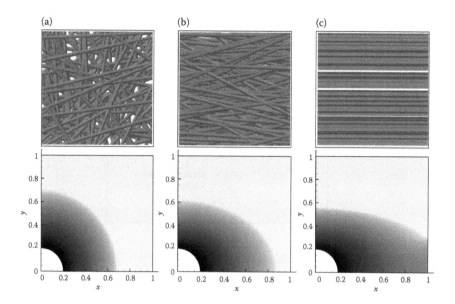

FIGURE 6.10
Comparison of water-absorption properties between thin fibrous sheets varying only in the orientation of their fibers. All have an SVF of 10% and a fiber diameter of 15 µm, surface tension of 0.7275 N/m, and contact angle of 60 degrees. Fiber orientations are as follows: (a) layered, with random in-plane fibers; (b) layered, with machine-directional fibers favoring the x-direction (standard deviation from x-axis of 22.5 degrees); and (c) unidirectional fibers, all in the x-direction. The simulation time for all three sheets is 0.91 s. (From Ashari, A et al. 2010a. *International Journal of Heat and Mass Transfer*, 53, 1750.)

orientation on capillary pressure and permeability as they are calculated through the FM and FM–Stokes simulation methods.

6.6 Summary

This chapter presents an overview for a modeling methodology for fluid absorption in fibrous porous media. Fibrous geometries obtained via imaging or modeling can be analyzed on the microscopic level to determine the effect that such parameters as porosity, fiber diameter and orientation anisotropy, and fiber-size distribution (in the case of bimodal media) have on the absorptive properties of a given fabric. This information can then be used to simulate performance in continuum domain on the scale of the actual fibrous product. There are a number of possible ways in which this methodology can be extended in future work to higher levels of complexity, for example, analyzing bicomponent fibrous media with different fluid contact angles therein, or heterogeneous materials whose properties fall into a distribution rather than a single value.

Nomenclature

b_1	Constant for Equation 6.2a
b_2	Exponent for Equation 6.2b
b_3	Exponent for Equation 6.2c
C_1	Constant for Equation 6.2a
C_2	Constant for Equation 6.2b
C_3	Constant for Equation 6.2c
d	Fiber diameter
d_c	Coarse fiber diameter
d_f	Fine fiber diameter
$D_g(S)$	Gravity diffusive coefficient
$D_{ij}(S)$	Capillary diffusive coefficient
h	Height
k	Permeability
K_0	Modified zero-order Bessel function of the second kind
K_1	Modified first-order Bessel function of the second kind
$k_{ij}(S)$	Total permeability tensor
k_{ij}^s	Single-phase (saturated) permeability tensor
$k^r(S)$	Relative permeability
k_{iso}^s	Single-phase permeability for 3-D isotropic medium
m_c	Coarse-fiber mass fraction
m_f	Fine-fiber mass fraction
n	Exponent in Equation 6.13
n_c	Coarse-fiber number fraction
n_f	Fine-fiber number fraction
$\langle p \rangle^f$	Volume-averaged wetting-phase pressure
$p_c(S)$	Capillary pressure
p_c^{Lan}	Capillary pressure based on Equation 6.2a
p_c^{Hav}	Capillary pressure based on Equation 6.2b
p_c^{Gen}	Capillary pressure based on Equation 6.2c
R	Wetted area radius
r	Fiber radius
R_0	Radius of fluid source
R_{cf}	Coarse-to-fine fiber-diameter ratio
r_c	Coarse fiber radius
r_{cap}	Capillary (pore) radius

Re	Reynolds number
r_{eq}	Equivalent fiber radius
r_{cap}^{eq}	Equivalent capillary radius
r_f	Fine fiber radius
S	Saturation
t	Time
$\bar{u}$	Point-wise velocity vector
$\langle u \rangle$	Volume-averaged fluid velocity in the x-direction
$\langle v \rangle$	Volume-averaged fluid velocity in the y-direction
$\langle w \rangle$	Volume-averaged fluid velocity in the z-direction

Greek Letters

β	Exponent in "weighted-average" bimodal media model (Equation 6.17)
ε	Medium's solid volume fraction
θ	Wetting phase contact angle
θ_w	Wetting phase contact angle when water is the wetting phase
λ	Pore-size distribution index
μ	Fluid viscosity
ρ	Fluid density
σ	Surface tension
σ_w	Surface tension of water
ϕ	Medium's porosity $(1-\varepsilon)$

Abbreviations

DVI	Digital volumetric Imaging
FM	Full morphology
MRI	Magnetic resonance imaging
NWP	Nonwetting phase
SVF	Solid volume fraction
VHR	Vertical height rise
WP	Wetting phase

References

Ashari, A and Tafreshi, HV. 2009a. A two-scale modeling of motion-induced fluid release from thin fibrous porous media, *Chemical Engineering Science*, 64, 2067.

Ashari, A and Tafreshi, HV. 2009b. General capillary pressure and relative permeability expressions for through-plane fluid transport in thin fibrous sheets, *Colloids and Surfaces A: Physicochemical and Engineering Aspects*, 346, 114.

Ashari, A, Bucher, TM, Tafreshi, HV, Tahir, TM, and Rahman, MSA. 2010a. Modeling fluid absorption in thin fibrous sheets: Effects of fiber orientation, *International Journal of Heat and Mass Transfer*, 53, 1750.

Ashari, A, Bucher, TM, and Tafreshi, HV. 2010b. A semi-analytical model for simulating fluid transport in multi-layered fibrous sheets made up of solid and porous fibers, *Computational Materials Science*, 50, 378–390.

Becker, J, Schulz, V, and Wiegmann, A. 2008. Numerical determination of two-phase material parameters of a gas diffusion layer using tomography images, *Journal of Fuel Cell Science and Technology*, 5(2), art. 021006.

Borhan, A and Rungta, KK. 1993. An experimental-study of the radial penetration of liquids in thin porous substrates, *Journal of Colloid Interface Science*, 158(2), 403–411.

Brooks, RH and Corey, AT. 1964. Hydraulic properties of porous media, Colorado State University, Hydrology Papers, Fort Collins, Colorado.

Brown, RC and Thorpe, A. 2001. Glass-fibre filters with bimodal fibre size distributions, *Powder Technology*, 118(1–2), 3–9.

Bucher, TM, Emami, B, Tafreshi, HV, Gad-el-Hak, M, and Tepper, GC. 2012. Modeling resistance of nanofibrous superhydrophobic coatings to hydrostatic pressures: The role of microstructure, *Physics of Fluids*, 24(2), 022109.

Clague, DS and Phillips, RJ. 1997. A numerical calculation of the hydraulic permeability of three-dimensional disordered fibrous media, *Physics of Fluids*, 9(6), 1562–1572.

Davies, CN. 1973. *Air Filtration*, Academic Press, London.

Danino, D and Marmur, A. 1994. Radial capillary penetration into paper—Limited and unlimited liquid reservoirs, *Journal of Colloid Interface Science*, 166(1), 245–250.

Dullien, FAL. 1992. *Porous Media Fluid Transport and Pore Structure*, 2nd edition, Academic Press, San Diego.

Emami, B, Bucher, TM, Tafreshi, HV, Gad-el-Hak, M, and Tepper, GC. 2011. Simulation of meniscus stability in superhydrophobic granular surfaces under hydrostatic pressures, *Colloids and Surfaces A: Physiochemical and Engineering Aspects*, 385(1–3), 95–103.

Fotovati, S, Tafreshi, HV, and Pourdeyhimi, B. 2010. Influence of fiber orientation distribution on performance of aerosol filtration media, *Chemical Engineering Science*, 65, 5285–5293.

Hazlett, RD. 1995. Simulation of capillary-dominated displacements in microtomographic images of reservoir rocks, *Transport in Porous Media*, 20(1–2), 21–35.

Haverkamp, R, Vauclin, M, Touma, J, Wierenga, PJ, and Vachaud, GA. 1977. Comparison of numerical simulation models for one-dimensional infiltration, *Soil Science Society of America Journal*, 124(3), 285–294.

Hilpert, M and Miller, CT. 2001. Pore-morphology-based simulation of drainage in totally wetting porous media, *Advances in Water Resources*, 24(3–4), 243–255.

Hodgson, KT and Berg, JC. 1987. The effect of surfactants on wicking flow in fiber networks, *Journal of Colloid Interface Science*, 121, 22–31.

Hoferer, J, Hardy, EH, Meyer, J, and Kasper, G. 2007. Measuring particle deposition within fibrous filter media by magnetic resonance imaging, *Filtration*, 7(2), 154–158.

Hollies, R, Kaessinger, M, Watson, B, and Bogaty, H. 1957. Water transport mechanisms in textile materials part II: Capillary-type penetration in yarns and fabrics, *Textile Research Journal*, 27, 8–13.

Hosseini, S and Tafreshi, HV. 2010. Modeling permeability of 3-D nanofiber media in slip flow regime, *Chemical Engineering Science*, 65, 2249–2254.

Hyvaluoma, J, Raiskinmaki, P, Jasberg, A, Koponen, A, Kataja, M, and Timonen, J. 2006. Simulation of liquid penetration in paper, *Physical Review E*, 73, art. 036705.

Jaganathan, S, Tafreshi, HV, and Pourdeyhimi, B. 2008a. Modeling liquid porosimetry in modeled and imaged 3-D fibrous microstructures, *Journal of Colloid and Interface Science*, 326, 166–175.

Jaganathan, S, Tafreshi, HV, and Pourdeyhimi, B. 2008b. A realistic approach for modeling permeability of fibrous media: 3-D imaging coupled with CFD simulation, *Chemical Engineering Science*, 63, 244–252.

Jaganathan, S, Tafreshi, HV, and Pourdeyhimi, B. 2008c. On the pressure drop prediction of filter media composed of fibers with bimodal diameter distributions, *Powder Technology*, 181(1), 89–95.

Jaganathan, S, Tafreshi, HV, and Pourdeyhimi, B. 2008d. A case study of realistic two-scale modeling of water permeability in fibrous media, *Separation and Science Technology*, 43(8), 1901–1916.

Jaganathan, S, Tafreshi, HV, and Pourdeyhimi, B. 2009a. A study on compression-induced morphological changes of nonwoven fibrous materials, *Colloids and Surfaces A—Physicochemical and Engineering Aspects*, 337(1–3), 173–179.

Jaganathan, S, Tafreshi, HV, and Pourdeyhimi, B. 2009b. A realistic modeling of fluid infiltration in thin fibrous sheets, *Journal of Applied Physics*, 105, art. 113522.

Koivu, V, Decain, M, Geindreau, C, Mattila, K, Alaraudanjoki, J, Bloch, JF, and Kataja, M. 2009. Flow permeability of fibrous porous materials: Micro-tomography and numerical simulations, *Advances in Pulp and Paper Research*, 1–3, 437–454.

Landeryou, M, Eames, I, and Cottenden, A. 2005. Infiltration into inclined fibrous sheets, *Journal of Fluid Mechanics*, 529, 173–193.

Lucas, R. 1918. Rate of capillary ascension of liquids, *Kolloid Z*, 23, 15.

Mao, N and Russell, SJ. 2003. Anisotropic liquid absorption in homogenous two dimensional nonwoven structure, *Journal of Applied Physics*, 94, 4135.

Mao, N. 2009. Unsteady-state liquid transport in engineered nonwoven fabrics having patterned structure, *Textile Research Journal*, 79(15), 1358–1363.

Marmur, A. 1988. The radial capillary, *Journal of Colloid and Interface Science*, 124(1), 301–308.

Mattern, KJ and Deen, WM. 2008. "Mixing Rules" for estimating the hydraulic permeability of fiber mixtures, *AICHE Journal*, 54(1), 32–41.

Miller, B and Jansen, SH. 1982. Wicking of liquids in nonwoven fibre assemblies, *Proceedings of 10th Technical Symposium on Advances in Nonwovens Technology*, pp. 216–226.

Pourdeyhimi, B, Ramanathan, R, and Dent, R. 1996. Measuring fiber orientation in nonwovens 1: Simulation, *Textile Research Journal*, 66(11), 713–722.

Richards, LA. 1931. Capillary conduction of liquids through porous medium, *Physics*, 1, 318–333.

Sarkar, S, Deevi, SC, and Tepper GC. 2007. Biased AC electrospinning of aligned polymer nanofibers, *Macromolecular Rapid Communications*, 28, 1034–1039.

Spielman, L and Goren, SL. 1968. Model for predicting pressure drop and filtration efficiency in fibrous media, *Environmental Science and Technology*, 2(4), 279–287.

Tafreshi, HV, Rahman, MSA, Jaganathan, S, Wang, Q, and Pourdeyhimi, B. 2009. Analytical expressions for predicting permeability of bimodal fibrous porous media, *Chemical Engineering Science*, 64, 1154–1159.

Tahir, MA and Tafreshi, HV. 2009. Influence of fiber orientation on the transverse permeability of fibrous media, *Physics of Fluids*, 21(8), no. 083604.

Thoemen, H, Walther, T, and Wiegmann, A. 2008. 3D simulation of macroscopic heat and mass transfer properties from the microstructure of wood fibre networks, *Composites Science and Technology*, 68(3&4), 608–616.

Van Genuchten, MT. 1980. A closed-form equation for predicting the hydraulic conductivity of unsaturated soils, *Soil Science Society of America Journal*, 44, 892–898.

Wang, Q, Maze, B, Tafreshi, HV, and Pourdeyhimi, B. 2006. A note on permeability simulation of multifilament woven fabrics, *Chemical Engineering Science*, 61(24), 8085–8088.

Washburn, E. 1921. The dynamics of capillary flow, *The Physical Review*, 17(3), 273–283.

Zhmud, BV, Tiberg, F, and Hallstensson, K. 2000. Dynamics of capillary rise, *Journal of Colloid and Interface Science*, 228(2), 263–269.

7

Wicking in Absorbent Swelling Porous Materials

Vladimir Mirnyy, Volker Clausnitzer, Hans-Jörg G. Diersch,
Rodrigo Rosati, Mattias Schmidt, and Holger Beruda

CONTENTS

7.1 Introduction

Progress in chemical engineering during the last few decades expanded the application of porous media modeling beyond environmental problems in soils to a large variety of artificial porous materials, such as industrial sponges, packed plastic granulates, hygiene products, and so on. Here, problem scales are reduced to centimeters or even millimeters. However, Darcy's law for flow in saturated porous media as well as Richards' equation for variably saturated media can still be usefully applied under certain known assumptions.

One important application area is the development of new materials for hygiene products with specific absorbent properties, where storage of liquid plays a significant role. Traditional absorbent materials like cotton wadding or cellulose fiber butts can absorb about 12 g of water or other aqueous solutions per gram of dry fiber (Buchholz and Graham, 1998). During the 1990s, they were mostly replaced by *absorbent gelling materials* (AGM) also known as *superabsorbent polymers* (SAP) with much better performance. Superabsorbers are able to store up to 1000 g of distilled water or 100 g of dissolved salt per gram of polymer. This astonishing property makes AGM very useful in various application areas such as composites and laminates, filtration, fire-retardant gels, waste stabilization and environmental remediation, fragrance carriers, hot and cold therapy packs, artificial snow, and so on. The largest use of AGM is found in diapers, feminine care, and adult incontinence products.

From the chemical point of view, AGMs are made from water-soluble polymers. As they consist of crosslinked chains, they do not dissolve in water, but absorb it in a diffusion process. The chain network expands until equilibrium between driving forces and retractive forces is achieved. A detailed chemical description of superabsorbers is given by Buchholz and Graham (1998).

Visually, AGMs look like white hard granular powders with particle size ranging from about 45 to 850 μm. When put into contact with water or water solution, the superabsorber granules turn into a soft, rubbery gel. Unlike the traditional absorbers, the superabsorbent polymer gel will not easily release the water under external force. It may squeeze out, but only very slowly. Another consequence of the transformation of AGM dry particles into gel is a substantial reduction of permeability. The so-called gel-blocking effect occurs, when soft gel particles with only little space to grow will close the adjacent void space, thereby prohibiting the penetration of water to the drier particles (Figure 7.1). This gel-blocking phenomenon is intentionally used for the protection of wires and cables against liquid. In the case of diapers and

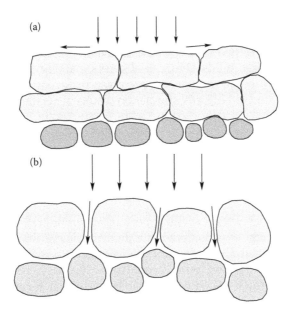

FIGURE 7.1
Schematic representation of the gel-blocking effect during liquid transport through AGM. Darker color corresponds to drier particles.

similar applications with fully opposite goals, gel blocking is to be avoided. Therefore, AGM is usually mixed with nonswelling fibers that distribute and deliver water to the dry AGM particles. On the one hand, an AGM-containing product must provide sufficient storage capacity as well as absorption performance, and on the other hand, it must observe certain shape requirements, for example, remain thin and comfortable. A number of modeling tasks arise:

- Prediction of flow and absorption processes in new composite fiber materials using two- and three-dimensional models
- Optimization of AGM properties for specific products
- Shape and structure optimization of material layers and overall product design

Successful realization of superabsorbent processes in a computational model may lead to new innovations in product development together with cost reductions and general economical benefits.

Section 7.2 describes a mathematical model based on Richards' equation, where most of the flow properties are dependent on swelling dynamics. How the resulting system of nonlinear partial differential equations (PDEs) can be solved numerically is explained in Section 7.3, where the finite-element method (FEM) is applied after temporal discretization and linearization. The numerical solution is implemented as a plug-in extension to the FEFLOW

(2010) simulation system. Among the simulation capabilities of FEFLOW is the solution of variably saturated flow and transport problems using the Galerkin FEM. The newly developed FEFLOW module makes possible management and simulation of swelling AGMs. Dynamic absorption and wicking experiments using AGM containing absorbent core are given in Section 7.4. Experimental results are compared to FEFLOW simulation results in Section 7.5.

7.2 Theoretical Development of Flow in Absorbent Swelling Porous Material

AGM containing absorbent swelling porous materials (referred further as AGM for brevity) can be considered as a deformable porous medium. Their theory relies on linear elasticity with extensions to cover poroelastoplastic-ity, poroviscoelasticity, and poroviscoplasticity (Coussy, 1995). This work extends the modeling basis with respect to large-scale deformation, changes in porous structure, reaction dependencies, and hysteretic behavior.

7.2.1 Balance Equations and Assumptions

Isothermal unsaturated flow in AGM is described as flow in a *porous medium* consisting of three phases: a solid phase s, a gas (air) phase g, and a mobile liquid phase l (e.g., Bear and Bachmat, 1991; Coussy, 1995; De Boer, 2000; Pinder and Gray, 2008). Any of the phases will be indicated as α ($\alpha = s,g,l$). The porous medium in *control space* $\Omega^s \subset \mathbb{R}^D (D = 2,3)$ is bounded by the solid surface Γ^s, while liquid and gas phases can enter or leave the control space Ω^s through Γ^s. The control space contains different α-phases and may swell or shrink in time during absorption. Let spatial point $x \in \Omega^s$ be the centroid of *representative elementary volume* (REV)* V, which is assumed to be constant in time. The *volume fraction* ε^α of the α-phase

$$\varepsilon^\alpha = \frac{V^\alpha}{V} = \frac{V^\alpha}{\sum_\alpha V^\alpha}, \quad \sum_\alpha \varepsilon^\alpha = 1 \tag{7.1}$$

defines *porosity* ε (fraction of void space) and *saturation* s of the dynamic liquid l, and the stagnant gas g phase as

$$\varepsilon^l = \varepsilon s^l, \quad \varepsilon^g = \varepsilon s^g, \quad \varepsilon^s = 1 - \varepsilon, \tag{7.2}$$

---

* For more information about representative elementary volume (REV), see Chapter 4.

$$s^l + s^g = 1, \quad 0 \le s^l \le 1, \quad 0 \le s^g \le 1. \tag{7.3}$$

The *density* of the liquid phase ρ^l is a function of pressure p^l. Its derivative can be derived from the definition of the liquid *compressibility* γ

$$\frac{\partial \rho^l}{\partial p^l} = \gamma \rho^l, \tag{7.4}$$

where we assume the compressibility to be constant.

The initial control space Ω_0^s at $t = t_0$ is changing in time due to absorption and resulting solid deformation. Since we apply continuum porous media formulation, the point x as well as the corresponding REV must contain material points X^α of all α phases at any time t. The movement of spatial point x can then be described as a function of reference material position vector X^α and time t (e.g., Coussy, 1995; De Boer, 2000):

$$x = u^\alpha(X^\alpha, t). \tag{7.5}$$

That states the *Lagrangian* or material *description of motion*. If the inverse to (7.5) exists, it represents the *Eulerian description of motion*:

$$X^\alpha = (u^\alpha)^{-1}(x, t). \tag{7.6}$$

The existence of the inverse can be expressed through the Jacobian J^α, which must be strictly positive:

$$J^\alpha = \det(\nabla u^\alpha) > 0. \tag{7.7}$$

From another point of view, the solid-phase Jacobian J^s represents deformation (swelling or shrinkage) of the porous media as the ratio between the current control-space volume at time t and the initial one:

$$J^s = \frac{|\Omega^s|(x, t)}{|\Omega_0^s|(X^s, 0)}. \tag{7.8}$$

We will call J^s the *volume dilatation function* of the porous solid.

The Eulerian formulation helps us to understand and describe the kinematics of deformation independently of any initial state. The *velocity of the solid phase* follows from (7.5):

$$v^s = \left.\frac{\partial u^s(X^s, t)}{\partial t}\right|_{X^s} = v^s(x, t), \tag{7.9}$$

where $|_{X^s}$ indicates that X^s is held constant.

7.2.1.1 Momentum Conservation

We consider momentum equations without inertial terms and assume sufficiently slow deformation of porous media.

- *Liquid phase:* We employ the well-known Darcy law for deformable porous media:

$$\varepsilon^l(v^l - v^s) = -\frac{k_r^l k}{\mu^l}(\nabla p^l - \rho^l g), \tag{7.10}$$

  where k_r^l is the relative permeability, k is the liquid-independent (saturated) permeability tensor, μ^l is the dynamic viscosity of the liquid, p^l is the liquid pressure, and g is the gravity vector. The left-hand side (LHS) of Equation 7.10 represents the volumetric flux or Darcy velocity:

$$q^l = \varepsilon^l(v^l - v^s). \tag{7.11}$$

- *Solid phase:* To avoid complicated and generally nonlinear dependencies of plasticity, we assume stress-free deformation (a *free-swelling* process) as a result of chemical reactions (cf. Coussy, 1995). Then the *solid strain* d^s is only a function of swelling solid volume $|\Omega^s|$, and then the equation for solid displacement can be reduced to

$$Lu^s = d^s(|\Omega^s|), \tag{7.12}$$

  where $d^s = [d_1^s, d_2^s, d_3^s, \gamma_{12}^s, \gamma_{23}^s, \gamma_{31}^s]^T$ and L is the symmetric gradient operator

$$L = \begin{bmatrix} \nabla_1 & 0 & 0 \\ 0 & \nabla_2 & 0 \\ 0 & 0 & \nabla_3 \\ \nabla_2 & \nabla_1 & 0 \\ 0 & \nabla_3 & \nabla_2 \\ \nabla_3 & 0 & \nabla_1 \end{bmatrix}, \quad \nabla_i = \frac{\partial}{\partial x_i} \quad (i = 1, 2, 3). \tag{7.13}$$

- *Gas phase:* We assume the gas phase g *stagnant* with respect to the solid phase and obtain the trivial equation

$$\varepsilon^g(v^g - v^s) \equiv 0. \tag{7.14}$$

  As a result, gas velocity is directly known from the solid velocity, $v^g = v^s$.

7.2.1.2 Mass Conservation

We restrict our model to *one* chemical component in the liquid and solid phases. The general multispecies mass conservation law in the α-phase system (taken from, e.g., Pinder and Gray, 2008; Diersch et al., 2010) can then be formulated in the following reduced form:

$$\frac{\partial}{\partial t}(\varepsilon^\alpha \rho^\alpha) + \nabla \cdot (\varepsilon^\alpha \rho^\alpha v^\alpha) = R^\alpha + Q^\alpha, \tag{7.15}$$

where ρ^α is the density and v^α is the velocity vector of the α-phase, R^α is a general reaction term containing homogeneous (intraphase) and heterogeneous (interphase) reactions. The bulk solute-mass source term Q^α is assumed to be zero.

- *Liquid phase:* Using the porosity definition (7.2), the density–pressure relationship (7.4), and the differentiation product rule, the temporal derivative becomes

$$\frac{\partial}{\partial t}(\varepsilon^l \rho^l) = \varepsilon^l \frac{\partial \rho^l}{\partial t} + \rho^l \frac{\partial \varepsilon^l}{\partial t} = \varepsilon s^l \gamma \rho^l \frac{\partial p^l}{\partial t} + \rho^l \left(\varepsilon \frac{\partial s^l}{\partial t} + s^l \frac{\partial \varepsilon}{\partial t} \right). \tag{7.16}$$

The divergence in the right-hand side (RHS) can also be developed using the product rule:

$$\nabla \cdot (\varepsilon^l \rho^l v^l) = \rho^l \nabla \cdot (\varepsilon^l v^l) + (\varepsilon^l v^l) \cdot \nabla \rho^l. \tag{7.17}$$

Note that the first term is much greater than the second term containing gradient of density (see, e.g., Freeze and Cherry, 1979), which can be eliminated. Further, we can substitute the divergence of Darcy velocity $\nabla \cdot q^l = \nabla \cdot (\varepsilon^l v^l) - \nabla \cdot (\varepsilon^l v^s)$ and the porosity ε^l from (7.2) and then apply the product rule again:

$$\rho^l \nabla \cdot (\varepsilon^l v^l) = \rho^l (\nabla \cdot q^l + \nabla \cdot (\varepsilon^l v^s)) = \rho^l (\nabla \cdot q^l + \varepsilon s^l (\nabla \cdot v^s) + v^s \cdot \nabla(\varepsilon s^l)). \tag{7.18}$$

If we additionally divide both sides by ρ^l, the mass-conservation equation for the liquid phase finally becomes

$$\varepsilon \frac{\partial s^l}{\partial t} + \varepsilon s^l \gamma \frac{\partial p^l}{\partial t} + \nabla \cdot q^l = \frac{R^l}{\rho^l} - s^l \varepsilon (\nabla \cdot v^s) - s^l \frac{\partial \varepsilon}{\partial t} - \phi^l, \tag{7.19}$$

where ϕ^l defines an additional *deformation term*: $\phi^l = v^s \cdot \nabla(\varepsilon s^l)$.

- *Solid phase:* Similarly, we apply the divergence product rule and substitute the definition of the velocity v^s (7.9)

$$\nabla \cdot (\varepsilon^s \rho^s v^s) = \varepsilon^s \rho^s (\nabla \cdot v^s) + v^s \cdot \nabla(\varepsilon^s \rho^s) = \varepsilon^s \rho^s (\nabla \cdot v^s) + \frac{\partial u^s}{\partial t} \cdot \nabla(\varepsilon^s \rho^s).$$

Then the conservation equation reads

$$\frac{\partial}{\partial t}(\varepsilon^s \rho^s) + \varepsilon^s \rho^s (\nabla \cdot v^s) = R^s - \frac{\partial u^s}{\partial t} \cdot \nabla(\varepsilon^s \rho^s). \tag{7.20}$$

7.2.1.3 Preliminary Equation System

From the conservation equations above with all the assumptions and simplifications we come to the following system of three equations with respect to three primary variables p^l, ρ^s, and u^s:

$$\varepsilon \frac{\partial s^l}{\partial t} + \varepsilon s^l \gamma \frac{\partial p^l}{\partial t} - \nabla \cdot \left[\frac{k_r^l k}{\mu^l} (\nabla p^l - \rho^l g) \right] = \frac{R^l}{\rho^l} - s^l \varepsilon (\nabla \cdot v^s) - s^l \frac{\partial \varepsilon}{\partial t} - \phi^l(u^s),$$

$$\frac{\partial}{\partial t}(\varepsilon^s \rho^s) + \varepsilon^s \rho^s (\nabla \cdot v^s) = R^s - \phi^s(u^s), \tag{7.21}$$

$$Lu^s = d^s(|\Omega^s|)$$

with the deformation terms

$$\phi^l(u^s) = \frac{\partial u^s}{\partial t} \cdot \nabla(\varepsilon s^l), \quad \phi^s(u^s) = \frac{\partial u^s}{\partial t} \cdot \nabla(\varepsilon^s \rho^s). \tag{7.22}$$

The first equation of the system is derived from the mass conservation for liquid phase (7.19), where flux density q^l (7.11) has been substituted from the momentum conservation equation for liquid phase (7.10). It can be recognized as a generalized Richards-type flow equation describing the liquid movement in unsaturated porous media extended to absorbing and swelling materials. The second equation represents the mass conservation for the solid phase (7.20), while the deformation term on the RHS, $\phi^s(u^s)$, has been defined in (7.22). The third equation of system (7.21) is the momentum conservation for the solid phase (7.13).

The system still requires additional relations for chemical reactions R^l and R^s as well as further quantities: porosity ε, saturation s^l, solid strain d^s, relative permeability k_r^l, and saturated permeability k. They are derived and discussed in the next section.

7.2.2 Constitutive and Closure Relations

7.2.2.1 Solid Components of AGM

The solid fraction of AGM can be divided into three components:

1. CM—inert *carrier material* such as airfelt, polymers, fibers
2. AGM—pure superabsorber*
3. L → S—liquid absorbed onto AGM and transformed into solid

These abbreviations are used throughout this section for quantity indices. The mass m^s and the volume V^s of the solid phase in an REV of constant volume V can then be expressed as

$$
\begin{aligned}
m^s &= m^s_{CM} + m^s_{AGM} + m^s_{L \to S}, \\
V^s &= V^s_{CM} + V^s_{AGM} + V^s_{L \to S},
\end{aligned}
\tag{7.23}
$$

where the latter sum of volumes implies that the volume of water absorbed into AGM is equal to the change in AGM volume. Accordingly, we define bulk concentrations:

$$
\bar{C}^s_{CM} = \frac{m^s_{CM}}{V}, \quad \bar{C}^s_{AGM} = \frac{m^s_{AGM}}{V}, \quad \bar{C}^s_{L \to S} = \frac{m^s_{L \to S}}{V}.
\tag{7.24}
$$

The mass of the first two components, CM and AGM, does not change in time in the deforming control space Ω^s. Therefore, we can formulate the mass conservation in the following integral form:

$$
\begin{aligned}
\int_{\Omega^s} \bar{C}^s_{CM} \, dV &= \int_{\Omega^s_0} \bar{C}^s_{CM_0} \, dV, \quad \int_{\Omega^s} \bar{C}^s_{AGM} \, dV = \int_{\Omega^s_0} \bar{C}^s_{AGM_0} \, dV, \\
\int_{\Omega^s} \bar{C}^s_{CM} \, dV &= \int_{\Omega^s_0} \bar{C}^s_{CM_0} \, dV, \quad \int_{\Omega^s} \bar{C}^s_{AGM} \, dV = \int_{\Omega^s_0} \bar{C}^s_{AGM_0} \, dV,
\end{aligned}
\tag{7.25}
$$

where subscript 0 denotes the values per initial reference control-space volume $|\Omega^s_0|$. Although the mass of absorbed liquid L → S is increasing, no liquid can leave the gel because of the known chemical properties. The concentration $\bar{C}^s_{L \to S}$ is monotonically increasing corresponding to the growing volume

---

* The abbreviation "AGM" is used for a subscript of the solid component here in contrast to the general notation for a composite absorbent material.

$|\Omega^s|$. Therefore, the mass conservation for the third component (L → S) can also be expressed as:

$$\int_{\Omega^s} \bar{C}^s_{L\to S}\,dV = \int_{\Omega^s_0} \bar{C}^s_{L\to S}\,dV.$$

$$\int_{\Omega^s} \bar{C}^s_{L\to S}\,dV = \int_{\Omega^s_0} \bar{C}^s_{L\to S_0}\,dV. \tag{7.26}$$

To transform integration set Ω^s into the reference set Ω^s_0, we use the Jacobian J^s defined in (7.8) and obtain for all three components:

$$\int_{\Omega^s_0} (\bar{C}^s_{I_0} - \bar{C}^s_I J^s)\,dV = 0 \quad (I = CM, AGM, L \to S).$$

$$\int_{\Omega^s_0} (\bar{C}^s_{I_0} - \bar{C}^s_I J^s)\,dV = 0 \quad (I = CM, AGM, L \to S). \tag{7.27}$$

The common index I has been introduced for the sake of brevity. After resolution of the integrals and applying the definition (7.24), concentration, mass, and volume can be expressed through their respective values in the reference control-space volume $|\ \Omega^s_0\ |$:

$$\bar{C}^s_I = \frac{1}{J^s}\bar{C}^s_{I_0}, \quad m^s_I = \frac{1}{J^s}\bar{C}^s_{I_0}V, \quad V^s_I = \frac{1}{J^s}\frac{\bar{C}^s_{I_0}}{\rho^s_{I_0}}V \quad (I = CM, AGM, L \to S). \tag{7.28}$$

where the density is assumed unchanged, that is, $\rho^s_I = \rho^s_{I_0}$.

We introduce the absorbent process quantity m^s_2 that describes how much liquid is absorbed per gram of dry AGM:

$$m^s_2 = \frac{m^s_{L\to S}}{m^s_{AGM}} = \left[\frac{g\ (\text{of absorbed liquid})}{g\ (\text{of AGM})}\right]. \tag{7.29}$$

Substituting $m^s_{L\to S}$ and m^s_{AGM} from (7.28) into the previous definition of m^s_2, we obtain the relation between absorbed liquid concentration and concentration of AGM:

$$\bar{C}^s_{L\to S_0} = m^s_2\bar{C}^s_{AGM_0}, \tag{7.30}$$

where $\bar{C}^s_{AGM_0}$ is a known material constant. Further, the sums in (7.23) can be rewritten using concentrations and relations (7.28) for particular mass and volume of the components, where mass and concentration of absorbed liquid (L $\rightarrow$ S) are expressed using corresponding quantities of AGM and m^s_2 together with (7.29) and (7.30). The second sum is also divided by the volume V to obtain the solid fraction ε^s:

$$m^s = \frac{V}{J^s}[\bar{C}^s_{CM_0} + \bar{C}^s_{AGM_0}(1 + m^s_2)],$$

$$\varepsilon^s = \frac{V^s}{V} = \frac{1}{J^s}\left[\frac{\bar{C}^s_{CM_0}}{\rho^s_{CM_0}} + \frac{\bar{C}^s_{AGM_0}}{\rho^s_{AGM_0}}\left(1 + m^s_2\frac{\rho^s_{AGM_0}}{\rho^s_{L \to S_0}}\right)\right].$$

(7.31)

Both the above expressions lead to the solid density:

$$\rho^s = \frac{m^s}{V^s} = \frac{m^s}{\varepsilon^s V} = \frac{\bar{C}^s_{CM_0} + \bar{C}^s_{AGM_0}(1 + m^s_2)}{\dfrac{\bar{C}^s_{CM_0}}{\rho^s_{CM_0}} + \dfrac{\bar{C}^s_{AGM_0}}{\rho^s_{AGM_0}}(1 + m^s_2(\rho^s_{AGM_0}/\rho^s_{L \to S_0}))}.$$

(7.32)

Note that the resulting relations depend only on initial properties of CM and AGM as well as on m^s_2.

In the initial state, it is assumed that $m^s_2 = 0$ and obviously $J^s = 1$. Defining the *initial solid fraction* as ε^s_0, it follows from (7.31) that

$$\varepsilon^s_0 = \varepsilon^s\big|_{m^s_2=0} = \frac{\bar{C}^s_{CM_0}}{\rho^s_{CM_0}} + \frac{\bar{C}^s_{AGM_0}}{\rho^s_{AGM_0}}.$$

(7.33)

The first summand of the previous equation can be substituted back into (7.31) to express the solid fraction ε^s using the initial solid fraction ε^s_0:

$$\varepsilon^s = \frac{1}{J^s}\left(\varepsilon^s_0 + m^s_2\frac{\bar{C}^s_{AGM_0}}{\rho^s_{L \to S_0}}\right).$$

(7.34)

Instead of using Equation 7.34 directly to compute porosity $\varepsilon = 1 - \varepsilon^s$, it is reformulated with respect to volume dilatation as a function of m^s_2,

$$J^s(m^s_2) = \frac{1}{\varepsilon^s(m^s_2)}\left(\varepsilon^s_0 + m^s_2\frac{\bar{C}^s_{AGM_0}}{\rho^s_{L \to S_0}}\right),$$

(7.35)

while the porosity dependency on m_2^s is determined empirically using experimental data. This dependency is given in (7.60).

7.2.2.2 Chemical Reactions

In this section, we enter the microscale world of AGM–liquid interaction in order to derive relations for reaction terms R^l and R^s of the equation system (7.21). We are especially interested in the surfaces of the AGM particles where this interaction occurs. Since the particles may have very different shapes and our model should not depend on particle geometrical properties, it will be sufficient to consider a surface area of AGM particles in an REV. We introduce the *liquid–solid interface area* and denote its maximum value as A_p.

The square brackets [...] will be used to define molar bulk concentrations and emphasize chemical activity. We further introduce

- $[H_2O(l)]$—the liquid concentration
- $[AGM_{raw}(s)]$—the *available* reacting AGM solid concentration
- $[AGM_{consumed}(s)]$—the *consumed* AGM solid concentration
- $[H_2O_{L\rightarrow S}(s)]$—the sorbed liquid concentration in the solid
- $[AGM(s)]$—total molar concentration of AGM

The liquid transfer over the interface area can be described as a nonequilibrium sorption process:

$$\text{excess}[H_2O(l)] + [AGM_{raw}(s)] \xrightarrow{\text{absorption } k^+} [H_2O_{L\rightarrow S}(s)] + [AGM_{consumed}(s)],$$

(7.36)

where k^+ is the rate constant. Total molar concentration of AGM in the REV is the sum of available and consumed concentrations at every time point due to equilibrium condition:

$$[AGM_{raw}(s)] + [AGM_{consumed}(s)] = [AGM(s)]. \qquad (7.37)$$

Moreover, every consumed absorption site corresponds directly to one absorbed molecule of $H_2O_{L\rightarrow S}(s)$; therefore

$$[AGM_{consumed}(s)] = [H_2O_{L\rightarrow S}(s)]. \qquad (7.38)$$

New concentration definitions also lead to another formulation for m_2^s

$$m_2^s = \frac{[H_2O_{L\rightarrow S}(s)]M_{H_2O}}{[AGM(s)]M_{AGM}}, \qquad (7.39)$$

where M_{H_2O} and M_{AGM} are molar masses of the liquid and AGM, respectively. For the sake of simplicity, we assume that all dry AGM particles can be reached by liquid over the pore space. This implies that no gel blocking occurs and the complete AGM is available for absorption in fully saturated state. Thus, once the AGM has absorbed the maximum possible amount of liquid, the AGM is considered *fully loaded* and no $[AGM_{raw}(s)]$ remains. Using Equations 7.37 and 7.39, we obtain

$$[AGM_{consumed}(s)]_{max} = [H_2O_{L \to S}(s)]_{max} = [AGM(s)],$$

$$m_{2\,max}^s = \frac{[H_2O_{L \to S}(s)]_{max} M_{H_2O}}{[AGM(s)]M_{AGM}} = \frac{M_{H_2O}}{M_{AGM}}. \tag{7.40}$$

The latter expression is a purely theoretical approximation for the maximum AGM load, while its real value also depends on external pressure and salt concentration in water which are not considered in this theory.

By assumption, AGM is always available via pores at the solid–liquid interface. Therefore, the density of reacting AGM must be proportional to its concentration $[AGM_{raw}(s)]$. The liquid density at the interface remains the same and approximately equal to the water density $\rho_{H_2O}^l$. The interface area involved in the reaction must be a function of liquid saturation s^l. It is generally material-specific and depends on the pore-size distribution. We define it as a fraction with respect to the total interface area A_p of the REV and denote it as $a^{sl}(s^l)$. Finally, we can formulate the reaction velocity v_r in the following form:

$$v_r = \frac{\partial}{\partial t}[H_2O_{L \to S}(s)] = k^+ \rho_{H_2O}^l[AGM_{raw}(s)]a^{sl}(s^l)A_p. \tag{7.41}$$

Substitution of the conditions (7.37) and (7.38) as well as definitions of m_2^s (7.39) and $m_{2\,max}^s$ (7.40) leads to

$$v_r = k^+ \rho_{H_2O}^l \frac{\overline{C}_{AGM}^s}{M_{AGM}}\left(1 - \frac{m_2^s}{m_{2\,max}^s}\right)a^{sl}(s^l)A_p, \tag{7.42}$$

where we employ the bulk concentration $\overline{C}_{AGM}^s = [AGM(s)]M_{AGM}$. It is also convenient to introduce the *normalized x-load*

$$\hat{m}_2^s = \frac{m_{2\,max}^s - m_2^s}{m_{2\,max}^s} = 1 - \frac{m_2^s}{m_{2\,max}^s}. \tag{7.43}$$

Its value ranges in interval [0, 1], where 0 corresponds to $m_{2\,\mathrm{max}}^s$. Replacing AGM concentration through its initial value $\bar{C}_{\mathrm{AGM}}^s = \dfrac{1}{J^s}\bar{C}_{\mathrm{AGM}_0}^s$ (7.28), we arrive at the final expressions for the reaction rates:

$$R^s = v_r M_{\mathrm{H_2O}} = \tau \frac{\bar{C}_{\mathrm{AGM}_0}^s}{J^s}\,\hat{m}_2^s a^{sl}(s^l), \tag{7.44}$$

$$R^l = -R^s,$$

where τ denotes the specific *AGM reaction constant* defined as

$$\tau = k^+ \rho_{\mathrm{H_2O}}^l \frac{M_{\mathrm{H_2O}}}{M_{\mathrm{AGM}}} A_p. \tag{7.45}$$

For the detailed derivation of the function $a^{sl}(s^l)$, we refer to the article by Diersch et al. (2010, Appendix B) and give only the result:

$$a^{sl}(\hat{s}(s_e^l)) = \frac{1 - e^{-\alpha_{\exp}\hat{s}}}{1 - e^{-\alpha_{\exp}}}, \quad \hat{s}(s_e^l) = \frac{s_e^l - s_{e,\mathrm{threshold}}^l}{1 - s_{e,\mathrm{threshold}}^l}, \quad s_e^l = \frac{s^l - s_r^l}{s_s^l - s_r^l}, \tag{7.46}$$

where s_e^l is the *effective liquid saturation*. The *maximum saturation* s_s^l is typically 1, while the *residual saturation* s_r^l is a small positive constant representing initial saturation of raw AGM at natural conditions. The *effective-saturation threshold* $s_{e,\mathrm{threshold}}^l$ can be interpreted as a start point for absorption process. Thus, this constant prevents simulation of absorption under near dry conditions.

7.2.2.3 AGM Swelling Process and Solid Strain

This section explains calculation of the solid displacement vector $u^s = (u_1^s, u_2^s, u_3^s)$ and the solid strain d^s, which appear in the third equation of the system (7.21). Generally, solid strain is a vector of six components:

$$d^s(m_2^s) = \left[d_1^s(m_2^s), d_2^s(m_2^s), d_3^s(m_2^s), \gamma_{12}^s, \gamma_{23}^s, \gamma_{31}^s \right]^T. \tag{7.47}$$

As shown in the previous sections, m_2^s is the primary measure for swelling and changing of the control space Ω^s. It follows that the solid strain is also dependent on m_2^s. We will consider the shear effects $\gamma_{12}^s, \gamma_{23}^s$, and γ_{31}^s as negligible. Under those assumptions, which are also proved experimentally, the third equation of system (7.21) is reduced to

$$\frac{\partial u_i^s}{\partial x_i} = d_i^s(m_2^s) \quad (i = 1, 2, 3). \tag{7.48}$$

Since in a real situation or experiment, AGM is placed on a nonmovable base, the swelling direction can be determined from the shape of this non-movable base and the shape of AGM. Therefore, we introduce a *scalar volumetric strain* d^s of the porous solid as a change of the control volume $|\Omega^s|$:

$$|\Omega^s| d^s = \Delta\Omega^s. \tag{7.49}$$

To consider the control-volume deformation in detail and establish a relation between solid strain components d_i^s and scalar d^s, we turn to the simplest case of a small cubic control space Ω^s and the swollen space $\Omega^s + \Delta\Omega^s$ (Figure 7.2) with corresponding side lengths l_i^s and $l_i^s + \Delta l_i^s$ ($i = 1,2,3$). The volumes of the cuboids are simply

$$|\Omega^s| = \prod_{i=1}^{3} l_i^s, \quad |\Omega^s| + \Delta\Omega^s = \prod_{i=1}^{3} (l_i^s + \Delta l_i^s). \tag{7.50}$$

Similar to (7.49), every dimension of the cuboid grows corresponding to solid strain components:

$$l_i^s d_i^s = \Delta l_i^s. \tag{7.51}$$

Substitution of the relations (7.50) and (7.51) into the definition (7.49) leads to

$$d^s = \frac{\Delta\Omega^s}{\Omega^s} = \prod_{i=1}^{3}(1 + d_i^s) - 1 \approx \sum_{i=1}^{3} d_i^s, \tag{7.52}$$

where the sum approximation is acceptable for very small values of d_i^s ($0 \le d_i^s \ll 1, i = 1,2,3$). Returning to (7.48), we can write the final relation between displacement vector and the scalar volumetric strain as

$$d^s = \sum_{i=1}^{3} \frac{\partial u_i^s}{\partial x_i} = \nabla \cdot u^s. \tag{7.53}$$

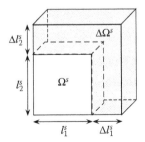

FIGURE 7.2
Incremental increase of a cubic volume Ω^s.

From the numerical point of view, the scalar volumetric strain d^s is related to $|\Omega^s|$ at every time step $n+1$ and the previous time step n as follows:

$$d^s_{n+1} = \frac{|\Omega^s_{n+1}| - |\Omega^s_n|}{|\Omega^s_n|} = \frac{|\Omega^s_{n+1}|}{|\Omega^s_n|} - 1. \tag{7.54}$$

This approach is appropriate for *large-scale deformations* (Lewis and Schrefler, 1998) as evident in superabsorbers, in contrast to small deformations in elasticity, where strain calculation refers always to the initial state Ω^s_0. Using the definition of the Jacobian (7.8) and its dependency on m^s_2, we can express d^s as a function of m^s_2 at every time step:

$$d^s_{n+1}(m^s_2) = \frac{J^s(m^s_{2(n+1)})}{J^s(m^s_{2(n)})} - 1. \tag{7.55}$$

7.2.2.4 Liquid Saturation on Capillary Pressure Dependency

Let us consider a wetting process of initially dry AGM. Liquid is penetrating first into AGM as into any nonswelling material. The amount of liquid phase corresponding to the nonwetting gas phase is reflected by the capillary pressure p_c. As is common for nonswelling media, the capillary pressure p_c is assumed to depend only on the liquid saturation s^l and is expressed using the pressure head ψ:

$$p_c(s^l) = -p^l(s^l) = -\psi^l \rho^l g, \tag{7.56}$$

where g is the gravitational acceleration. We apply the relation for effective saturation introduced by van Genuchten (1980):

$$s^l_e = \begin{cases} \dfrac{1}{[1+ |\alpha \psi^l|^n]^m} & \psi^l < 0 \\[2mm] 1 & \psi^l \geq 0 \end{cases}, \tag{7.57}$$

where α, n, and m are fitting parameters. Experiments with superabsorbers have shown strong hysteretic behavior. That means that the pressure–saturation relation should have different parameters for wetting and for drying processes. It has been experimentally established that those parameters also depend on porosity ε and therefore on m^s_2 according to (7.34). We found that it is sufficient to consider only α as a function of m^s_2, avoiding the need for additional empirical functions and computational effort.

7.2.2.5 Empirical Material Relations

In this section, we present all AGM parameters whose dependencies on other physical quantities could not be derived theoretically, but were fitted from experimental data measured on real superabsorbent materials.

- The α parameter of the saturation–pressure dependency (7.57) strongly decreases with increasing m_2^s:

$$\alpha(m_2^s) = \frac{\alpha_{max}}{[1 + \alpha_{scale} m_2^s]^{\alpha_{exp}}},\qquad (7.58)$$

where we also note the existence of two different values for wetting and drying processes:

$$\alpha_{max} = \begin{cases} \alpha_{max,wetting} & \dfrac{\partial \psi^l}{\partial t} > 0 \\[2ex] \alpha_{max,drying} & \dfrac{\partial \psi^l}{\partial t} < 0 \end{cases}. \qquad (7.59)$$

- Porosity is influenced by m_2^s:

$$\varepsilon(m_2^s) = \frac{2\varepsilon_{max}}{1 + (m_2^s \varepsilon_{scale} + 1)^{\varepsilon_{exp}}},\qquad (7.60)$$

where ε_{max}, ε_{scale}, and ε_{exp} are fitting parameters.

Relative permeability is described by an exponential function in contrast to the classical van Genuchten–Mualem relation:

$$k_r^l(s^l) = (s_e^l)^\delta, \qquad (7.61)$$

where δ is a fitting parameter.

- The following function provides sufficient flexibility in modeling saturated permeability as an isotropic material $k = kI$:

$$k(m_2^s) = k_{base}[1 + k_{coeff} \exp(k_{expcoeff} m_2^s) \sin(2\pi k_{sincoeff} m_2^s + k_{sinphase})], \qquad (7.62)$$

where k_{base}, k_{coeff}, $k_{expcoeff}$, $k_{sincoeff}$, and $k_{sinphase}$ are fitting parameters.

7.2.3 Final Equation System

The preliminary system of PDEs (7.21) can now be reformulated according to the theoretical and empirical relations in Section 7.2.2. Anticipating the subsequent numerical solution using FEM, we select the following primary unknowns of the system:

1. Pressure head $\psi^l = p^l/\rho^l g$.
2. Concentration of absorbed liquid per initial bulk volume $\bar{C}^s_{L \to S_0}$. It is directly related to the characteristic quantity of the absorption process $m_2^s = \bar{C}^s_{L \to S_0}/\bar{C}^s_{AGM_0}$, where $\bar{C}^s_{AGM_0}$ is a constant parameter of AGM. The concentration $\bar{C}^s_{L \to S_0}$ can be volume-integrated and is therefore a better choice for the numerical FEM solution than m_2^s.
3. Solid displacement u^s.

All other quantities, such as s^l, q^l, v^s, ε, and ρ^s, can be computed from the primary variables.

From the assumption (7.53), we can derive a simple relationship for the divergence of the solid phase velocity v^s. We differentiate (7.53) by time, use the symmetry of second derivatives and substitute the definition of the solid phase velocity (7.9):

$$\frac{\partial}{\partial t} d^s = \frac{\partial}{\partial t} \nabla \cdot u^s = \nabla \cdot \frac{\partial u^s}{\partial t} = \nabla \cdot v^s \cdot \frac{\partial}{\partial t} d^s = \frac{\partial}{\partial t} \nabla \cdot u^s = \nabla \cdot \frac{\partial u^s}{\partial t} = \nabla \cdot v^s.$$

Comparing the property of the Jacobian (7.8) with (7.49), we obtain that $J d^s = \Delta J$ and therefore

$$\nabla \cdot v^s = \frac{1}{J^s} \frac{d J^s}{dt}. \tag{7.63}$$

This relation will be substituted in both balance equations of the system (7.21).

First, we simplify the balance equation for solid phase, the second in the system (7.21), in order to express it with respect to $\bar{C}^s_{L \to S_0}$. Moreover, it will deliver a short relation for reaction terms $R^s = -R^l$, which will then be substituted into the first equation of the system. We explain the simplification of the LHS in five steps:

$$\frac{\partial}{\partial t}(\varepsilon^s \rho^s) + \varepsilon^s \rho^s (\nabla \cdot v^s) \overset{(1)}{=\!=} \frac{\partial}{\partial t}(\varepsilon^s \rho^s) + \varepsilon^s \rho^s \frac{1}{J^s} \frac{d J^s}{dt}$$

$$\overset{(2)}{=\!=} \frac{\partial(\varepsilon^s \rho^s)}{\partial m_2^s} \frac{\partial m_2^s}{\partial t} + \frac{\varepsilon^s \rho^s}{J^s} \frac{\partial J^s}{\partial m_2^s} \frac{\partial m_2^s}{\partial t} \tag{7.64}$$

$$\overset{(3)}{=\!=} \frac{1}{J^s} \frac{\partial(\varepsilon^s \rho^s J^s)}{\partial m_2^s} \frac{\partial m_2^s}{\partial t} \overset{(4)}{=} \frac{\bar{C}^s_{AGM_0}}{J^s} \frac{\partial m_2^s}{\partial t} \overset{(5)}{=} \frac{1}{J^s} \frac{\partial \bar{C}^s_{L \to S_0}}{\partial t}.$$

1. Relation (7.63) is substituted.
2. The differentiation chain rule is applied because the physical quantities depend on m_2^s.
3. The derivative $\partial m_2^s / \partial t$ can be factored out resulting in a sum that represents the differentiation product rule for the derivative $\partial(\varepsilon^s \rho^s J^s)/\partial m_2^s$.
4. From relation (7.31), we obtain

$$\varepsilon^s \rho^s J^s = \frac{m^s}{V} J^s = [\bar{C}_{CM_0}^s + \bar{C}_{AGM_0}^s (1 + m_2^s)], \quad \frac{\partial(\varepsilon^s \rho^s J^s)}{\partial m_2^s} = \bar{C}_{AGM_0}^s.$$

5. Concentration of absorbed liquid $\bar{C}_{L \to S_0}^s$ can be substituted due to differentiation of (7.30)

$$\frac{\partial \bar{C}_{L \to S_0}^s}{\partial t} = \bar{C}_{AGM_0}^s \frac{\partial m_2^s}{\partial t},$$

where $\bar{C}_{AGM_0}^s$ is constant over time.

The RHS receives the reaction term according to (7.44) and the second balance equation takes the following form:

$$\frac{1}{J^s} \frac{\partial \bar{C}_{L \to S_0}^s}{\partial t} = R^s - \phi^s(u^s),$$

$$\frac{\partial \bar{C}_{L \to S_0}^s}{\partial t} = \tau a^{sl}(s_e^l) \hat{m}_2^s \bar{C}_{AGM_0}^s - J^s \phi^s(u^s). \tag{7.65}$$

On the other hand, we can use the latter equation to express the reaction term

$$R^s = -R^l = \frac{\bar{C}_{AGM_0}^s}{J^s} \frac{\partial m_2^s}{\partial t} + \phi^s(u^s) \tag{7.66}$$

and then substitute it into the first equation.

In the balance equation for the liquid phase, the first in the system (7.21), the RHS is modified according to the following two steps:

$$\frac{R^l}{\rho^l} - s^l \varepsilon(\nabla \cdot v^s) - s^l \frac{\partial \varepsilon}{\partial t} - \phi^l(u^s) \overset{(1)}{=} \frac{R^l}{\rho^l} - \frac{s^l \varepsilon}{J^s} \frac{\partial J^s}{\partial m_2^s} \frac{\partial m_2^s}{\partial t} - s^l \frac{\partial \varepsilon}{\partial m_2^s} \frac{\partial m_2^s}{\partial t} - \phi^l(u^s)$$

$$\overset{(2)}{=} - \left[\frac{1}{\rho^l} \frac{\bar{C}_{AGM_0}^s}{J^s} + s^l \left(\frac{\partial \varepsilon}{\partial m_2^s} + \frac{\varepsilon}{J^s} \frac{\partial J^s}{\partial m_2^s} \right) \right] \frac{\partial m_2^s}{\partial t} - \frac{1}{\rho^l} \phi^s(u^s) - \phi^l(u^s).$$

$$\tag{7.67}$$

1. Relation (7.63) is substituted and the differentiation chain rule is applied, introducing derivatives with respect to m_2^s.

2. Reaction term (7.66) is substituted and the derivative $(\partial m_2^s / \partial t)$ is factored out.

Finally, the liquid pressure p^l in the LHS is substituted by the hydraulic head ψ^l using (7.56). The complete system of three PDEs reads

$$\varepsilon \frac{\partial s^l(\psi^l)}{\partial t} + \varepsilon s^l \bar{\gamma} \frac{\partial \psi^l}{\partial t} - \nabla \cdot [k_r^l K(\nabla \psi^l + e)] = - \left[\frac{1}{\rho^l} \frac{\bar{C}_{AGM_0}^s}{J^s} + \varepsilon \frac{\partial s^l}{\partial m_2^s} \right.$$

$$\left. + s^l \left(\frac{\partial \varepsilon}{\partial m_2^s} + \frac{\varepsilon}{J^s} \frac{\partial J^s}{\partial m_2^s} \right) \right] \frac{\partial m_2^s}{\partial t}$$

$$- \frac{1}{\rho^l} \phi^s(u^s) - \phi^l(u^s), \qquad (7.68)$$

$$\frac{\partial \bar{C}_{L \to S_0}^s}{\partial t} = \tau a^{sl} (\hat{s}(s_e^l)) \, \hat{m}_2^s \, \bar{C}_{AGM_0}^s - J^s \phi^s(u^s),$$

$$\nabla \cdot u^s = d^s.$$

For simplicity, we introduced the *gravitational unit vector* e, which fulfills $g = -ge$, the hydraulic conductivity $K = k \frac{\rho^l g}{\mu^l} K = k \rho^l g / \mu^l$ and the specific liquid compressibility $\bar{\gamma} = \gamma \rho^l g$. Further quantities must be already known from this section and can be computed with the help of empirical relations (7.58) through (7.62) or using dependencies on other known parameters. Note that the first equation represents a generalized Richards-type flow equation written in a mixed $(\psi - s)$-form (e.g., Diersch and Perrochet, 1999), where both pressure head ψ and saturation s are employed. Its discrete formulation can provide better mass conservation and accuracy than the original pressure-based Richards' equation. The pressure head ψ is chosen as the primary variable even in the present $(\psi - s)$-formulation, which is capable of simulating both saturated and unsaturated conditions.

7.3 Finite-Element Solution

The system of PDEs (7.68) developed in the previous section contains nonlinear terms due to complicated dependencies of physical quantities on m_2^s. For arbitrary 2D and 3D domain geometries of realistic problems, such a system can only be solved numerically. We apply an adaptive predictor–corrector iterative method for time discretization, either the Picard or the Newton method for linearization of nonlinear terms, and FEM for spatial discretization.

First, we rewrite the equation system in simplified notation without phase superscripts and abbreviated RHSs:

$$\varepsilon \frac{\partial s(\psi)}{\partial t} + \varepsilon s \gamma \frac{\partial \psi}{\partial t} - \nabla \cdot [k_r K \cdot (\nabla \psi + e)] = R_\psi(\psi, C, u)$$

$$\frac{\partial C}{\partial t} = R_c(\psi, C, u) \qquad (7.69)$$

$$\nabla \cdot u = d(C)$$

System (7.69) must be solved for the liquid pressure head ψ, the sorbed liquid concentration $C \equiv m_2^s \bar{C}_{AGM_0}^s$, and the solid displacement u on the bounded domain Ω^s with the boundary Γ^s.

The displacement vector u^s can be written as a product of a scalar displacement norm u^s and a displacement–direction vector a^s of unit size: $u^s = u^s a^s$. The divergence of the displacement can be then decomposed into two summands

$$\nabla \cdot u^s = a^s \cdot \nabla u^s + u^s \nabla \cdot a^s. \qquad (7.70)$$

The second term will be zero if all points of the swelling domain move in the same direction. In most simulations of real experiments, the investigated superabsorbent material has normally at least one straight nonmovable base. As a result, the swelling directions differ only slightly. Thus, we can neglect the last term in (7.70) under the condition

$$a^s \cdot \nabla u^s \gg u^s \nabla \cdot a^s. \qquad (7.71)$$

More details and discussion on this assumption can be found in Diersch et al. (2010, Section 4).

The resulting PDE $\nabla \cdot (au) = d$ must be stabilized due to numerical reasons (see, e.g., Donea and Huerta, 2003). We apply a streamline upwinding method introducing an additional solid stress in the form of artificial diffusion of displacement. Therefore, the displacement equation is considered numerically in a stabilized form:

$$\kappa \frac{\partial u}{\partial t} + a \cdot \nabla u - \nabla \cdot (\sigma^\circ \cdot \nabla u) = d, \quad \sigma^\circ = \beta_{upwind}(a \otimes a), \qquad (7.72)$$

where κ is a small artificial compression factor and the assumption (7.71) is implied. The upwind parameter β_{upwind} denotes *numerical dispersivity.* It can be estimated as a half of characteristic finite-element length l: $\beta_{upwind} \approx l/2$.

To be uniquely solvable, system (7.69) requires appropriate initial and boundary conditions. We set the following initial conditions:

$$\psi(x,0) = \psi_0(x) \text{ in } \Omega^s \cup \Gamma^s,$$
$$C(x,0) = C_0(x) \text{ in } \Omega^s \cup \Gamma^s, \qquad (7.73)$$
$$u(x,0) = 0 \text{ in } \Omega^s \cup \Gamma^s$$

and boundary conditions (BC)

$$\psi(x,t) = \psi_D(t) \text{ on } \Gamma^s_{D_\psi} \text{ Dirichlet BC,}$$
$$u(x,t) = 0 \text{ on } \Gamma^s_{D_u} \text{ Dirichlet BC,}$$
$$-[k_r K \cdot (\nabla \psi + e)] \cdot n = q_n(t) \text{ on } \Gamma^s_{N_\psi} \text{ Neumann BC,} \qquad (7.74)$$
$$- [\sigma^\circ \cdot \nabla(au)] \cdot n = 0 \text{ on } \Gamma^s_{N_u} \text{ Neumann BC,}$$

where $\Gamma^s_{D_\psi} \cup \Gamma^s_{N_\psi} = \Gamma^s_{D_u} \cup \Gamma^s_{N_u} = \Gamma^s$ is composed of the two nonoverlapping boundary segments on which Dirichlet-type and Neumann-type BCs are specified for the pressure head ψ and the displacement u, respectively. The latter BC in (7.74) represents the natural boundary condition including the artificial stress σ°. Note that no BC is required for the concentration C.

7.3.1 Discretization in Space and Time

The FEM was preferred over other possible numerical approaches such as finite-difference or finite-volume methods due to its rigorous theoretical background as well as arbitrary solution domain geometry and general robustness. Since a full FEM discretization of the system (7.69) would exceed the scope of this chapter, we refer to Zienkiewicz and Taylor (2000) for a comprehensive description of FEM. After a discretization of the control space Ω^s_0 into finite elements and assembling of matrices according to the system (7.69) over all the elements, we can rewrite the system in the following matrix form:

$$B(C)\dot{S} + O(C)\dot{\Psi} + K(S,C)\Psi = F(S,C,U) \text{ in } \Omega^s_{n+1}$$
$$A\dot{C} = Z(S,C,U) \text{ in } \Omega^s_0 , \qquad (7.75)$$
$$P\dot{U} + DU = Q(C) \text{ in } \Omega^s_{n+1}$$

which is solved for vectors Ψ, C, and U. A dot over a vector indicates differentiation with respect to time t. Parentheses show the dependencies between equation terms. The vectors Ψ, C, and U return the resulting nodal values of the liquid pressure head for the deformed (swollen) volume Ω^s_{n+1}, the concentration of absorbed liquid per initial undeformed bulk volume Ω_0^s, and the solid displacement for the deformed volume Ω_{n+1}^s, respectively. The saturation vector S can be evaluated from known Ψ. In contrast to the other two equations, the second equation in (7.75) is solved on Ω_0^s because it involves no transport in the domain and its primary variable C is defined with respect to the undeformed geometry. The matrices B, O, A, and P are symmetric. The conductance matrix K is nonsymmetric if a Newton iteration technique is employed for the solution procedure; otherwise it is symmetric. The displacement matrix D is always nonsymmetric. The vectors of the RHS F, Z, and Q represent liquid sink, kinetic absorption reaction, and solid-strain source, respectively.

The physical domain Ω^s is deforming due to the swelling process and is therefore time dependent. We start with the time t_0 and define time-step increment $\Delta t_n = t_{n+1} - t_n$, where n is the time-level index. The increment is automatically adjusted by the employed adaptive strategy, which is explained below. For the sake of simplicity, we define $\Omega_n^s = \Omega^s(t_n)$ and $\Omega_{n+1}^s = \Omega^s(t_n + \Delta t_n)$. The same rule will be used for matrix and vector notations on time levels n and $n+1$. The solution process requires computation of unknown vectors Ψ, C, and U at time level $(n+1)$ assuming that they are known at the previous time step n. The connection between the next $(n+1)$ and the current time level n is traditionally established by a discretization method of ordinary differential equations. We apply a first-order fully implicit predictor-corrector (forward Euler/backward Euler) time-stepping scheme with a *residual control*. This is a two-step method, where the two steps correspond to the forward and backward Euler method, respectively. A detailed description of the scheme can be found in Diersch and Perrochet (1999) and FEFLOW (2010). It results in the following system of three equations:

1. $\displaystyle R_{n+1}^\tau = \frac{B(C_{n+1}^\tau)}{\Delta t_n} S_{n+1}^\tau + \left(\frac{O(S_{n+1}^\tau, C_{n+1}^\tau)}{\Delta t_n} + K(S_{n+1}^\tau, C_{n+1}^\tau) \right) \Psi_{n+1}^\tau$

$\displaystyle \qquad - \frac{B(C_{n+1}^\tau)}{\Delta t_n} S_n - \frac{O(S_{n+1}^\tau, C_{n+1}^\tau)}{\Delta t_n} \Psi_n - F(S_{n+1}^\tau, C_{n+1}^\tau, U_n) \quad \text{in } \Psi\Omega_{n+1}^s,$

2. $\displaystyle \frac{A}{\Delta t_n} C_{n+1} = \frac{A}{\Delta t_n} C_n + Z(S_{n+1}^\tau, C_{n+1}^\tau, U_n) \quad \text{in } \Omega_0^s,$

3. $\displaystyle \left(\frac{P}{\Delta t_n} + D \right) U_{n+1} = \frac{P}{\Delta t_n} U_n + Q(C_{n+1}^\tau) \quad \text{in } \Omega_{n+1}^s.$

The first equation defines vector R_{n+1}^τ, which is the residual of the discretized Richards' equation. The new superscript τ denotes the iteration counter for the additional linearization procedure. The first iteration starts with $\tau = 0$ at the new time level $(n + 1)$ using predictor values Ψ_{n+1}^P and C_{n+1}^P, which are computed from

$$\Psi_{n+1}^0 = \Psi_{n+1}^P = \Psi_n + \Delta t_n \dot{\Psi}_n,$$

$$C_{n+1}^0 = C_{n+1}^P = C_n + \Delta t_n \dot{C}_n,$$

(7.76)

where $\dot{\Psi}_n$ and $\dot{C}_n$ are derivative vectors. Further, the next solution approach is computed according to

$$J(S_{n+1}^\tau, C_{n+1}^\tau) \Delta \Psi_{n+1}^\tau = -R_{n+1}^\tau$$

(7.77)

with the solution refinement

$$\Delta \Psi_{n+1}^\tau = \Psi_{n+1}^{\tau+1} - \Psi_{n+1}^\tau.$$

(7.78)

The Jacobian of the linearization procedure is by definition the derivative of the residual vector with respect to the pressure

$$J(S_{n+1}^\tau, C_{n+1}^\tau) = \frac{\partial R_{n+1}^\tau(S_{n+1}^\tau, C_{n+1}^\tau)}{\partial \Psi_{n+1}^\tau}.$$

(7.79)

FEFLOW provides two methods to linearize nonlinear terms. They differ in the Jacobian definition.

1. The *Picard method* has at most a linear convergence rate, but a relatively large convergence radius. Even if the initial estimate Ψ_{n+1}^0 lies far from the actual solution, the method can still converge, where the Newton method would fail. The Jacobian represents a symmetric matrix in the form:

$$J = \frac{B}{\Delta t_n} \frac{\partial S_{n+1}^\tau}{\partial \Psi_{n+1}^\tau} + \frac{O}{\Delta t_n} + K.$$

2. The *Newton method* has quadratic convergence rate but a smaller convergence radius than the Picard method. The Jacobian is a nonsymmetric matrix and has the following more complicated form:

$$J = \frac{\partial O}{\partial \Psi_{n+1}^\tau} \left(\frac{\Psi_{n+1}^\tau - \Psi_n}{\Delta t_n} \right) + \frac{B}{\Delta t_n} \frac{\partial S_{n+1}^\tau}{\partial \Psi_{n+1}^\tau} + \Psi_{n+1}^\tau \frac{\partial K}{\partial \Psi_{n+1}^\tau} + \frac{O}{\Delta t_n} + K - \frac{\partial F}{\partial \Psi_{n+1}^\tau}.$$

The linearization iterations (7.77) and (7.78) must be continued until the residual norm will be smaller than predefined error tolerance η_l:

$$\| R_{n+1}^{\tau} \| < \eta_l, \quad \| R_{n+1}^{\tau} \| = \sqrt{\sum_{i=1}^{N} (R_{n+1,i}^{\tau})^2 / N}. \tag{7.80}$$

Practical simulation experience shows that the linearization requires only a few iterations. The maximum possible number of iterations was set to 12.

After the linearization procedure has converged, the last determined pressure head vector $\mathbf{\Psi}_{n+1}$ is the numerical solution for the pressure head at time level $n + 1$. It allows the computation of the remaining unknowns C_{n+1} and U_{n+1}. The complete solution process is summarized in Table 7.1. The linear equation systems arising in steps 2, 4, and 5 possess sparse system matrices. Therefore, iterative solvers from the family of Krylov subspace methods would be beneficial. We apply the preconditioned conjugate-gradient method (PCG) for the solution of the symmetric systems that appear in step 2 when using Picard linearization. For nonsymmetric systems, BiCGStab and GMRES methods have been found to be most efficient from the iterative solvers available in FEFLOW.

Once the solution at time level $(n + 1)$ has been computed, the new time-step length is estimated from the relation between predictor and corrector values according to

$$\Delta t_{n+1} = \Delta t_n \min \left[\sqrt{\frac{\eta_t}{\| d_\psi^{n+1} \|}}, \sqrt{\frac{\eta_t}{\| d_c^{n+1} \|}}, \sqrt{\frac{\eta_t}{\| d_u^{n+1} \|}} \right] \tag{7.81}$$

with the error vectors for pressure head, concentration, and solid displacement

TABLE 7.1

Numerical Solution Algorithm

1.	Initialize	$\mathbf{\Psi}_{n+1}^0 = \mathbf{\Psi}_{n+1}^P, \quad C_{n+1}^0 = C_{n+1}^P$	
2.	Solve	$J\Delta\mathbf{\Psi}_{n+1}^{\tau} = -R_{n+1}^{\tau} \quad$ until $\| R_{n+1}^{\tau} \| < \eta_l$	in Ω_{n+1}^s
3.	Evaluate	$S_{n+1}^{\tau} = S_{n+1}^{\tau}(\mathbf{\Psi}_{n+1}^{\tau}), \quad C_{n+1}^{\tau} = C_{n+1}^{\tau}(\mathbf{\Psi}_{n+1}^{\tau})$	
4.	Solve	$\dfrac{A}{\Delta t_n} C_{n+1} = \dfrac{A}{\Delta t_n} C_n + Z(S_{n+1}^{\tau}, C_{n+1}^{\tau}, U_n)$	in Ω_0^s
5.	Solve	$\left(\dfrac{P}{\Delta t_n} + D \right) U_{n+1} = \dfrac{P}{\Delta t_n} U_n + Q(C_{n+1}^{\tau})$	in Ω_{n+1}^s

$$d_\psi^{n+1} = \frac{1}{2}(\mathbf{\Psi}_{n+1} - \mathbf{\Psi}_{n+1}^P), \quad d_c^{n+1} = \frac{1}{2}(\mathbf{C}_{n+1} - \mathbf{C}_{n+1}^P), \quad d_u^{n+1} = \frac{1}{2}(\mathbf{U}_{n+1} - \mathbf{U}_{n+1}^P),$$

$$(7.82)$$

where $\|...\|$ is the root mean square (RMS) error norm defined as in (7.80). Parameter η_t is a constant predefined time error tolerance. As a result, the time-step size is adapted at every time level according to changes in the solution vectors. The time-step size is increased whenever possible and decreased only if necessary. This adaptive strategy is very computationally efficient and has been practically proven to be robust (see Diersch et al., 2011, for more details).

7.3.2 Hysteresis Implementation

Hysteretic behavior of the pressure–saturation relationship $s(\psi)$ means that it depends on the direction of pressure head change. If it increases, the saturation also increases that represents a wetting process. Otherwise, a drying process occurs. In case of hysteresis, there are actually two main dependencies: $s_w(\psi)$ for wetting and $s_d(\psi)$ for drying. Usually, wetting curve lies under drying curve. They meet at point ($\psi = 0$) and must not cross but instead converge exponentially as ψ approaches minus infinity. By default, the simulations considered here start assuming a wetting process. A special point ψ^* denotes a change of the process from wetting to drying or vice versa. According to Scott et al. (1983), instead of jumping from one main curve to another at point ψ^*, a new scanning curve $s^*(\psi)$ is obtained by scaling the main curve through the current reversal point $s(\psi^*)$ (Figure 7.3). We consider the effective saturation $s_e(\psi)$ rewriting it here from (7.46), omitting the phase index:

$$s_e(\psi) = \frac{s(\psi) - s_r}{s_s - s_r}, \tag{7.83}$$

where the constants s_r and s_s correspond to the residual (minimum) and fully saturated (maximum) saturation values, respectively. These constants

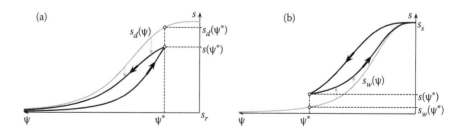

FIGURE 7.3
Scanning curves in a hysteresis loop. (a) From wetting to drying and (b) from drying to wetting.

are assumed to be the same for both wetting and drying curves. Figure 7.3a illustrates the case where the process changes from wetting to drying and the drying scanning curve must be derived. If $s(\psi)$ denotes the current wetting curve that turns into a drying scanning curve $s_d^*(\psi)$ in the point $s(\psi^*) = s_d^*(\psi^*)$, then the coefficient c_d for the scaling through $s(\psi^*)$ of the main drying curve is obtained from

$$c_d(\psi) = \frac{s_d^*(\psi) - s_r}{s_d(\psi) - s_r}, \quad c_d(\psi^*) = \frac{s(\psi^*) - s_r}{s_d(\psi^*) - s_r}, \tag{7.84}$$

where the latter gives the value of c_d and the former returns the desired drying scanning curve $s_d^*(\psi)$. Similarly, the wetting scanning curve $s_w^*(\psi)$ can be derived with the help of the Figure 7.3b. The results are summarized below:

- Drying process ($\psi < \psi^*$)

$$s_d^*(\psi) = c_d s_d(\psi) + (1 - c_d)s_r, \quad c_d = \frac{s(\psi^*) - s_r}{s_d(\psi^*) - s_r}. \tag{7.85}$$

- Wetting process ($\psi^* < \psi < 0$)

$$s_w^*(\psi) = c_w s_w(\psi) + (1 - c_w)s_s, \quad c_w = \frac{s_s - s(\psi^*)}{s_s - s_w(\psi^*)}. \tag{7.86}$$

7.4 Wicking Experiments in Absorbent Swelling Materials

The model described in the previous paragraphs of this chapter has been tested versus wicking experiments with swelling materials. The wicking experiments described in Sections 7.4.1 and 7.4.2 were chosen to test the model across a range of different conditions. A commercial diaper core, comprising about 12.6 g of AGM and 11.5 g of cellulose fibers (also known as airfelt or pulp or fluff fibers), was used in all the experiments and simulations shown in the following paragraphs; it will be referred to as "core sample" in the rest of the chapter.

7.4.1 Dynamic Absorption Test

This test method measures the amount of liquid being absorbed by the absorbent swelling material at a fixed capillary pressure. In this test, the flow and swelling directions are both normal to the material plane, leading to a homogeneous swelling of the sample.

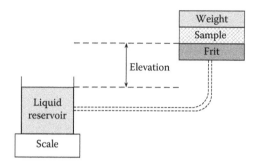

FIGURE 7.4
Schematic construction of dynamic absorption test.

An initially dry core sample of 6 cm diameter and of 3.5 g mass was loaded into a sample holder and placed on a porous glass frit that was connected via tubing to a liquid reservoir. A weight was applied onto the core sample corresponding to a confining pressure of 0.3 psi (see Figure 7.4). The core sample elevation above the liquid level of the reservoir was set at the beginning of each test. Three tests were performed, using sample elevations of 0, 0.2, and 0.4 m, respectively, thereby increasing the capillary pressure and correspondingly decreasing the saturation before the onset of AGM absorption and related swelling. The choice of the pore size of the porous glass is such that it remains saturated within the elevation range of interest, preventing air from entering the tubings. The reservoir resided on a scale connected to a computer for dynamic data acquisition. For each test, the amount of 0.9% NaCl–water solution* absorbed per gram of AGM was measured continuously over a period of 1 h. The tests were performed at 23°C and 50% relative humidity (RH). The 0.9% NaCl–water solution test liquid has a surface tension of 72.5 mN/m, a density of 1.005 g/cm³, and a viscosity of 1.019 mPa · s. Additionally, a correction of the measured uptake was needed to account for the evaporation and the absorption of the sample holder. To account for test variability, two replicates were executed for each experiment, based on which standard deviations were added as error bars on the plot: it can be seen that the method is quite reproducible, only the experiments at 20 cm showed a relative standard deviation of up to 8% (see Figure 7.5).

7.4.2 Vertical Wicking Experiment

This method measures the liquid uptake and wicking-front height in a material hanging vertically, with its lower edge slightly submerged in the liquid reservoir. Here, the direction of the vertical wicking flow against gravity is

---

* This solution is also known as normal saline (NS) or physiological saline or isotonic saline. It is often used in EDANA (European Disposables and Nonwovens Association) methods.

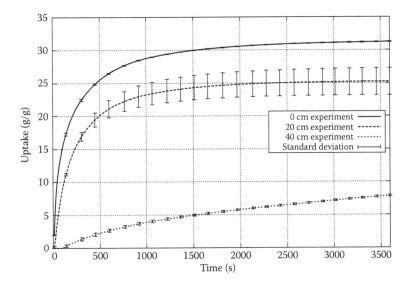

FIGURE 7.5
Experimental results of the dynamic absorption test.

perpendicular to the swelling direction which is leading to heterogeneous swelling. As in Section 7.4.1, 0.9% NaCl–water solution was used as test liquid. Indigo Carmine was added as dye to help recognize the wicking-front position.

The core sample analyzed here was taken directly from a diaper. The length and average width of the absorbent material were 419 mm and 88.3 mm, respectively.

The initially dry core sample was placed between two plexiglas plates with dimensions sufficiently larger than the sample core. An inflatable rubber pillow was placed between the two plates and maintained at a constant pressure of 0.3 psi (see Figure 7.6). After the pillow and the core sample were placed between the two plates, the plates were fixed via screws placed at the four corners and in the center of the plates. The thus assembled device was then hung vertically and dipped into the liquid reservoir which resided on a scale connected to a computer for dynamic data acquisition. While the liquid uptake was measured via the scale at 5 s intervals, the liquid-front height was measured via a calibrated ruler and recorded at 5, 10, 15, 20, 25, and 30 min elapsed time from the beginning of the test. The wicking-front position was assessed visually as average across the width of the core sample. The total duration of the experiment was 30 min. The tests were performed at 23°C and 50% RH. To account for test variability, three replicates were executed for each experiment, reflected by the standard deviation error bars in the plots of Figure 7.7. A correction of the measured uptake was needed to account for the evaporation.

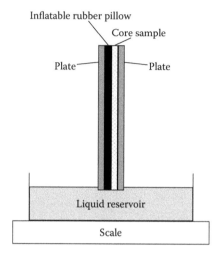

FIGURE 7.6
Schematic construction of vertical wicking experiment.

7.5 Simulation of Wicking Experiments

The physical experiments described in Section 7.4 were simulated in FEFLOW to provide quantitative validation by comparing the predicted and measured values for liquid uptake and wicking-front height.

7.5.1 Determination of Input Parameters for the Core Sample

The simulation of absorbent swelling materials requires the determination of 30 material input parameters, including fitting parameters for the empirical constitutive relationships as described in Section 7.2.2. A comprehensive description of the test methods and parameter determination would go beyond the scope of this chapter. Instead, we give a brief description of the key approaches for determining such input parameters experimentally. All measurements were performed at 23°C and 50% RH under a confining pressure of 0.3 psi. All parameters were determined from experiments that were independent from the subsequent validation tests. One notable exception was the dynamic absorption test at 20 cm height which was needed to estimate one parameter as explained in detail in this section.

The *porosity* ε of the swelling material depends on the extent of swelling m_2^s. Therefore, measurements were performed with NaCl–water solutions at different concentrations, ranging from 0.9% to 25% of NaCl by weight. In each experiment, the porosity is obtained by dividing the measured void volume by the measured total volume of the sample, while m_2^s by dividing

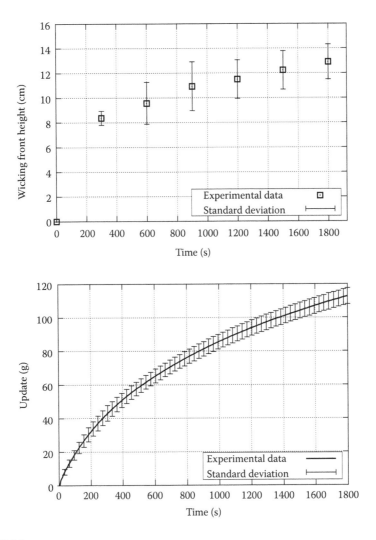

FIGURE 7.7
Results of the vertical wicking in a core sample.

the immobile liquid weight after removing the liquid from the interstitials by the amount of AGM in the core sample. Empirical relation (7.60) was then fitted to the data using ε_{max}, ε_{scale}, and ε_{exp} as fitting parameters that have to be determined using standard fitting methods. Figure 7.8 shows the experimental porosity values and the corresponding fitting curve.

The *hydraulic conductivity* K is defined as

$$K = \frac{k\rho^l g}{\mu^l},$$

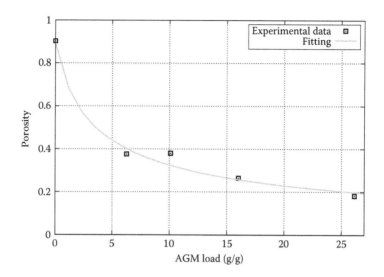

FIGURE 7.8
Fitting of the porosity ε dependent on AGM load m_2^s.

where k is the saturated permeability, ρ^l is the liquid density, g is the gravitational acceleration, and μ^l is the liquid viscosity. As has been reported elsewhere (e.g., Buchholz, 2006), permeability for a swelling material is generally a function of m_2^s and saturation s_l^e

$$k = f(s_l^e, m_2^s) = K(m_2^s) \cdot k_r^l(s_l^e, m_2^s),$$

where k_r^l is the relative permeability.

In general, permeability may be anisotropic as reflected by the model equations. This potential complexity was simplified in our model by assuming that the dependence of conductivity on fluid-swellable composite material load m_2^s and saturation s_l^e satisfies the relations (7.61) and (7.62) for the relative and the saturated permeability, respectively.

The hydraulic conductivity K of an absorbent material can be directly measured in the laboratory. The in-plane and cross-plane saturated permeability can be measured using a constant hydrostatic pressure head method. The saturated swelling material, under the confining pressure of 0.3 psi, is subjected to a flow rate under a constant hydrostatic pressure head. The mass flow rate of fluid through the layer is recorded for a fixed period of time and then used to calculate the saturated permeability of the absorbent, knowing the area of the absorbent orthogonal to the direction of flow, the thickness of the absorbent in a direction in-plane with the direction of flow, and the change in total pressure head across the absorbent.

Permeability k as function of m_2^s can be obtained by performing measurements with water solution at different NaCl concentrations, ranging from

0.9% to 25% NaCl by weight. The value of m_2^s corresponding to a certain salt concentration can be measured as in the porosity method described above.

In relation (7.61), the relative permeability is described through a power-law model where the coefficient δ can be estimated from capillary pressure–saturation relationships (van Genuchten, 1980) or obtained via inverse fitting. In the present study, a coefficient δ of 3.5 was used, based on literature data available for fibrous materials (Landeryou et al., 2005; Ashari and Tafreshi, 2009) and soils (Honarpour et al., 1986).

Parameters τ (7.45) and $m_{2\max}^s$ (7.43) were obtained by fitting *uptake-kinetics test* data by relation (7.44). The uptake-kinetics test is executed under near-saturated conditions (i.e., s^l close to 1 everywhere) to simplify the relation.

The *capillary pressure–saturation relationship* was obtained using the liquid porosimetry as described in the literature (e.g., Miller and Tyomkin, 1994). The method has been adapted to a swelling material as follows: the core sample is preswollen to equilibrium under the confining pressure of 0.3 psi, by leaving the specimen for at least 2 h in contact with liquid, then followed by drainage to dewater pores, then the standard measurement of wetting and drainage curves is performed. The measurements have been performed with water solution at different NaCl concentrations, as previously described for porosity and permeability. Data were fitted by the constitutive relationships (7.57) and (7.58).

The *density of dry AGM* was assumed to be homogeneous at 1.6 g/cm^3. *AGM concentration* was calculated as the ratio between the amount of AGM and dry volume of core sample. Finally, for the effective phase–interface area relationship (7.46) the threshold parameter $s_{e,\text{threshold}}^l$ was assumed to be 0.01, while the exponent parameter α_{exp} was estimated at 0.4 via inverse fitting of experimental data from the dynamic absorption test at 20 cm height.

7.5.2 Simulation of Dynamic Absorption Test

The three experimental tests presented in Section 7.4.1 were simulated by FEFLOW using a rectangular domain 2.83 mm width and 4.1 mm height. The initial hydraulic head was set equal to −0.8 m everywhere to represent initially dry conditions. At the lower horizontal boundary, a constant hydraulic head was applied while the remaining boundaries were implied to be impermeable in all cases. The value of the constant-head condition was set equal to 0, 0.2, and 0.4 m, corresponding to the respective elevation of the sample relative to the liquid level in the reservoir. The finite-element mesh consists of 140 quadrilateral elements and 168 nodes with an element size of 8.32×10^{-8} m^2 (Figure 7.9).

The results of the simulations are reported in Figure 7.10 and show reasonable agreement compared to the laboratory data over the tested time and elevation range.

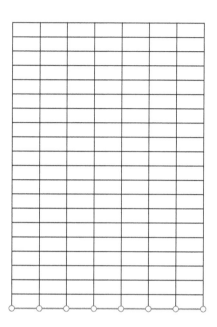

FIGURE 7.9
Finite-element mesh with location of boundary condition.

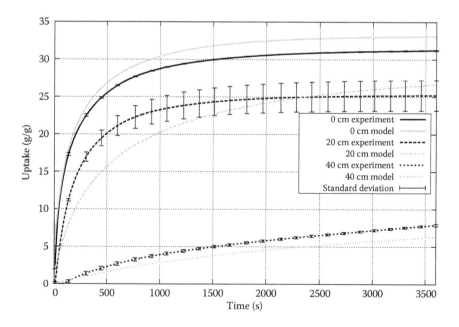

FIGURE 7.10
Simulation results of dynamic absorption test compared to experimental data.

7.5.3 Simulation of Vertical Wicking Experiment

The vertical wicking experiment described in Section 7.4.2 was simulated by FEFLOW. The core sample was represented by a 200 mm high rectangular domain. The finite-element mesh consists of 5688 quadrilateral elements and 6408 nodes. The elements were generated with varying width along the machine direction of the core sample (here, the vertical direction) to reproduce the dry caliper profile of the original core sample along the machine direction. The initial hydraulic head was set equal to −0.8 m everywhere to represent initially dry conditions. At the lower horizontal boundary, a constant hydraulic head of 0 m was applied, while the remaining boundaries were implied to be impermeable.

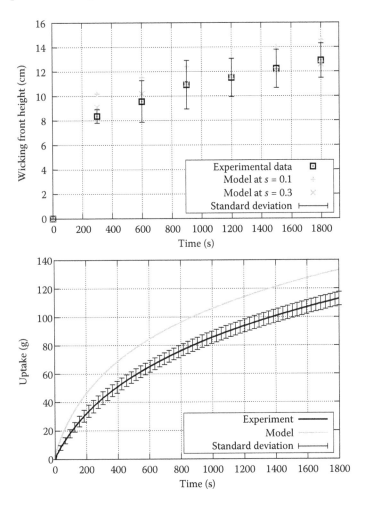

FIGURE 7.11
Simulation results of vertical wicking compared to experimental data.

The predicted uptake curve has a similar shape as the experimental curve, with a maximum deviation of 18% at the end of the experiment.

Comparison of wicking height is less straightforward. As described in the literature (Jaganathan et al., 2009), at a certain time, wicking height will be different depending on the chosen saturation value. Here, it was decided to report predicted wicking height at two saturation levels, 0.1 and 0.3, to cover a reasonable range when comparing to experimental values. Also in this case the comparison is considered acceptable (see Figure 7.11).

7.6 Summary

This chapter presented a theoretical model development for the fluid movement in variably saturated swelling porous media. The resulted system of nonlinear PDEs (7.68) consists of three equations with the pressure head ψ^l, the absorbed liquid concentration $\bar{C}^s_{L \rightarrow S_0}$, and the solid displacement u^s as the primary variables. The system involves a number of empirical relations with fitting parameters that were estimated with the help of specific experiments.

The system of PDEs was solved numerically in FEFLOW for two- and three-dimensional problems using the FEM, linearized by the Picard or the Newton method, and discretized in time using a predictor–corrector scheme including an adaptive time-stepping strategy. The solution process contains large-scale mesh deformation according to the swelling of absorbent materials.

Dynamic absorption and wicking experiments were presented in detail. Graphical results of those experiments show a good agreement with the numerical simulations.

Nomenclature

Roman Letters

A		Symmetric matrix in the numerical solution
a	1	Solid displacement direction vector
A_p	L^2	Maximum solid–liquid interface area per REV
$a^{sl}(s^l)$	1	Saturation-dependent fraction of the solid–liquid interface area
B		Symmetric matrix in the numerical solution
C	ML^{-3}	Solution vector of concentration
c	1	Scaling factor
$\bar{c}$	ML^{-3}	Bulk concentration

D		Displacement matrix in the numerical solution
d		Error vector
d^s	1	Solid strain vector
d^s	1	Volumetric solid strain
e	1	$= -g/\lvert g \rvert$, gravitational unit vector
F		Liquid sink vector in the numerical solution
g	LT^{-2}	Gravity vector
g	LT^{-2}	$= \lvert g \rvert$, gravitational acceleration
J^s	1	Jacobian of solid domain, volume dilatation function
K	LT^{-1}	Hydraulic conductivity tensor
K		Conductance matrix in the numerical solution
k	L^2	Permeability tensor
k_r	1	Relative permeability
k^+	$M^{-1}LT^{-1}$	Reaction rate constant
L	L^{-1}	Gradient operator
l	L	Characteristic element length
l_i^s	L	Side lengths of a small solid cuboid
M	M	Molar mass
m	M	Mass
m	1	Van Genuchten curve-fitting parameter
m_2^s	1	AGM x-load
$\hat{m}_2^s$	1	$= (m_{2\max}^s - m_2^s)/m_{2\max}^s$, normalized AGM x-load
$m_{2\max}^s$	1	Maximum AGM x-load
n	1	Van Genuchten curve-fitting parameter
O		Symmetric matrix in the numerical solution
P		Symmetric matrix in the numerical solution
p	$ML^{-1}T^{-2}$	Pressure
Q		Solid-strain source vector in the numerical solution
Q	$ML^{-3}T^{-1}$	Mass supply
q	LT^{-1}	Volumetric Darcy flux
q_n	LT^{-1}	Normal component of Darcy flux
R^τ	L^3T^{-1}	Residual vector
R	$ML^{-3}T^{-1}$	Chemical reaction term
R_c	$ML^{-3}T^{-1}$	Kinetic reaction term for immobile liquid
R_ψ	T^{-1}	Kinetic reaction term for mobile liquid
S	1	Solution vector of saturation
s	1	Saturation
t	T	Time
U	L	Solution vector of solid displacement
u	L	Solid displacement vector
u	L	Scalar solid displacement norm
V	L^3	REV volume
v	LT^{-1}	Velocity vector
x	L	Eulerian spatial coordinates

x_i	L	Components of x
Z		Kinetic absorption reaction vector in the numerical solution

Greek Letters

α	L^{-1}	Van Genuchten curve-fitting parameter
β_{upwind}	L	Numerical dispersivity
Γ^s	L^2	Closed boundary of solid control space Ω^s
γ	$M^{-1}LT^2$	Liquid compressibility
$\bar{\gamma}$	L^{-1}	$= \gamma\rho^l g$, specific liquid compressibility
γ_{ij}	1	Shear strain component
Δ		Increment or difference
δ	1	Exponential fitting parameter
ε	1	Porosity, void space
ε^α	1	Volume fraction of α-phase
η_l	1	Linearization error tolerance
η_t	1	Time error tolerance
κ	$L^{-1}T$	Artificial compression of solid
μ	$ML^{-1}T^{-1}$	Dynamic viscosity
ρ	ML^{-3}	Density
σ	L	Artificial (dampening) "diffusive" stress of solid
τ	T^{-1}	AGM reaction (speed) rate constant
φ^l, φ^s	$T^{-1}, ML^{-1}T^{-1}$	Deformation (sink/source) terms for liquid and solid
Ψ	L	Solution vector of pressure head
ψ^l	L	Pressure head of liquid phase l
Ω^s	L^3	Control space of porous solid or domain
$|\Omega^s|$	L^3	Volume of the control space
∇	L^{-1}	Nabla operator
∇_i		$= \partial/\partial x_i$, partial differentiation with respect to x_i.

Subscripts

0	Reference, initial or dry
AGM	AGM
AGM_0	AGM at initial state
$AGM_{consumed}$	Consumed AGM in reaction
AGM_{raw}	Available AGM in reaction
C	Capillary
D	Dirichlet-type BC
D	Drying
E	Effective or elemental
H_2O	Water
I	Index of solid component ($I = CM, AGM, L \rightarrow S$)
I	Spatial Eulerian coordinate ranging from 1 to 3 or nodal indices

L $\to$ S	AGM absorbed liquid
N	Neumann-type BC
N	Numerical time level index
R	Residual, reactive, or relative
W	Wetting

Superscripts

α	Phase indicator
D	Number of space dimensions
g	Gas phase
l	Liquid phase
P	Predictor
s	Solid phase
T	Transpose
τ	Iteration counter

Abbreviations

AGM	Absorbent gelling material
BC	Boundary condition
CM	Inert carrier material
LHS	Left-hand side
REV	Representative elementary volume
RH	Relative humidity
RHS	Right-hand side
RMS	root-mean square
SAP	Superabsorbent polymer

References

A. Ashari and H.V. Tafreshi. A two-scale modeling of motion-induced fluid release from thin fibreous media. *Chemical Engineering Science*, 64:2067–2075, 2009.

J. Bear and Y. Bachmat. *Introduction to Modeling of Transport Phenomena in Porous Media*. Springer Netherland, Berlin, 1st edition, 1991.

F.L. Buchholz. Model of liquid permeability in swollen composites of superabsorbent polymer and fiber. *Journal of Applied Polymer Science*, 102:4075–4084, 2006.

F.L. Buchholz and A.T. Graham. *Modern Superabsorbent Polymer Technology*. Wiley-VCH, New York, 1998.

O. Coussy. *Mechanics of Porous Continua*. Wiley, Chichester, 1995.

R. De Boer. *Theory of Porous Media: Highlights in Historical Development and Current State*. Springer, Berlin, 2000.

H.-J.G. Diersch and P. Perrochet. On the primary variable switching technique for simulating unsaturated-saturated flows. *Advances in Water Resources*, 23:271–301, 1999.

H.-J.G. Diersch, V. Clausnitzer, V. Myrnyy, R. Rosati, M. Schmidt, H. Beruda, B.J. Ehrnsperger, and R. Virgilio. Modeling unsaturated flow in absorbent porous media: 1. Theory. *Transport in Porous Media*, 83:437–464, 2010.

H.-J.G. Diersch, V. Clausnitzer, V. Myrnyy, R. Rosati, M. Schmidt, H. Beruda, B.J. Ehrnsperger, and R. Virgilio. Modeling unsaturated flow in absorbent porous media: 2. Numerical simulation. *Transport in Porous Media*, 86:753–776, 2011.

J. Donea and A. Huerta. *Finite Element Methods for Flow Problems*. John Wiley and Sons, Chichester, 2003.

FEFLOW. FEFLOW finite element subsurface flow and transport simulation system—Users Manual/Reference Manual/White Papers. Recent release 6.0. Technical report, DHI-WASY GmbH, Berlin, 2010. http://www.feflow.info.

R.A. Freeze and J.A. Cherry. *Groundwater*. Prentice Hall, Englewood Cliffs, N.J., 1979.

M. Honarpour, L. Koederitz, and A.H. Harvey. *Relative Permeability of Petroleum Reservoirs*. CRC Press, Boca Raton, FL, 1986.

S. Jaganathan, H.V. Tafreshi, and B. Pourdeyhimi. Realistic modeling of fluid infiltration in thin fibreous sheets. *Journal of Applied Physics*, 105(113522):2009.

M. Landeryou, I. Eames, and A. Cottenden. Infiltration into inclined fibreous sheets. *Journal of Fluid Mechanics*, 529:173–193, 2005.

R.W. Lewis and B.A. Schrefler. *The Finite Element Method in the Static and Dynamic Deformation and Consolidation of Porous Media*. Wiley, Chichester, 2nd edition, 1998.

B. Miller and I. Tyomkin. Liquid porosimetry: New methodology and applications. *Journal of Colloid and Interface Science*, 162:163–170, 1994.

G.F. Pinder and W.G. Gray. *Essentials of Multiphase Flow and Transport in Porous Media*. Wiley, Hoboken, NJ, 2008.

P.S. Scott, G.J. Farquhar, and N. Kouwen. Hysteretic effects on net infiltration. In *Advances in Infiltration*, pp. 163–170, St Joseph, 1983. Am Soc Agric Eng.

M.Th. van Genuchten. A closed-form equation for predicting the hydraulic conductivity of unsaturated soils. *Soil Science Society of America Journal*, 44:892–898, 1980.

O.C. Zienkiewicz and R.L. Taylor. *The Finite Element Method*. Elsevier Butterworth-Heinemann, Oxford, 5th edition, 2000.

8

<hr>

Evaporation and Wicking

Stéphanie Veran-Tissoires, Sandrine Geoffroy, Manuel Marcoux,
and Marc Prat

CONTENTS

Wicking and evaporation occur simultaneously in a variety of situations. In this chapter, we present and study several aspects of wicking/evaporation when evaporation is external mass transfer driven and unconfined or weakly confined. These include the influence of evaporation on the final impregnation height and the impregnation dynamics, the structure and the influence of the evaporation flux distribution along the wick surface, and the transport of a dissolved species. Other effects, still poorly studied, such as the influence of liquid films or the shape of the internal evaporation front in the wick are also addressed. Despite the practical importance of wicking/evaporation situations, this chapter also shows that further work is needed to increase our understanding of the combined action of wicking and evaporation and our prediction capabilities.

8.1 Introduction

The combined action of wicking and evaporation can be encountered in a variety of natural as well as industrial situations. Although not typically referred to as wicking and in fact more complicated than pure wicking due to capillary actions, the transport of sap in trees or plants (Perez-Diaz et al. 2010), is an obvious example where wicking and evaporation are combined in natural systems.

Evaporation and wicking are also combined issues in fabric companies, especially for the design of sportswear. The ability to remove moisture and enhance speed evaporation is crucial in keeping the body dry and comfortable. It is important that if the body works up a sweat, the moisture is transported away from the skin to the surface of the garment where it can evaporate quickly.

Wicking with evaporation is frequently encountered in the studies related to salt weathering issues (see, e.g., Goudie and Viles, 1997). As discussed for example in Scherer et al. (2001), salts can enter stone and masonry by several routes, the simplest of which is the capillary rise of groundwater. As the water wicks up into a wall, it also evaporates from the free surface. Near the bottom of the wall (i.e., the source of the water), the flow through the pores is likely to be able to keep up with the rate of evaporation, so the liquid/vapor interface remains at the exterior surface (i.e., the outer surface of the porous medium). In that region, if the solution becomes supersaturated with salt, crystals will precipitate at the surface; this is called *efflorescence*, and is responsible for the whitish stain often seen on new brick walls. Higher up the wall, the rate of capillary rise is slower, so that evaporation may dry out the surface and drive the liquid/vapor interface into the body. In that case, supersaturation leads to precipitation inside the pores, or *subflorescence*, which can cause severe damage (Figure 8.1). Hence, here evaporation and wicking are combined with the transport of a dissolved species (salt in this example), see Puyate and Lawrence (2000) for a simple case.

This is also the case in the simple commercial devices designed for dispersing of a volatile liquid in a room. The liquid can be a perfume or an insect repellent, for example (see, for instance, www.toscochem.com/template/product_03_01.htm). Another interesting example presented in Fries et al. (2008) is related to propellant management devices (PMD) used in spacecraft tanks. In all these examples, though heat transfers are not necessarily always negligible, evaporation is essentially driven by mass transfer.

In this respect, the two-phase cooling devices based on wick action in which evaporation is driven by heat transfer form a rather different but very important category. These include mainly the heat pipes and capillary loops, for example (Peterson 1994, Faghri 1995), developed notably

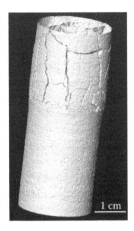

FIGURE 8.1
X-ray tomography image of a rock sample submitted to a wicking evaporation process in the presence of a dissolved salt. The intense fracturing in the top region of the sample is due to the salt crystallization resulting from the wicking/evaporation process. (Reproduced from Noiriel, C. et al. Intense fracturing and fracture sealing induced by mineral growth in porous rocks. *Chemical Geology* 269(3–4):197–209. Copyright 2010, with permission of Elsevier.)

in relation with spacecraft thermal management issues or the cooling of electronic devices.

This leads us to distinguish two main subtopics depending on the nature of the transfer inducing the liquid–vapor phase change; namely mass-transfer-driven evaporation and heat-transfer-driven evaporation. Again, both transfers can be important in some situations but we believe that this distinction makes sense and is helpful.

In this chapter, we only consider wicking in the presence of unconfined (or weakly confined) mass-transfer-driven evaporation. For simplicity, we focus on slow evaporation situations so that the process can be considered as isothermal. Reviews of results obtained in relation with the study of the vaporization process occurring in the porous wick of cooling devices (wicking in the presence of confined heat-transfer-driven evaporation), loop heat pipes in particular, are notably available in Maydanik (2005) and Prat (2010).

8.2 Wicking and Unconfined Mass Transfer Induced Evaporation

There are many different geometrical configurations in which evaporation and wicking can occur simultaneously. Thus, the idea is not to describe all these situations but to focus on a few generic aspects, present or at least frequently encountered in the mass transfer-induced evaporation problems. It can be noticed beforehand that the wicking/evaporation problems and the field of drying of porous media are not unrelated. Among the recent references covering this field, one can refer for example to the book series *Modern Drying Technology* edited by Wiley. The works based on pore-network models are of more direct interest to this chapter. One can then refer to a few review articles (Prat 2002, Metzger et al. 2007, Prat 2011). A first ingredient, which makes the study of wicking/evaporation somewhat difficult, is that evaporation takes place usually in an unconfined environment. The consequence is that the evaporation rate is sensitive to external factors, such as the external air flow or the radiative transfers that can be difficult to evaluate in practical situations. As a result, we only consider hereafter some basic aspects.

In many mass-transfer-controlled evaporation situations, the external gas phase is a multi-component gas phase typically formed by air, which is generally the main component, and the vapors of the volatile species present in the porous medium. In this chapter, we only consider the basic but important case where the gas phase can be considered as a binary mixture made of air and the vapor of the liquid present in the wick. Neglecting heat transfer, the equations governing the transport of the vapor within the external gas phase can then be expressed as

$$\frac{\partial \rho}{\partial t} + \nabla \cdot (\rho \mathbf{u}) = 0. \tag{8.1}$$

$$\rho \frac{\partial \mathbf{u}}{\partial t} + \rho \mathbf{u} \cdot \nabla \mathbf{u} = -\nabla P + \mu \nabla^2 \mathbf{u} + \rho \mathbf{g} \tag{8.2}$$

$$\frac{\partial \rho X_v}{\partial t} + \nabla \cdot (\rho X_v \mathbf{u}) = \nabla \cdot (\rho D \nabla X_v) \tag{8.3}$$

where ρ is the density of the gas mixture, μ is the gas viscosity, P the gas pressure, $\mathbf{u}$ the gas velocity vector, $\mathbf{g}$ the gravitational acceleration, D the binary molecular diffusion coefficient, and X_v is the vapor mass fraction.

Equations 8.1 through 8.3 are the gas-phase continuity equation, momentum conservation equation (Navier–Stokes equations) and vapor transport equation, respectively. It should be noted that ρ and μ are functions of X_v, see for instance, Bird et al. (2002). Hence, there is *a priori* a coupling between the vapor transport and the flow. This will be illustrated in Section 8.4. To solve the above equations and determine the evaporation flux along the liquid saturated surface of the porous wick, boundary conditions should be supplemented. To this end, we first need to discuss the boundary condition to impose at an evaporative porous surface.

8.3 Evaporation from a Porous Surface

Consider a saturated porous surface that is a porous surface where all surface pores are occupied by liquid. The basic question is whether the evaporation rate from a porous surface, that is a surface which is partially solid and partially liquid, is different (and *a priori* smaller) from the same surface fully covered by liquid. This problem was first addressed by Suzuki and Maeda (1968), who showed that the evaporation rate from a free surface liquid, provided the porous surface was sufficiently and finely divided. In other terms, the pore sizes must be sufficiently small compared to the overall size of the surface for the rate being essentially identical to that of a liquid surface. As an example we consider evaporation from a model porous surface when the external vapor transport is due only to diffusion (no free or forced convective effects). This is illustrated in Figures 8.2 and 8.3. The porous surface is of size L and is formed by a simple square array of identical cylindrical pores of diameter d. The size L of the porous surface as well as the porosity of the surface (~0.5) are kept constant. Flat menisci are assumed at the entrance of surface pores and the influence of pore size d is studied by solving numerically the equations governing the external mass transfer (supposed here purely diffusive) (Geoffroy et al. 2011). The evaporation rate from the porous surface is denoted by J whereas the evaporation rate from the same surface but assumed fully liquid is J_{ref}. As can be seen in Figure 8.3, $J / J_{ref} \approx 1$ when $d \ll L$. Hence, the remarkable result is that a porous surface behaves as a fully wetted surface provided that the pore size d (or more exactly the mean distance between two pores) is sufficiently small compared to its size L. Hence, the evaporation rate from a partially wet surface is essentially identical to the evaporation rate of a free surface of liquid provided that the size of wet spots on the partially wet surface is sufficiently small. This phenomenon can be analyzed as a screening phenomenon, that is, the system does not see the solid fraction once the surface is sufficiently finely divided. Screening effects are common when a diffusion process takes place from a heterogeneous surface (Chraïbi et al. 2009). The constraint $d \ll L$ is of course generally well satisfied in the wicking/evaporation situations. The external diffusional screening phenomenon also explains why a change in the

FIGURE 8.2
Model porous surface made of a regular array of cylindrical pores of diameter d. Owing to symmetry, only 1/4 of the surface is shown. (Adapted from Geoffroy, S. et al. 2011. *To be submitted.*)

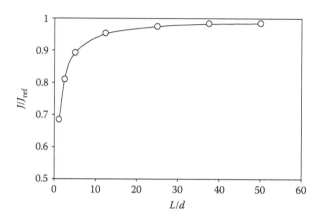

FIGURE 8.3
Evaporation rate as a function of porous surface fineness L/d (see Figure 8.2). The surface poros-ity is $\Phi = 0.5$. The number of surface pores is $4\varphi/\pi$ $(L/d)^2$. J is the evaporation rate from the porous surface, J_{ref} is the evaporation rate from a liquid free surface. (Adapted from Geoffroy, S. et al. 2011. *To be submitted.*)

pore occupancy by the liquid at the porous surface does not affect the evapora-tion rate significantly as long as there is a sufficient fraction of liquid saturated pores at the porous surface.

In the example presented in Figures 8.2 and 8.3, the external transfer of the vapor was governed by diffusion only. Thus, the question arises as to whether the screening effect still holds when convective effects are present. Convective effects were considered by Suzuki and Maeda (1968), who showed that the screening effect still holds provided that the external boundary layer thickness (a flow parallel to the porous surface was considered) is sufficiently large compared to the pore sizes. Under these circumstances, vapor transfer is in fact still governed near the surface (i.e., at distances from the surface of the order of the grain or pore size) by diffusion.

The conclusion from this section is therefore that the boundary condition to impose at the surface of a saturated porous medium in order to compute the vapor concentration field in the external gas is the same as for a liquid surface, namely

$$X_v = X_{ve} \tag{8.4}$$

where X_{ve} is the equilibrium vapor mass fraction at a liquid/gas interface. The evaporation flux is given by

$$\dot{j}_e = -\frac{\rho D}{1 - X_{ve}} \nabla X_v \cdot \mathbf{n} \tag{8.5}$$

where $\mathbf{n}$ is the unit normal vector.

For solving the flow equation, Equation 8.2, a no-slip boundary condition is imposed as regards the component of the velocity vector tangential to the surface whereas the normal component is expressed as

$$\mathbf{u} \cdot \mathbf{n} = -\frac{D}{(1 - X_{ve})} \nabla X_v \cdot \mathbf{n} \qquad (8.6)$$

Equation 8.6 reflects the fact that the evaporation process induces a flow within the gas phase. One can refer for instance to Carey (2008) for more details. Notice that the problem of vapor transport was formulated in this section using the vapor mass fraction X_v. As described for example in Bird et al. (2002), the problem can also be formulated as well by using the vapor concentration or the vapor partial pressure.

8.4 External Mass Transfer

In the simplest situation, the external mass transfer is governed by diffusion. This is a correct assumption when the liquid is not too volatile and the evaporative surface is small (a typical example is the evaporation of a small liquid droplet lying on a solid surface, e.g., Deegan et al. 1997). When the size of the evaporative surface is not small, convective effects in the gas phase cannot be ignored. This can be understood from the expression of the Grashof number, which characterizes the competition between buoyancy effects and viscous effects as

$$Gr = \frac{g\beta(\rho_{vsat} - \rho_{v\infty})L^3}{v^2} \qquad (8.7)$$

where $\beta = -(1/\rho)(\partial\rho/\partial\rho_v)_{T,P}$, v is the gas kinematic viscosity, ρ_{vsat} is the vapor concentration at the surface and $\rho_{v\infty}$ is the vapor concentration in ambient (surrounding) medium. As can be seen the Grashof number varies as L^3, where L is the characteristic length of the surface. The high value of the exponent (3) explains why free convection effects cannot generally be neglected as soon as the size of the surface is typically greater than a few millimetres. This is illustrated from the simple experiment sketched in Figure 8.4. A cylindrical wick, 38 mm in diameter, made of a random packing of 1 mm glass beads is in contact at the bottom with a water reservoir. As sketched in Figure 8.4, the wick is in a cylindrical container and the distance between the wick top and the hollow cylinder top is denoted by δ. This system is set in an enclosure of controlled temperature ($T = 22°C$) and relative humidity (RH). The RH in the enclosure is imposed by a LiBr

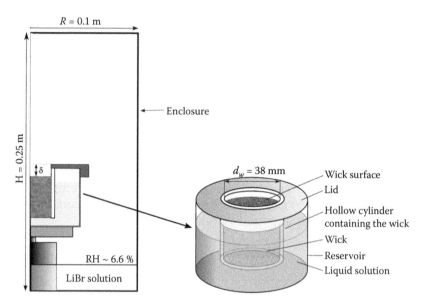

FIGURE 8.4
Cylindrical wick contained in a hollow cylinder placed in a cylindrical enclosure.

saturated aqueous solution (RH ≈ 6.6%) located at the bottom of the enclosure as shown in Figure 8.4. Note that the cylinder containing the wick is open at the top. Thus the wick surface is in contact with the dryer air in the enclosure. The wick being fully saturated with water, evaporation takes place at the surface of the wick. The evaporation rate J is deduced from the evolution of the mass $m(t)$ of the system wick—water reservoir, which is measured by continuous weighting. Hence, $J = dm/dt$. Then the evaporation rate is computed from the numerical solution of Equations 8.1 through 8.3 supplemented by the boundary condition $X_v = X_{ve}$ at the porous medium surface where X_{ve} is the equilibrium mass fraction for pure water at 22°C and the condition $X_v = 0.066X_{ve}$ at the surface of the LiBr solution. A zero-flux condition is imposed at other surfaces bounding the gas phase within the enclosure. The same computation was also made assuming a purely diffusive transport in the gas phase, that is, by solving Equation 8.3 with $\mathbf{u} = 0$ and ρ = constant. In both cases the evaporation rate is computed from

$$J = \int_{A_w} j_e \, dS = -\int_{A_w} \frac{\rho D}{1 - X_{ve}} \nabla X_v \cdot \mathbf{n} \, dA \tag{8.8}$$

where A_w is the evaporative surface of the wick (i.e., the top disk-like surface of the wick).

TABLE 8.1

Comparison between Computed and Measured Evaporation Rates

δ (mm)	2.5	7.5	15
$J_{e,exp}$ (kg/s)	3.95 10^{-8}	3.35 10^{-8}	2.85 10^{-8}
$J_{e,num}$ (kg/s) pure diffusion	2.46 10^{-8}	1.99 10^{-8}	1.56 10^{-8}
Relative error (%)	37.7	40.4	45.3
$J_{e,num}$ (kg/s) free convection + diffusion	3.63 10^{-8}	2.82 10^{-8}	2.11 10^{-8}
Relative error (%)	8.1	15.6	26

Note: The relative error is between experimental results and computed ones.

This was done for three different values of δ so as to vary the evaporation rate. Reducing δ increases the overall evaporation rate (the external mass transfer resistance due to transfers between the porous medium surface and the hollow cylinder entrance is reduced) and modifies the local flux evaporation distribution over the surface (Veran-Tissoires et al. 2012). Increasing δ reduces the evaporation rate (the resistance due to transfers between the porous medium surface and the hollow cylinder entrance is increased) and makes the local evaporation flux distribution more and more uniform. The situation is somewhat similar to evaporation from a (flat) meniscus in an open capillary tube. When the meniscus is right at the tube entrance the evaporation rate is maximum and the flux is much greater near the contact line (perimeter of the meniscus) whereas the evaporation flux is lower and uniform over the meniscus when the meniscus is located within the tube (as soon as the meniscus is about one tube diameter away from the entrance). The results are summarized in Table 8.1. As can be seen, the purely diffusive computation leads to noticeable discrepancies with the experimental results. Although the agreement is not perfect, the computation including the free convection effect leads to significantly better results. More details can be found in Veran-Tissoires (2011).

8.5 A Reference Situation

A basic situation sketched in Figure 8.5 in the context of wicking/evaporation is wicking in a porous sheet. The thickness of the porous sheet is typically of the order of a few millimeters or less and a typical height is typically of the order of a few centimetres. In addition to engineering applications, such as water or ink/dye transport in paper and textiles, body fluids adsorption by medical dressings and sanitary pads or moisture migration in tiles and porous building cladding and surfaces (see Lockington et al. (2007) and references therein), this situation has also attracted the interest of physicists (e.g., Barabási and Stanley 1995), in relation with the problem of interface

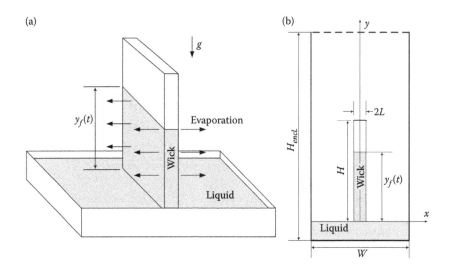

FIGURE 8.5
(a) A porous sheet in contact with liquid at the bottom. (b) Two-dimensional sketch of the situation studied in Fries et al. (2008).

roughening in disordered media (as illustrated in Figure 8.6a the front is not flat but rough owing to the disordered nature of porous microstructures). As shown in Figure 8.6b, interface roughening can be captured by the simple pore network model described in Section 8.7. In what follows, we briefly present a simple model taking into account the influence of evaporation on

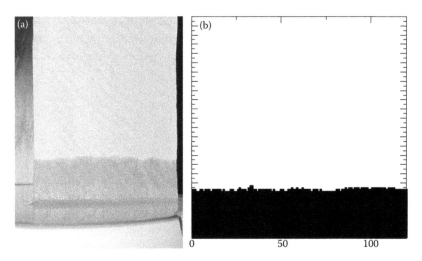

FIGURE 8.6
(a) Imbibition front in a vertical toilet paper sheet in contact with a water tank. The front is not perfectly flat and presents an interface roughening; (b) rough invasion front obtained with the pore network model described in Section 8.7.

wicking and discussed various aspects such as the imbibition dynamics in the presence of evaporation and the condition for the wick to be fully saturated under steady-state conditions.

The wick is fully dry initially and put in contact with a wetting liquid reservoir at its base at $t = 0$. The liquid is volatile and the surrounding gas (air typically) is not saturated so that mass transfer evaporation can take place during the imbibition process and will also possibly affect the final position of the top menisci within the wick. The system is thin (its height H is much greater than its width $2L$, $H \gg L$) so that it is reasonable to assume a flat travelling front (see Figure 8.6) during the imbibition process. The final distribution of the liquid at the top of the liquid region at the end of the imbibition will be, however discussed in Section 8.7. We also assume that the wick is made of a random packing of sintered monodisperse beads of radius R_b. According to Masoodi et al. (2007) the capillary pressure threshold or maximum suction pressure can be expressed in this case as

$$P_{cth} = \frac{3(1 - \varepsilon)\gamma \cos \theta}{\varepsilon R_b} \tag{8.9}$$

where γ, θ, and ε are the surface tension, the contact angle, and the porous medium porosity, respectively.

Using a two-dimensional Cartesian coordinate system (as sketched in Figure 8.5b), the position of the imbibition front in the wick is denoted by $y_f(t)$. The wick is supposed to be saturated below $y_f(t)$ so that the flow is governed by Darcy's law which will be combined with the continuity equation,

$$\frac{\partial U_x}{\partial x} + \frac{\partial U_y}{\partial y} = 0 \quad \text{in } V_\ell \tag{8.10}$$

where U_x and U_y are the components of the volume-averaged Darcy velocity, and V_ℓ is the domain occupied by the liquid in the wick. At the lateral surface of the wick where evaporation takes place (evaporation from the moving internal front is neglected), we have,

$$U_x = j_e/\rho_\ell \quad \text{at } \partial V_\ell \ 0 \leq y \leq y_f \tag{8.11}$$

where ∂V_ℓ is the wick surface. For simplicity, it is assumed that the evaporation flux j_e is constant along the wick. To obtain an approximate but simple solution we introduce the spatial average $\bar{\phi} = (1/L)\int_0^L \phi \, dx$ that we apply to Equation 8.10. This gives

$$\frac{\partial \bar{U}_y}{\partial y} + \frac{j_e}{\rho_\ell L} = 0. \tag{8.12}$$

Equation 8.12 is combined with Darcy's law ($\bar{U}_y = -(K/\mu_\ell)((\partial \bar{P}_\ell/\partial y) + \rho g)$) to obtain

$$\frac{\partial^2 \bar{P}_\ell}{\partial y^2} - \frac{\mu_\ell j_e}{\rho_\ell L K} = 0 \quad 0 \le y \le y_f. \tag{8.13}$$

At the front (using the atmospheric pressure as the reference pressure), we have

$$\bar{P}_\ell = -P_{cth} \quad \text{at } y = y_f \tag{8.14}$$

whereas at the wick bottom

$$\bar{P}_\ell = 0 \quad \text{at } y = 0. \tag{8.15}$$

For simplicity, we assume a constant evaporation flux at the wick surface. Solving Equations 8.13 through 8.15 becomes straightforward. We are interested in the pressure gradient at $y = y_f$, which reads

$$\frac{\partial \bar{P}_\ell}{\partial y} = -\frac{P_{cth}}{y_f} + 0.5\frac{\mu_\ell j_e}{\rho_\ell L K} y_f \tag{8.16}$$

The front velocity is given by

$$\varepsilon \frac{dy_f}{dt} = \bar{U}_y \quad \text{at } y = y_f \tag{8.17}$$

which combined with Darcy's law and Equation 8.16 yields

$$\frac{K}{\mu_\ell}\left(\frac{P_{cth}}{y_f} - 0.5\frac{\mu_\ell j_e}{\rho_\ell L K} y_f - \rho_\ell g\right) \tag{8.18}$$

This equation can be expressed as

$$\frac{dy_f}{dt} = \frac{a}{y_f} - b - c y_f \tag{8.19}$$

with $a = (K/\mu_\ell)(P_{cth}/\varepsilon)$, $b = \rho_\ell g K/\mu_\ell \varepsilon$, $c = 0.5(j_e/\rho_\ell L \varepsilon)$.

As pointed out in Fries et al. (2008), one can distinguish different cases from Equation 8.19 as regards the final height of the liquid within the wick, which is given by Equation 8.19 setting $dy_f/dt = 0$:

(i) No evaporation occurs ($c = 0$). In this case, the competition between gravity and capillary forces governs the maximum reachable height given by (Jurin's law)

$$y_{max1} = \frac{a}{b} = \frac{P_{cth}}{\rho_\ell g} \tag{8.20}$$

(ii) Negligible gravity effects ($b = 0$). The competition between capillary effects and evaporation sets the maximum reachable height, which is given by

$$y_{max2} = \sqrt{\frac{a}{c}} = \sqrt{\frac{2KP_{cth}\rho_\ell L}{\mu j_e}} \tag{8.21}$$

(iii) Both gravity and evaporation must be considered, which leads to

$$y_{max3} = \frac{-b}{2c} + \sqrt{\frac{b^2}{4c^2} + \frac{a}{c}} \tag{8.22}$$

Naturally in the absence of gravity and evaporation, there is no effect balancing the capillary effects and the liquid will always reach the wick top.

The permeability K of the wick (for, as mentioned before, a packing of sintered monodisperse beads) can be expressed using the classical Kozeny–Carman model relationship as

$$K = C \frac{\varepsilon^3 R_b^2}{(1 - \varepsilon)^2} \tag{8.23}$$

with $C \approx 0.022$.

As an example we consider a 10 cm high wick ($H = 10$ cm) and of thickness $2L = 200$ μm. The working fluid is FC77 ($\gamma = 15 \times 10^{-3}$ N/m, $\rho_\ell = 1780$ kg/m³, $\mu = 1.28 \times 10^{-3}$ Pa s at $T = 25°C$). The fluid is supposed to be perfectly wetting ($\theta = 0$). The experiments with this fluid reported in Fries et al. (2008) lead to values of the evaporation flux in the range $[10^{-6}, 10^{-3}]$ (kg/m²/s). For the computations shown below we have taken an intermediate value in this range, namely $j_e = 10^{-4}$ kg/m²/s. One simple and nice way to study the impact of the wick properties is to vary the bead radius R_b, which has a direct impact on both the permeability (Equation 8.23) and the capillary suction (Equation 8.9). We have plotted in Figure 8.7 the evolution of y_{max1}, y_{max2}, and y_{max3} as a function of the bead radius R_b, varying in the range [0.1–50] μm. The porosity is constant and equal to 0.36.

As can be seen from Figure 8.7, evaporation here has a strong influence on the height reached by the liquid. The evaporation effect is dominant in limiting the rise for beads of radius lower than 5 μm (as also shown in the inset in Figure 8.7) whereas both gravity effects and evaporation limit the rise for a greater bead size. As a consequence, there is a maximum in the impregnation height when both limiting effects, that is, evaporation and gravity, are comparable (see Figure 8.7). As can be also seen, the wick would be fully saturated in the absence of evaporation for bead radii lower than about 45 μm.

Let us now look at the effect of evaporation on the imbibition dynamics. Notice that we suppose that the wick top is impervious so that evaporation from the wick top can be neglected. After neglecting evaporation from the internal front, only evaporation from the lateral sides is thus taken into

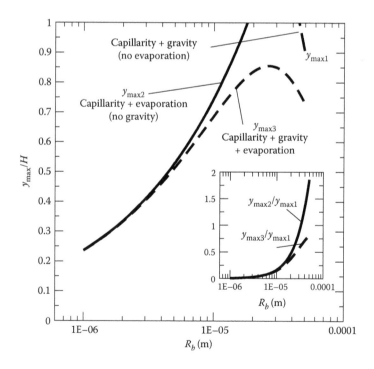

FIGURE 8.7
Evolution of final maximum impregnation height as a function of bead diameter R_b.

account. As reported in Fries et al. (2008), Equation 8.19 can be solved analytically. The solution reads

$$t = \frac{1}{2c}\left[-\ln\left(\frac{-c\,y_f^2 - b\,y_f + a}{a}\right)\right]$$

$$-\frac{b}{2c\sqrt{-\psi}}\ln\left[\frac{(-2c\,y_f - b - \sqrt{-\psi})(-b + \sqrt{-\psi})}{(-2c\,y_f - b + \sqrt{-\psi})(-b - \sqrt{-\psi})}\right] \tag{8.24}$$

where $\psi = -4ac - b^2$.

The time $t_{0.99}$ needed for reaching 99% of the wick height (when the evaporation rate is sufficiently low) or 99% of the final imbibition height (when the evaporation rate or gravity effects are such that the wick is not fully saturated at the end of impregnation) can therefore be computed using Equation 8.24. Time $t_{0.99}$ is plotted as a function of R_b in Figure 8.8a (for the same set of data as before). The time needed to invade the wick in the absence of gravity and evaporation as well as the time needed to reach the height = min (y_{max1}, H) when the rise is controlled only by capillary and gravity effects (no evaporation) are also shown in Figure 8.8a. The shorter time obtained in the presence of evaporation when the bead size is lower than about 5 μm is due to the fact that the final rise height is significantly lower than the height of the wick for the evaporation rate considered (see Figure 8.7). For greater bead size, the size of the invaded zone in the presence of evaporation becomes of the order of the wick height (see Figure 8.7) and evaporation slows down the impregnation as expected compared to the cases where evaporation is negligible (except for the larger beads when gravity effects lead to larger imbibition times owing to the fact that the final imbibition height is lower in the presence of evaporation (compared y_{max1} and y_{max3} in Figure 8.7)). As can be seen also in this example, the impregnation time is practically constant over a decade in bead size (between $R_b = 1$ μm and about $R_b = 10$ μm in Figure 8.8a). As can be seen from Figure 8.7, the final impregnation height increases with the bead size in this range. Hence the increase in the average invasion velocity with R_b compensates almost exactly here the increase in the final imbibition height so that the impregnation time is almost constant.

The evolution of the imbibition time is plotted in Figure 8.8b as a function of evaporation flux (for $R_b = 10$ μm) keeping all other data as before. The inset shows the imbibition height (which is equal to the wick height when y_{max3} as computed from Equation 8.22 is larger than the wick height H). For sufficiently low evaporation rates, evaporation does not affect the impregnation time as expected, which is then controlled by the capillary and gravity effects (evaporation fluxes lower than 10^{-5} kg/m^2/s in Figure 8.8b). Then one can observe a sharp increase in the impregnation time up to a maximum which corresponds to the evaporation flux for which $y_{max3} = H$. For greater

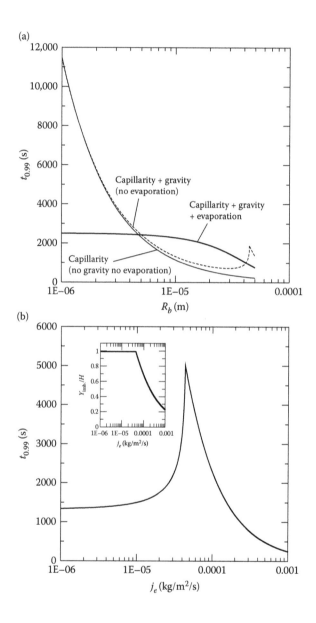

FIGURE 8.8
(a) Evolution of time $t_{0.99}$ (time needed to reach 99% of the final impregnation height) as a function of bead diameter ($j_e = 10^{-4}$ kg/m²/s); (b) evolution of time $t_{0.99}$ as a function of evaporation flux j_e ($R_b = 10$ μm). The inset shows the final impregnation height Y_{imb} as a function of evaporation flux.

evaporation fluxes, the final impregnation height significantly decreases with j_e (see inset in Figure 8.8b) and the time needed for impregnating this height decreases with j_e.

The simple example considered in this section illustrates that evaporation can have a very significant effect on the final imbibition of the wick as well as on the imbibition dynamics. Note that on the right-hand side of Equation 8.18, the term containing the evaporation flux varies as L^{-1}, where L is the half-thickness of the porous medium. Therefore, the influence of evaporation becomes negligible for a sufficiently thick porous material (except possibly near the surface of the wick and in the upper part of the impregnated region, see Section 8.7).

8.6 Distribution of External Mass Transfer Coefficient

The simple model presented in Section 8.5 is based on the assumption that the evaporation flux is constant along the wick surface. This can also be discussed in terms of mass transfer coefficient (denoted by α) when the evaporation flux is expressed using relationship of the form

$$j_e = \alpha (\rho_{vi} - \rho_{v\infty}) \tag{8.25}$$

where ρ_{vi} is the vapor concentration at the wick surface (here we assume that $\rho_{vi} = \rho_{vsat}$ where ρ_{vsat} is the vapor saturation concentration) and $\rho_{v\infty}$ is the vapor concentration in the far field in the gas phase. Hence, assuming a constant evaporation flux is equivalent to assuming a constant mass transfer coefficient.

Interestingly, the constant evaporation flux assumption was questioned in Fries et al. (2008) in order to explain why the impregnation height computed with their model (similar to Equation 8.18) overestimates the height measured in their experiment. The observed discrepancies are *a priori* not sufficient to modify qualitatively the results presented in Section 8.5 but the question is worth considering. To shed some light on this problem, we have computed the evaporation flux distribution for the situation depicted in Figure 8.5b (with $W = 50$ mm, $H = 50$ mm, $2L = 150$ μm, $H_{encl} = 100$ mm, a 10 mm liquid layer at the bottom of the enclosure), which is similar to the one considered in Fries et al. (2008), assuming the wick fully saturated and considering only the steady-state evaporation regime observed once the imbibition process is over. We have also assumed that no evaporation can take place from the wick horizontal top surface. Only evaporation from the lateral sides of the wick is thus taken into account. Thus we impose $\rho_v = \rho_{vsat}$ on the surface of the liquid layer and at the lateral surface of the wick, a zero-flux condition at the top wick surface and at the lateral internal surface of the enclosure.

Vapor escapes from the open top of enclosure ($y = H_{encl}$) where we impose $\rho_v = \rho_{v\infty} < \rho_{vsat}$. The result, obtained solving Equation 8.3 with the commercial code Comsol and expressed in terms of dimensionless evaporation flux $j_e^* = j_e / j_{ref}$ using the mean evaporation flux as reference, $j_{ref} = E/A$, where E is the overall evaporation rate and A is the evaporative surface of the wick, is shown in Figure 8.9. This distribution is for the pure diffusion limit (free convection effect neglected) (a more thorough study of the situation studied in Fries et al. (2008) would probably require 3D free convection computations owing to the possible influence of edge effects and the relatively high evaporation rate measured but this is left for future work and would not change the main conclusion of this section).

As can be seen from Figure 8.9, the evaporation flux distribution is far from uniform in this type of configuration. The evaporation flux is much higher in the top region of the wick. It decreases rapidly in the first 5 mm from the wick top. Then it still decreases along the wick surface but much more slowly. Notice that Figure 8.9 is a semi-log plot. Hence, the variations in the flux along the wick evaporative surface are quite significant. This result suggests that the evaporation flux is much higher in the zone of the imbibition front during the impregnation dynamics. As guessed by Fries et al. (2008), this result indicates that the assumption of spatially constant evaporation flux along the wick must be abandoned. A proper computation of the evaporation flux along the evaporative surface of the wick should significantly improve the comparison between their model and their experimental results. This is left for future work. It can also be noted that the side walls in Figure 8.5b

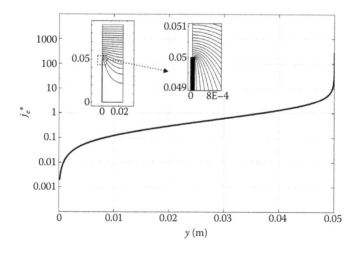

FIGURE 8.9
Evaporation flux distribution along the wick lateral surface for the situation sketched in Figure 8.5b and a fully saturated wick. The insets show the vapor partial pressure isocontour lines in the gas phase. (Wick in black, all figures expressed in m.)

induce a confinement effect, which limits the evaporation from the lower lateral surfaces. The situation depicted in Figure 8.5a for instance (no side walls) should lead to a different evaporation flux distribution along the wick but still not spatially constant.

The fact that the flux distribution is an important factor in wicking/evaporation can be illustrated nicely as follows. To this end, we return to the situation depicted in Figure 8.4. As shown in Figure 8.10 for the case $\delta = 2.5$ mm, the evaporation flux j_e along the radius of the wick at the wick surface is also not spatially uniform. In this case, the evaporation flux is greater in the periphery of the wick surface, especially when δ is small (we recall that δ is the distance between the wick surface and the top of the hollow cylinder containing the wick, see Figure 8.4). As shown in Figure 8.10, this holds for the pure diffusion limit as well as when free convection effects are taken into account. This has important consequences as illustrated in Figure 8.11 for $\delta = 2.5$ mm, which shows a direct correlation between the higher evaporation fluxes and the salt crystallization spots visible in Figure 8.11. This indicates that the evaporation flux spatial distribution has a direct impact

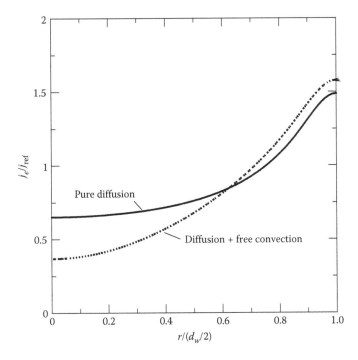

FIGURE 8.10
Evaporation flux distributions along a radius at the wick surface for the wicking/evaporation situation shown in Figure 8.4 ($\delta = 2.5$ mm). The evaporation flux is significantly higher at the periphery of wick surface; $j_{ref} = E/A$, where E is the overall evaporation rate and A is the surface area of the wick.

FIGURE 8.11
View of the surface of the wick shown in the experimental set-up sketched in Figure 8.4. The white patches are NaCl crystallization spots at the surface of the wick. The crystallization spots are located in the peripheral region of the wick surface where the evaporation fluxes are higher as shown in Figure 8.10.

on the transport of dissolved species (the crystallization spots in Figure 8.11 correspond to local maxima in the concentration of the dissolved salt at the surface of the wick). One can refer to Veran-Tissoires et al. (2012) for more details.

8.7 Pore Network Study of Wicking/Evaporation

8.7.1 Introduction

One of the assumptions made to develop the model presented in Section 8.5 was to assume a flat imbibition front. This is a reasonable assumption for materials made of relatively narrow pore size distributions when the imbibition height is large compared to the system thickness ($2L$) (this will be discussed in this section). As shown in Figure 8.6, however, the front is rarely perfectly flat. As explained for instance in Barabási and Stanley (1995), this is due to the disordered nature of the porous microstructure. The roughening of the interface can be captured using a simple pore network model as shown later in Section 8.7.2.

Even if the flat front assumption can be sufficient to estimate the imbibition dynamics and the final imbibition height as discussed in Section 8.5, the final distribution of the liquid in the top region of the impregnated zone does not necessarily correspond to a flat front owing to the effect of evaporation. The second objective of this section is therefore to discuss the shape of the evaporation front in the wick. We consider only the steady-state distribution (the shape of the frontal region during the impregnation dynamics is not considered). As many other problems in porous media, this problem can also certainly be tackled using continuum (Darcy's scale) models. The main problem is to determine the position of the front and several techniques, such as the VOF (volume of fluid) method proposed in Carciofi et al. (2011), could be used in this case within the framework of continuum models. Here, we use, however, a pore network (PN) approach, see the chapter by Joekar-Niasar and Hassanizadeh in this book for more details on pore-network modeling. The reasons in favor of a PN approach are numerous. First, a porous sheet can be very thin. For example, the gas diffusion layer, which is one of the porous layers forming a PEMFC (proton exchange membrane fuel cell) is only a few pore sizes thick, for example, Rebai and Prat (2009) and Ceballos et al. (2011). As a result, the length scale separation between the REV (representative elementary volume) and the porous sheet width is poor in this example. This is also the case with many paper sheets. It is well known that a sufficient length scale separation is needed for the continuum models to represent satisfactory models. This length scale separation requirement is not needed for the PN models. Second, tracking the interface position is not very difficult using PN models. Third, the PN approach can be particularly well adapted for analyzing what happens at the evaporation front when the wicking/evaporation process induces the transport of dissolved species, for example, Veran-Tissoires (2011). As an example, a PN model can explain the occurrence of the discrete crystallization spots visible in Figure 8.11. This is not possible with a traditional continuum model. It should be mentioned that the studies of wicking/evaporation based on PNM (Pore Network Models) are scarce. Except from the works presented in this section and in Veran-Tissoires (2011), one can only cite the related work on evaporation of a sessile droplet on a porous medium presented in Markicevic and Nawaz (2010b). We believe, however, that this is a promising approach which is going to be used much more in future works.

As discussed in many works, for example, Prat (2010, 2011), Joekar-Niasar and Hassanizadeh (2011) and references therein, a pore network model is based on a representation of the pore space as a network of sites (pores) connected by bonds (throats). This is sketched in Figure 8.12. For simplicity, we use a simple square network in this chapter. The pores (sites) are spheres of diameter d_p placed on a regular square grid. The distance d_ℓ between two pores (called the lattice spacing) is constant. As shown in Figure 8.12, each pore is connected to four neighbor pores. The connecting bonds (throats) are tubes of diameter d_t. The diameter d_p (d_t, respectively) are randomly

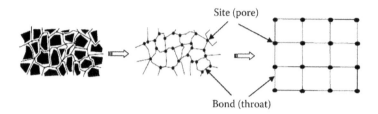

Site (pore)

Bond (throat)

FIGURE 8.12
Representation of the pore space as a network of interconnected pores (sites) and bonds (throats). The solid phase is in black. Modeling of pore space by a square network of sites (pores) and bonds (throats).

distributed in a given range of values $[d_{pmin}, d_{pmax}]$ ($[d_{tmin}, d_{tmax}]$, respectively) according to some prescribed probability density function (p.d.f.) with the constraint that the diameter of a bond is smaller than the smallest size of the two adjacent pores. The length of a bond is then given by $d_\ell - 0.5 (d_{pi} + d_{pj})$ where i and j refer to the two pores adjacent to the considered bond.

8.7.2 A Simple Pore-Network Model of Wicking

The PN approach has been used to study wicking in several previous works. These notably include Ridgway et al. (2001), Ridgway and Gane (2002), Ghassemzadeh et al. (2001), Ghassemzadeh and Sahimi (2004), Markicevic and Navaz (2010a,b). An approximate but simple version of this type of model can be described as follows.

The porous medium is represented by a square network of pores connected by throats (see Section 8.7.1). Initially the network is dry and put in contact with a liquid reservoir at time $t = 0$ (Figure 8.5). The pores are storage elements of volume V_p. The pressure field P_ℓ in the part of the network occupied by the liquid phase is computed by expressing the mass conservation of the liquid at each liquid node (pore) of network,

$$\sum_i Q_{ij} = 0. \tag{8.26}$$

where the volumetric flow rate between two liquid pores i and j is expressed as

$$Q_{ij} = \frac{g_{ij}}{\mu_\ell}(P_{\ell i} - P_{\ell j}) \tag{8.27}$$

The local hydraulic throat conductance g_{ij} (viscous resistance is neglected in the pores) is expressed considering the throat as a circular tube (Poiseuille's law),

$$g_{ij} = \frac{\pi d_{tij}{}^4}{128\, d_\ell} \tag{8.28}$$

At the bottom of the network, the pressure is known ($P_\ell = P_{ref}$ where P_{ref} is the atmospheric pressure). At a bond (throat) belonging to the invasion front (liquid–gas interface), the flow rate is expressed as

$$Q_{ij} = \frac{g_{ij}}{\mu_\ell}(P_{\ell i} + P_{cthij}) \tag{8.29}$$

where i is the adjacent liquid pore and $P_{cthij} = 4\gamma \cos\theta/d_{tij}$ is the capillary pressure associated with the throat (capillary pressure responsible for the invasion of the empty or partially empty adjacent pore j).

The time for filling an interfacial pore numbered j is computed from the equation,

$$t_{fillj} = V_{pj}(1 - S_j)\sum_i Q_{ij} \tag{8.30}$$

where the flow rate into the considered pore is obtained by summing up all the flow rates from neighboring liquid pores i; S_j is the pore liquid saturation.

The invasion algorithm can be finally summarized as follows: (1) compute pressure field in the liquid by solving numerically the system of Equations 8.26; (2) compute flow rates in interfacial bonds from Equation 8.29; (3) compute the time step from Equation 8.30, $t_{step} = \min (t_{fillj})$ for all interfacial pores; (4) fill fully the interfacial pore corresponding to $\min (t_{fillj})$ and update saturation in other interfacial pores; and (5) return to (1) or stop invasion at a desired stage.

As can be seen from Figure 8.13, this type of PN model leads to invasion dynamics in full agreement with the Washburn scaling $y_f \propto \sqrt{t}$ (note that gravity or evaporation effects are not considered in this subsection). Note that we have plotted in Figure 8.13 the position of the least advanced point of the front (and not the average position of the front) as a function of $\sqrt{t}$. This explains the staircases in Figure 8.13. Two simulations for a 120×120 network are actually shown in Figure 8.13. The difference lies in the throat size distribution. In the first simulation, the microstructure is perfectly ordered (all the throats have the same size $0.5d_\ell$ and all the pores have the same volume V_p). As a result, the front is flat (see the inset on the left-hand side in Figure 8.13) and there is no fluctuation in the steps of the staircase describing the dynamics of the least advanced point of the front. By contrast, the second simulation was performed for a disordered microstructure d_t varying in the range $[0.3d_\ell, 0.7d_\ell]$. This leads to fluctuations in the size of the steps of the corresponding staircase in Figure 8.13 as well as, interestingly, a rough invasion

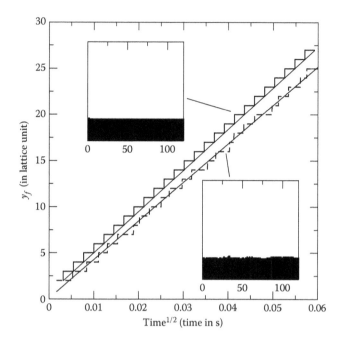

FIGURE 8.13

Pore network computation of evolution of least advanced point y_f (expressed in lattice unit) of imbibition front as a function of the square root of time (expressed in seconds). The straight lines are linear fits of the staircase lines. Comparison between a perfectly ordered network and a disordered network (see text). The insets show an example of front shape for both cases (imbibed region in black).

front (see the inset on the right-hand side in Figure 8.13 and a magnified view of the inset in Figure 8.6b).

However, evaluating whether this simple model can lead to a correct roughening dynamics, for example, Barabási and Stanley (1995), would require work which is much beyond the scope of the present chapter. Also imbibition can be considerably more complicated than considered here, for example, Blunt et al. (1992) and Blunt and Scher (1995). Phenomena associated with liquid films (see Section 8.8) such as snap-off for instance, can be important, especially to determine the final steady-state distribution (when gravity effects limit the invasion height, for example). Also, no capillary pressure is associated with the pores (capillary pressures are associated with throats only). This is clearly a shortcoming of the simple model presented in this section. Pore filling can be significantly more complex than considered here owing to cooperative effects between adjacent menisci, for example, Blunt et al. (1992), Blunt and Scher (1995), Martys et al. (1991). In brief, we believe that this simple model is reasonable as long as viscous effects are dominant (early stage of wicking) and should be refined for later stages (when the imbibition front approaches its final position in particular).

8.7.3 Steady-State Shape of Internal Evaporation Front

We now consider the problem of the steady-state shape of the internal invasion front in the presence of evaporation. From the experimental results presented in Noiriel et al. (2010), this shape can be expected to be roughly parabolic. This is of course in contrast with the flat front (neglecting the small scale front roughness), which is expected when the stabilization of the front results from the competition between capillary and gravity effects in the absence of evaporation. It is interesting to evaluate first the conditions for which the internal evaporation can be expected to have a significant effect on the evaporation front shape. Setting $dy_f/dt = 0$, Equation 8.18 gives us the position y_{f1} of the front when the internal evaporation is neglected (y_{f1} is thus given by Equation 8.22). Taking into account the internal evaporation should lead therefore to a lower (mean) position. Assume that the distance between these two positions is $O(L)$ (we recall that L is the half width of porous wick, see Figure 8.5) so as to have a significant effect on internal evaporation. According to Equation 8.18 this corresponds to an additional mass flow rate δm given by

$$\delta m = 2L\rho_\ell \frac{K}{\mu_\ell}\left(\frac{P_{cth}}{y_{f1}-L} - 0.5\frac{\mu_\ell j_e}{\rho_\ell LK}(y_{f1}-L) - \rho_\ell g\right) \tag{8.31}$$

An order of magnitude of the internal evaporation is given by

$$J_{int} \approx 2L\varepsilon D^* \frac{\Delta\rho_v}{L} \tag{8.32}$$

where D^* is the effective diffusion coefficient of the porous medium and $\Delta\rho_v = \rho_{vsat} - \rho_{v\infty}$ is the vapor concentration difference between a meniscus and the surrounding air. Expressing that δm and J_{int} must be of the same order of magnitude (for again expecting a significant effect of the internal evaporation on the front shape) gives

$$\rho_\ell \frac{K}{\mu_\ell}\left(\frac{P_{cth}}{y_{f1}-L} - 0.5\frac{\mu_\ell j_e}{\rho_\ell LK}(y_{f1}-L) - \rho_\ell g\right) = \varepsilon D^* \frac{\Delta\rho_v}{L} \tag{8.33}$$

Then consider again a random packing of monodisperse beads of radius R_b; $K = C(\varepsilon^3 R_b^2/(1-\varepsilon)^2)$; $P_{cth} = 3(1-\varepsilon)\gamma \cos\theta/\varepsilon R_b$; $D^* \approx \varepsilon^{0.4}D$ and as in Noiriel et al. (2010), water at 30°C as working fluid ($\rho_{vsat} \approx 0.0264$ kg/m³, $\mu_\ell = 10^{-3}$ Pa s, $\gamma = 72 \cdot 10^{-3}$ N/m) and express the lateral evaporation flux as $j_e = (D/\delta_{ext})\Delta\rho_v$ where δ_{ext} represents an external mass transfer length scale (a reasonable value is $\delta_{ext} = 10$ mm). For these conditions and with $L = 0.5$ cm (as in Noiriel et al. (2010)), the two sides of Equation 8.33 are about equal for

$R_b \approx 75$ μm. For $R_b \gg 75$ μm, gravity effects should lead to a flat front (the internal shape is then controlled by the competition between gravity and capillary effects) whereas for $R_b \ll 75$ μm, it is expected that the internal front shape depends only on the competition between viscous and capillary effects. Hence, a nonflat shape is expected here for $R_b \approx 75$ μm or smaller (again for greater values gravity effects should lead to a flat front). In terms of network representation, this represents a network of width $= 2L/2R_b \approx 65$ (in lattice unit). For these conditions, $y_{f1} \approx 38$ cm, which expressed in lattice unit gives $y_{f1}/2R_b \approx 2500$. Thus a two-dimensional 2500×65 network would be necessary to represent the whole wick. However, we are only interested in the top region of the wick imbibed pack. Accordingly, a 120×65 network should therefore be sufficient to determine the internal front shape at the top of the imbibed region.

To compute the shape of the interface we need to be able to compute the pressure field in the liquid for a given position of the interface. The pressure is first expressed as

$$P_\ell = P_{\ell h} + \tilde{P}_\ell \tag{8.34}$$

where $P_{\ell h} = -\rho_\ell \, gy + P_{\text{atm}}$ is the hydrostatic pressure and $\tilde{P}_\ell$ is the pressure due to the viscous flow induced by the evaporation process (the system of coordinates is similar to the one shown in Figure 8.5b).

To compute $\tilde{P}_\ell$, the method is as described in Section 8.7.2. A local hydraulic conductance is assigned to each throat, see Equation 8.28 (viscous resistance is neglected in the pores). As in Section 8.7.2, a throat is considered as a circular tube. The local flow rate between two pores is expressed using Equation 8.27. Expressing the mass conservation equation over each pore, see Equation 8.26, yields a system of linear equations that is solved numerically to obtain the pressure $\tilde{P}_\ell$ in each pore of network. The boundary conditions are,

$$\tilde{P}_\ell = P_{\text{bottom}} \quad \text{at } y = y_{\text{bottom}} \tag{8.35}$$

where y_{bottom} and P_{bottom} are the position and the pressure at the bottom of the region of the wick represented by the pore network (top of the wick imbibed region). Hence, $y_{\text{bottom}} \approx y_{f1} - 2L$ and $P_{\text{bottom}} \approx -y_{\text{bottom}}\rho_\ell g$. At the interfacial bonds (throats containing a meniscus), we express that the liquid flow rate in the bond should balance the evaporation rate,

$$Q_l = J_e/\rho_\ell \tag{8.36}$$

An interfacial throat is a throat with a meniscus at its entrance. One can distinguish the interfacial throats located at the surface of the wick from the ones located within the wick. The evaporation rate J_{ext} for a throat located at the surface of the wick is computed as

$$J_{ext} = d_\ell^2 j_e \tag{8.37}$$

where the evaporation flux j_e is expressed as before as $j_e = (D/\delta_{ext})\Delta\rho_v$. As explained in Section 8.3, the porous surface behaves as a saturated surface when sufficiently divided. This explains why J_{ext} is computed using Equation 8.37 and not from an expression of the form $(\pi d_{tij}^2/4)j_e$, for example.

To determine J_e at the surface of a meniscus located within the wick, one needs to solve the diffusive transport of the vapor within the region of the wick occupied by the gas phase. The method is similar to the one used for the flow problem. A diffusive conductance is assigned to each "gaseous" throat,

$$g_D = \frac{\pi d_t^2}{4 d_\ell} D \tag{8.38}$$

The local mass flow rate F_{ij} between two pores (pores i and j) is expressed as

$$F_{ij} = g_{Dij}(\rho_{vi} - \rho_{vj}) \tag{8.39}$$

Expressing the mass conservation equation over each pore

$$\sum_i F_{ij} = 0. \tag{8.40}$$

yields a system of linear equations that is solved numerically to obtain the vapor concentration in each gas pore of network. The boundary conditions are

$$\rho_v = \rho_{vsat} \quad \text{at a meniscus} \tag{8.41}$$

$$F_{ij} = J_{ext} \quad \text{at surface throats} \tag{8.42}$$

with $J_{ext} = d_\ell^2(D/\delta_{ext})(\rho_{vi} - \rho_{v\infty})$ (a surface throat is a throat located at the external surface of the wick). The method to compute the vapor concentration in the gas pore is actually quite similar to the one used in previous works on pore network modeling of drying, for example, Prat (1993, 1995). One can thus refer to these works for further details.

The last necessary ingredient is the condition for a meniscus to be in a stable position at the entrance of a throat. This condition can be expressed as

$$P_{cth} \geq \rho_\ell g \, y - \tilde{P}_\ell \tag{8.43}$$

where P_{cth} is the invasion capillary pressure threshold of the throat ($P_{cth} = 4\gamma \cos \theta / d_t$ for a circular throat).

To determine the position of the evaporation front within the wick under steady-state evaporation condition the procedure is finally as follows: (1) start from an almost liquid saturated network, (2) compute vapor concentration in the part of the network occupied by gas phase, (3) compute evaporation rate at menisci and compute the viscous pressure field $\tilde{P}_\ell$ in the liquid part of the network, (4) test menisci stability (Equation 8.43) and dry each pore adjacent to a nonstable meniscus (i.e., a meniscus located in a throat and such that $\bar{P}_{cth} \leq \rho_\ell g \, y - \tilde{P}_\ell$), (5) completely invade the interfacial pore associated with the smallest drying time and update liquid saturation in other drying pores accordingly, and (6) return to (2) until all menisci are stable.

The results obtained with this model are presented in Figure 8.14. The shape obtained for the reference situation (network permeability and throat capillary pressure threshold corresponding to $R_b = 75$ μm, $\delta_{ext} = 10$ mm, see right after Equation 8.33 for the other data) is shown in Figure 8.14a. As can be seen, the shape is roughly parabolic in good agreement with the schematic shape presented in several previous papers, for example, Scherer (2004) and references therein, and the experimental results presented in Noiriel et al. (2010). Note that this shape has been obtained considering a perfectly ordered square network (all throats have the same conductance and same capillary pressure threshold). When some disorder is introduced (by distributing randomly the throat size in the range [0.95 $\bar{d}_t$, 1.05 $\bar{d}_t$], where $\bar{d}_t$ is the throat size corresponding to the perfectly ordered reference situation), the front is rough as expected in a drying process (the final steady-state position is obtained through a drying process with the algorithm presented in this section), see Prat and Bouleux (1999) for more details on this type of irregular fractal front.

Hence one interesting question is whether the steady-state front should be rough as shown in Figure 8.14b or much smoother as suggested in Scherer (2004) and indicated by the experimental results of Noiriel et al. (2010). Actually, this should depend on how the steady state is reached. Suppose the front reaches some steady-state position within the wick and is then submitted to a greater evaporation. The front must then recede into the wick. The process is then analogous to a drainage process (displacement of a wetting fluid (the liquid here) by a nonwetting one (the gas phase)) as in drying. This case leads to rough fronts as illustrated in Figure 8.4b (especially in a two-dimensional pore network). A markedly different case

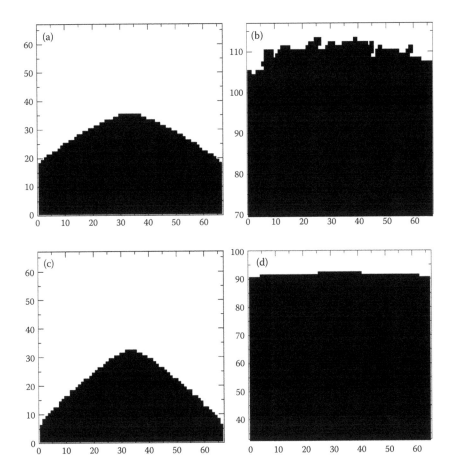

FIGURE 8.14
Steady-state shape of internal invasion front (top of imbibed region) within a porous sheet submitted to evaporation computed using the pore network model described in Section 8.7.3. Imbibed region in black. (a) Reference situation (perfectly ordered pore network, (b) reference situation (disordered pore network), (c) lower permeability (see text) for a perfectly ordered pore network, (d) greater permeability (see text) for a perfectly ordered pore network.

is when the steady state is reached through an imbibition process (displacement of a nonwetting fluid by a wetting one). This is the situation expected in wicking for given (fixed) evaporation conditions (in wicking, the wick is often dry initially and therefore the final distribution is reached through an imbibition process). It is well known, for example, Blunt et al. (1992), Martys et al. (1991), that imbibition leads to much smoother fronts (see Figure 8.6) due to cooperative mechanisms between adjacent menisci. It would be possible to define pore local capillary pressures taking into account the cooperative effects, see Chapuis et al. (2007) or Chraïbi et al. (2009) and references therein, and to develop an algorithm somewhat analogous to the one presented in this section but reaching the state through an imbibition/evaporation process rather than a drying process. This will be presented in a future work. The consideration of a perfectly ordered pore network with the present much simpler algorithm should therefore be understood here as just a trick to obtain smooth fronts as expected in the wicking situation.

Using this trick (a perfectly ordered network), we can also get some insights into the influence of pore size (or equivalently R_b or equivalently the porous medium permeability) on the results. Figure 8.14c and 8.14d shows the shapes obtained imposing in the network the permeability and capillary pressure threshold corresponding to $R_{bref}/5$ and $5R_{bref}$, respectively ($R_{bref} = 75$ μm). As expected (see discussion leading to Equation 8.33), the height of the "parabolic" shape is greater with the lower R_b, that is a lower permeability (Figure 8.14c), whereas an almost flat front is obtained with the greater R_b, that is a greater permeability (as discussed below in Equation 8.33, this is due to the fact that the front shape becomes essentially controlled by the competition between capillary and gravity effects, the viscous pressure drop induced in the liquid by the evaporation process becoming negligible compared to hydrostatic variations). Note that the "parabolic" shape observed when viscous effects are sufficiently important is of course due to the presence of a lateral evaporation, which induces two-dimensional pressure drops within the wick. Increasing the strength of the vapor flux from the sides increases the induced viscous pressure drop within the wick and thus should have an effect similar to the effect obtained in diminishing the bead radius in Figure 8.14.

8.8 Liquid Film Effects

It is now widely admitted that liquid films can play a major role in the drying kinetics of a porous medium, for example, Yiotis et al. (2004) and Prat (2007). Thus, it is quite likely that they can also play a major role in wicking/evaporation. In this respect, it is convenient to distinguish the thick films from the thin films. The liquid films forming in the corners or others geometrical singularities of the pore space are referred to as capillary liquid films or thick films. They can be distinguished from the very thin films driven by the disjoining pressure that forms on a flat surface. Contrary to capillary (thick) films, which are important in the analysis of evaporation phenomena in microporous media, the thin films are not expected to extend over large regions of the system during an evaporation process and are therefore generally ignored in mass-transfer-driven evaporation process.

The effect of thick films can be well illustrated with the consideration of evaporation from a single capillary tube of square cross section. Whereas evaporation in a capillary tube of circular cross section is well understood since the nineteenth century with the works of Stefan (1971), it is somewhat surprising that we had to wait until very recently for a complete description of evaporation in a single tube of square cross section, Chauvet et al. (2009). One reason is that the evaporation process in a square tube is considerably more complex due to the effect of liquid films trapped by capillarity in the tube internal corners, see Figure 8.15.

The corner films provide paths for liquid transport between the bulk meniscus and the entrance of the tube. The liquid is transported within the films under the action of the pressure gradient induced by the corner meniscus curvature variation along the films. This effect is therefore termed capillary pumping. As a result, the phase change occurs preferentially at the entrance of the tube as long as the corner liquid films remain attached at the tube entrance. This very efficient transport mechanism of liquid by the films is naturally absent in a circular tube, in which the transport mechanism between the bulk meniscus and the tube entrance is the poorly efficient molecular diffusion in the gas phase. This explains why evaporation can be several orders of magnitude greater in a tube with corners compared to a circular tube (Prat 2007, Chauvet et al. 2009).

The effect of thick liquid films when a liquid evaporates from a porous medium has been the subject of several studies in the framework of pore network models, for example, Yiotis et al. (2004), Prat (2007), Prat (2011), and

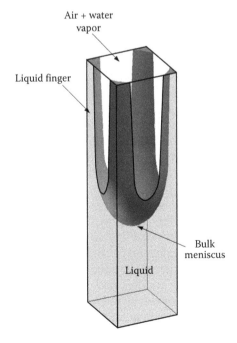

FIGURE 8.15
Sketch of the thick liquid films during evaporation in a capillary tube of square cross section.

references therein. The models presented in these studies could be valuably adapted to study the effects of film on wicking/evaporation process. This is a completely open research area. It is also worth mentioning that the films are likely to greatly influence the transport and therefore the spatial and temporal distribution of dissolved species during wicking/evaporation. As discussed for instance in Ponomarenko et al. (2011), the influence of thick films on wicking is also interesting to consider in the absence of evaporation, see also Blunt et al. (1992) and Blunt and Scher (1995).

8.9 Conclusion

Wicking occurs in the presence of evaporation in many practical situations of interest. A major area not reviewed in this article is wicking with heat-transfer-driven evaporation occurring in two-phase cooling devices, for example, Maydanik (2005) and Prat (2010). In this chapter, we focussed instead on wicking in the presence of unconfined mass-transfer-driven evaporation. The first conclusion is that this field, albeit quite important in relation with several applications (salt weathering issues, volatile species diffusers, PMD, clothing designs, etc.), has been the subject of relatively few scientific studies as regards the prediction of evaporation rate and the influence of evaporation on the impregnation dynamics and the final steady-state distribution of the liquid. As pointed out in this chapter, this is in part due to the fact that the evaporation flux distribution along the gas–liquid interface (located at the surface of the wick or within the wick) is not well documented (even for the steady-state regime). As a consequence there remains lots to be done in this area.

The results reported in this chapter clearly show that evaporation can have a major impact on both impregnation dynamics and the final fluid distribution within the wick. Evaporation induces a flow driven by capillarity even when the final fluid distribution has been reached within the wick and this has important consequences for the transport of dissolved species (as illustrated by the transport of a dissolved salt in this article).

As illustrated in this chapter, the modeling of wicking/evaporation can be performed according to various techniques. Wick cross section averaged continuum (Darcy's scale) models are well adapted to capture the main features of the impregnation dynamics and the approximate final impregnation height whereas pore network models provide detailed information on the fluid distribution or on the effect of phenomena difficult to analyze with continuum models, such as for instance the influence of liquid films or the structure of the concentration field of dissolved species at the scale of the pore network.

Nomenclature

Roman Symbols

A	Evaporative surface area of the wick (m²)
D	Binary molecular diffusion coefficient (m²/s)
d	Pore diameter (m)
E	Evaporation rate (kg/s)
F_{ij}	Local mass flow rate between pores i and j (kg/s)
g	Gravitational acceleration (m/s²)
g_{ij}	Throat hydraulic conductance (kg/Pa/s)
H	Height (m)
HR	Relative humidity (-)
J	Evaporation rate (kg/s)
j_e	Evaporation flux (kg/m²/s)
K	Permeability (m²)
L	Length (m)
m	Mass (kg)
n	Unit normal vector (-)
P	Pressure (Pa)
$\tilde{P}_l$	Pressure due to the viscous flow induced by the evaporation process (Pa)
Q	Volumetric flow rate (m³/s)
R_b	Beads radius (M)
S	Pore liquid saturation (-)
t	Time (s)
U	Volume-averaged Darcy velocity (m/s)
u	Gas velocity vector (m/s)
V	Volume (m³)
V_e	Domain occupied by the liquid in the wick (m³)
W	Width of the enclosure (m)
x, y	Cartesian coordinates (m)
X_v	Vapor mass fraction (-)
y_f	Position of the imbibition front (m)

Greek Symbols

α	Mass transfer coefficient (kg/m²/s)
β	Mass expansion coefficient (-)
δ	Distance between the wick surface and the top of the hollow cylinder containing the wick (m)
δ_{ext}	External mass transfer length scale (m)
δ_m	Mass flow rate (kg/s)

ε	Porosity (-)
Φ	Surface porosity (-)
γ	Surface tension (N/m)
μ	Gas-phase dynamic viscosity (Pa s)
θ	Contact angle (rad)
ρ	Gas-phase density (kg/m³)
ρ_v	Vapor concentration (kg/m³)
υ	Gas-phase kinematic viscosity (m²/s)

Subscripts

∞	Ambient
*	Effective coefficient for the porous media
atm	Atmosphere
c	Capillary
e	Equilibrium
encl	Enclosure
exp	Experimental
ext	External
f	Front
fill	Filling
int	Internal
l	Liquid phase
num	Numerical
p	Pores
sat	Saturation
th	Threshold
v	Vapor
w	Wick
x, y	Cartesian components

References

Barabási, A.-L. and H. E. Stanley. 1995. *Fractal Concepts in Surface Growth*. New York: Cambridge University Press.

Bird, R. B., W. E. Stewart, and E. N. Lightfoot. 2002. *Transport Phenomena*. 2nd ed. New York: John Wiley and Sons.

Blunt, M., M. J. King, and H. Scher. 1992. Simulation and theory of two-phase flow in porous media. *Physical Review A* 46(12):7680–7699.

Blunt, M. J. and H. Scher. 1995. Pore-level modeling of wetting. *Physical Review E* 52(6):6387–6403.

Carciofi, B. A. M., M. Prat, and J. B. Laurindo. 2011. Homogeneous volume-of-fluid (VOF) model for simulating the imbibition in porous media saturated by gas. *Energy & Fuels* 25(5):2267–2273.

Carey, V. P. 2008. *Liquid–Vapor Phase Change Phenomena*. 2nd ed. New York: Taylor & Francis.

Ceballos, L., M. Prat, and P. Duru. 2011. Slow invasion of a nonwetting fluid from multiple inlet sources in a thin porous layer. *Physical Review E* 84:056311.

Chapuis, O. and M. Prat. 2007. Influence of wettability conditions on slow evaporation in two-dimensional porous media. *Physical Review E* 75(4):046311.

Chauvet, F., P. Duru, S. Geoffroy, and M. Prat. 2009. Three periods of drying of a single square capillary tube. *Physical Review Letters* 103(12):124502.

Chraïbi, H., M. Prat, and O. Chapuis. 2009. Influence of contact angle on slow evaporation in two-dimensional porous media. *Physical Review E* 79(2):026313.

Deegan, R. D., O. Bakajin, T. F. Dupont, G. Huber, S. R. Nagel, and T. A. Witten. 1997. Capillary flow as the cause of ring stains from dried liquid drops. *Nature* 389(6653):827–829.

Faghri, A. 1995. *Heat Pipe Science and Technology*. Washington: Taylor & Francis.

Fries, N., K. Odic, M. Conrath, and M. Dreyer. 2008. The effect of evaporation on the wicking of liquids into a metallic weave. *Journal of Colloid and Interface Science* 321(1):118–129.

Geoffroy, S., S. Beyhaghi, M. Prat, and K. M. Pillai. 2011. Evaporation of a single-component liquid from porous wicks. *To be Submitted*.

Ghassemzadeh, J. and M. Sahimi. 2004. Pore network simulation of fluid imbibition into paper during coating—III: Modelling of the two-phase flow. *Chemical Engineering Science* 59(11):2281–2296.

Ghassemzadeh, J., M. Hashemi, L. Sartor, and M. Sahimi. 2001. Pore network simulation of imbibition into paper during coating: I. Model development. *AIChE Journal* 47(3):519–535.

Goudie, A. and H. Viles. 1997. *Salt Weathering Hazards*. Chichester: Wiley.

Lockington, D. A., J. Y. Parlange, and M. Lenkopane. 2007. Capillary absorption in porous sheets and surfaces subject to evaporation. *Transport in Porous Media* 68(1):29–36.

Markicevic, B. and H. K. Navaz. 2010a. Primary and secondary infiltration of wetting liquid sessile droplet into porous medium. *Transport in Porous Media* 85(3):953–974.

Markicevic, B. and H. K. Navaz. 2010b. The influence of capillary flow on the fate of evaporating wetted imprint of the sessile droplet in porous medium. *Physics of Fluids* 22:122103.

Martys, N., M. Cieplak, and M. O. Robbins. 1991. Critical phenomena in fluid invasion of porous-media. *Physical Review Letters* 66(8):1058–1061.

Masoodi, R., K. M. Pillai, and P. P. Varanasi. 2007. Darcy's law-based models for liquid absorption in polymer wicks. *AIChE Journal* 53(11):2769–2782.

Maydanik, Y. 2005. Loop heat pipes. *Applied Thermal Engineering* 25:635–657.

Metzger, T., E. Tsotsas, and M. Prat. 2007. Chapter 2: Pore network models: A powerful tool to study drying at the pore level and understand the influence of structure on drying kinetics. In *Modern Drying Technology, Volume 1, Computational Tools at Different Scales*, A. Mujumdar and E.T. Tsotsas (Eds). Germany: Wiley VCH Weinheim.

Noiriel, C., F. Renard, M. L. Doan, and J. P. Gratier. 2010. Intense fracturing and fracture sealing induced by mineral growth in porous rocks. *Chemical Geology* 269(3–4):197–209.

Perez-Diaz, J. L., J. C. Garcia-Prada, F. Romera-Juarez, and E. Diez-Jimenez. 2010. Mechanical behaviour analyses of sap ascent in vascular plants. *Journal of Biological Physics* 36(4):355–363.

Peterson, G. P. 1994. *An Introduction to Heat Pipes, Modeling, Testing and Applications.* New York: Wiley Interscience.

Ponomarenko, A., D. Quere, and C. Clanet. 2011. A universal law for capillary rise in corners. *Journal of Fluid Mechanics* 666:146–154.

Prat, M. 1993. Percolation model of drying under isothermal conditions in porous-media. *International Journal of Multiphase Flow* 19(4):691–704.

Prat, M. 1995. Isothermal drying of nonhygroscopic capillary-porous materials as an invasion percolation process. *International Journal of Multiphase Flow* 21(5):875–892.

Prat, M. 2002. Recent advances in pore-scale models for drying of porous media. *Chemical Engineering Journal* 86(1–2):153–164.

Prat, M. 2007. On the influence of pore shape, contact angle and film flows on drying of capillary porous media. *International Journal of Heat and Mass Transfer* 50(7–8):1455–1468.

Prat, M. 2010. Application of pore network models for the analysis of heat and mass transfer with phase change in the porous wick of loop heat pipes. *Heat Pipe Science and Technology* 1(2):129–149.

Prat, M. 2011. Pore network models of drying, contact angle and films flows. *Chemical Engineering & Technology* 34(7):1029–1038.

Prat, M. and F. Bouleux. 1999. Drying of capillary porous media with a stabilized front in two dimensions. *Physical Review E* 60(5):5647.

Puyate, Y. T. and C. J. Lawrence. 2000. *Steady State Solutions for Chloride Distribution Due to Wick Action in Concrete.* Vol. 55. Kidlington, ROYAUME-UNI: Elsevier.

Rebai, M. and M. Prat. 2009. Scale effect and two-phase flow in a thin hydrophobic porous layer. Application to water transport in gas diffusion layers of proton exchange membrane fuel cells. *Journal of Power Sources* 192(2):534–543.

Ridgway, C. J. and P. A. C. Gane. 2002. Dynamic absorption into simulated porous structures. *Colloids and Surfaces a-Physicochemical and Engineering Aspects* 206(1–3):217–239.

Ridgway, C. J., J. Schoelkopf, G. P. Matthews, P. A. C. Gane, and P. W. James. 2001. The effects of void geometry and contact angle on the absorption of liquids into porous calcium carbonate structures. *Journal of Colloid and Interface Science* 239(2):417–431.

Scherer, G. W. 2004. Stress from crystallization of salt. *Cement and Concrete Research* 34:1613–1624.

Scherer, G. W., R. Flatt, and G. Wheeler. 2001. Materials science research for the conservation of sculpture and monuments. *MRS Bulletin* 26(1):44–50.

Stefan, J. 1871. Uber das gleichgewicht und die bewegung in besondere die diffusion von gasgemengen. *Sitzungsber Math-Naturwiss. Akad. Wiss. Wien* 63:63–124.

Suzuki, M. and S. Maeda. 1968. On the mechanism of drying of granulars beds. *Journal of. Chemical Engineering of Japan* 1(1):26–31.

Veran-Tissoires, S. 2011. PhD Thesis, University of Toulouse.

Veran-Tissoires, S., M. Marcoux, and M. Prat. 2012. Discrete salt crystallization at the surface of a porous medium. *Physical Review Letters* 108, 054502.

Yiotis, A. G., A. G. Boudouvis, A. K. Stubos, I. N. Tsimpanogiannis, and Y. C. Yortsos. 2004. Effect of liquid films on the drying of porous media. *AIChE Journal* 50(11):2721–2737.

9

Pore-Network Modeling of Wicking:
A Two-Phase Flow Approach

Vahid Joekar-Niasar and S. Majid Hassanizadeh

CONTENTS

The wicking process, or capillary rise in a dry porous medium, is a two-phase phenomenon. The dynamics of wicking in porous media has not been extensively studied compared to that of capillary tubes. In a porous medium, the interconnected pore channels create a complex topology and geometries that influence the dynamics of the wicking significantly. Furthermore, the initial conditions of the wetting phase for wicking play an important role.

In this study, we have developed an advanced dynamic pore-network model for the wicking process based on two-phase flow. Thus, properties of both fluid phases (density and viscosity) have been incorporated in the model. Many pore-scale invasion mechanisms such as capillary pinning, snap-off, mobilization of the nonwetting phase, pinch-off of the wetting phase, pore-level countercurrent flow, piston-like movement, corner flow, and capillary diffusion have been included in the model. The presented model can be utilized for a better understanding of the fundamental physics of multiphase flow at the pore level and its consequence at the Darcy scale.

We investigate how the macroscopic wicking front velocity can depend on the initial wetting phase saturation and its connectivity. Our analysis provides a possible answer to the experimental observations that illustrate the influence of the initial saturation on the macroscopic front velocity.

9.1 Introduction

9.1.1 Capillary Rise in Porous Media

The wicking process or capillary rise in a dry porous medium is a two-phase phenomenon that can be found in many industrial and biological applications such as paper and filter industry, powder technology, vadose zone hydrogeology, and so on. In this phenomenon, capillary forces are the major driving forces that cause invasion of the wetting fluid in the porous medium. The capillary forces in a capillary rise problem are counteracted by the gravitational forces acting on the moving fluid body. The height at which the equilibrium occurs depends on pore size distribution, interfacial tension, as well as the densities of the fluids.

In 1906, Bell and Cameron found that the height of imbibition/wicking front in soil is almost proportional to the square root of time. Capillary rise in a single tube was first theoretically and experimentally studied in 1918 by Lucas and later in 1921 in an independent work by Washburn. The Lucas–Washburn (LW) equation was then extended by several researchers for application in porous media, although flow in a capillary tube and flow in a porous medium are very different. In several works, the LW equation for short-time and long-time events has been modified and partially verified in capillary tube experiments (e.g., Schoelkopf et al. 2002; Chebbi 2007).

Lago and Araujo (2001) found that the capillary rise versus time in glass beads can be defined by two regimes. They showed that the early stage can be explained by the LW equation, while the second stage can be explained by a simple analytical-empirical expression. This model has been compared to the experimental data obtained from a glass-bead column. However, such a distinct behavior in capillary rise was not found in Berea sandstone. Finally, Lago and Araujo (2001) conjectured that invasion in glass beads starts with a piston-like movement and changes to a capillary-dominated regime, where the LW equation does not hold anymore. Fries (2008) developed an analytical solution that is claimed to be valid for cylindrical tubes as well as for a porous medium. Although there is a good agreement between the analytical solution and the experiments performed in different capillary tubes, the validity of the equation for a porous medium has not been shown.

Marmur (1997) used the LW relation to calculate the effective radius and effective contact angle using the capillary rise data versus time. In a cylindrical capillary with a radius of r and in contact with a fluid with viscosity μ, the Washburn relation can be written as follows (Marmur 1997):

$$At = -Bh - \ln(1 - Bh), \quad A = \frac{\rho^2 g^2 r^3}{16\sigma^{nw}\mu\cos\theta}, \quad B = \frac{\rho g r}{2\sigma^{nw}\cos\theta} \quad (9.1)$$

where ρ is the density, θ is the contact angle between wetting and nonwetting fluids at the solid wall, σ^{nw} is the interfacial tension, and h is the height of the capillary rise in time t. Marmur (1997) showed the variation of $-Bh - \ln(1 - Bh)$ versus time for paper filter and sand porous media. Interestingly, they did find a linear relation for sand and in the major part of the graph, the slope of the graph increased with time. Gombia (2008) measured the capillary rise height (imbibition front) in Lecce stone for two different wettability conditions using MRI. They conjectured that invasion of the wetting fluid into a porous medium may be delayed and a delay time should be introduced into the LW equation that can be well fitted to their experimental data.

Schoelkopf et al. (2002) conjectured that the original LW equation cannot be applied to a porous medium since there exist many different sizes of pore channels with variable lengths in a network. Based on pore-network modeling, they stated that with the inclusion of inertia effects in LW equation, the agreement between their pore-network model and experiments would improve. Later, Ridgway et al. (2002) conjectured that deviation from the LW equation due to neglecting the inertia effect is stronger in fine materials with short pore throats. However, there are major shortcomings in their pore-network analysis (explained in the next paragraph) that does not let one generalize their statement.

To model the wicking process in a porous medium, it is necessary to include the essential features of a natural porous medium. We develop our model considering the following features:

(i) A porous medium is an interconnected network of capillary channels and not a single capillary tube. The pressure field in the wetting phase is continuous and a change in the capillary pressure (due to variable pore sizes) at a moving interface may lead to the receding of the wetting phase from the other pores. This relaxation in the pressure field will lead to an oscillation in capillary interfaces referred to as "capillary pinning." Capillary pinning has been discussed in empirical-analytical analysis by Lago and Araujo (2001). This phenomenon has a major impact on the residential time of pore filling, which is absent in LW relation as well as the pore-network study of Schoelkopf et al. (2002) and Ridgway et al. (2002).

(ii) A natural porous medium (e.g., soils and rocks) is made of large pores (referred to as "pore bodies") connected to each other through narrow pores (referred to as "pore throats"). Pore bodies have smaller capillary pressures compared to the pore throats. It means that during the wicking process, where capillary forces are the driving force, there will be a strong drop in capillary pressure as soon as the wetting phase enters the pore body. This will lead to a change of wetting phase pressure. As the wetting phase is continuous, it will flow due to changes in its pressure field. This variation of capillary forces is absent in the LW equation and in the studies performed with this analogy. In the dynamic pore-network model of Schoelkopf et al. (2002) and Ridgway et al. (2002), pore bodies do not contribute to flow at all.

(iii) Roughness of the solid phase and angularity of the void space are important characteristics of any natural porous medium. If a porous medium is not fully dry, the wetting phase will exist in the corners along the edges of pores. Thus, there will be a hydraulically connected path in the wetting phase along the edges. If the wetting phase can accumulate in the corners, eventually there will be snap-off, that is, the interfaces will coalesce and the nonwetting phase can be trapped. Trapping of the nonwetting (air) phase will change the topology of the available void for the wetting phase to flow.

To consider all these complexities in our model, we employ a two-phase dynamic pore-network model to investigate the capillary rise problem in a porous medium. Since the simulation of capillary rise problem requires a relatively large domain, other pore-scale modeling techniques such as lattice Boltzmann or smoothed particle hydrodynamics are not computationally feasible. In dynamic pore-network modeling, based on equations of volume conservation and flow, the pressure field and fluid volume change in time are computed.

Dynamic pore-network modeling was initiated by Koplik and Lasseter (1985) to simulate the imbibition process in a two-dimensional network. Later, several dynamic pore-network models were developed for various applications (see, e.g., Dias and Payatakes 1986; Nordhaug et al. 2003; Joekar-Niasar and Hassanizadeh 2011b). One of the important drawbacks of the existing dynamic pore-network models of wicking process is the use of circular cross sections for network elements, which does not allow local countercurrent flow (e.g., Ridgway et al. 2002). Since, in porous media, the connectivity of pores plays a very important role, the use of angular cross sections is essential for having simultaneous flow of both fluids. In a circular cross section, a gradual change of hydraulic conductance of each phase is not possible and it behaves in an "on–off" fashion. Furthermore, as it is not possible to have simultaneous existence of both fluids, a continuous interface between pores cannot exist. Thus, it will not be possible to simulate the continuous and gradual change of capillary forces along the pores, that is, the capillary diffusion through the wetting corners will be excluded.

Another drawback of existing dynamic pore-network models is that many of them do not include any pore body in the network (e.g., Dias and Payatakes 1986; Ridgway et al. 2002). Instead, all pore throats are assumed to be virtually connected to a node, in which pressure is calculated. Based on this assumption, not only all the physics associated with the variable capillary pressure in a pore body and capillary pinning will be ignored, but also a serious computational instability will appear (Koplik and Lasseter 1985; Dias and Payatakes 1986).

Finally, many existing dynamic pore-network models use the single-pressure algorithm for a two-phase problem. This means that pore-scale phase permeabilities are averaged and the pressure field for an average phase is calculated. This does not allow for countercurrent flow and the consequent snap-off to be included in the modeling. Except for the model developed by Dias and Payatakes (1986), mobilization of a nonwetting phase blob is absent in all dynamic pore-network models. This means that, as soon as a part of the nonwetting phase gets disconnected, it will be trapped regardless of the local flow velocity. But it is known that the mobilization of the trapped phase can occur with the increase of pore-scale viscous forces. For the imbibition of water in a porous medium, due to the high local wetting phase velocity and very favorable viscosity ratio (almost 100), mobilization of the disconnected air bubble is a relevant issue.

In this study, we used the DYPOSIT model for simulating the wicking process (Joekar-Niasar and Hassanizadeh 2011c). This model alleviates the shortcomings discussed earlier and includes major pore-scale mechanisms, such as pore-scale countercurrent flow, capillary pinning, snap-off, pinch-off, piston-like movement, variable corner flow, variable capillary pressure,

(a)   (b)   (c)

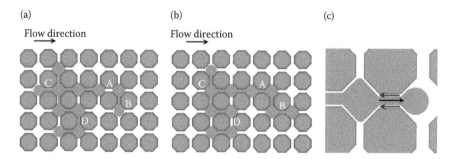

FIGURE 9.1
Schematic presentation of various mechanisms that can be captured by the two-pressure algorithm of the DYPOSIT model. (a) and (b) Two different snapshots illustrating coalescences at site A, the piston-like movement of the nonwetting phase at site B, gradual receding of the nonwetting phase at site C (variable capillary pressure), break-up (snap-off) at site D due to the local nonwetting phase pressure drop or wetting phase local pressure increases. Countercurrent flow can be easily simulated as shown in (c).

and mobilization of a blob. These mechanisms have been schematically illustrated in Figure 9.1. To show the ability of the model in simulating wicking in porous media, capillary rise in a medium is simulated. Then, the influences of the connectivity of the wetting phase on the saturation profile, wetting phase fingering, and macroscopic velocity of the wicking interface are studied. Numerical results are qualitatively compared to the column experiments reported in Bauters et al. (2000) and explanations for the observations are provided.

9.2 Model Description

9.2.1 Structure and Geometry

Our pore network is a regular three-dimensional lattice with a fixed coordination number of six. Pore bodies and pore throats are presented by "truncated octahedrons" and "parallelepipeds," respectively. This allows the simultaneous existence of both phases in a single pore element. The octahedron pore bodies can be unequally truncated since they are connected to pore throats of different sizes, as shown in Figure 9.2. Truncated sections of the octahedron have the shape of square pyramids with a base width of $2r_{ij}$ (which is equal to the size of the pore throat ij) as shown in Figure 9.2.

The size distribution of pore bodies (radius R_i of the inscribed spheres) is specified by a truncated log-normal distribution, with no spatial correlation.

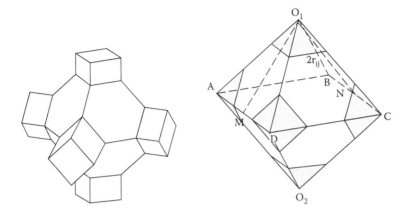

FIGURE 9.2
Schematic presentation of a pore body and its connected pore throats in a pore-network shown in Figure 9.1. Truncated parts of the pore body are square pyramids with the base width of $2r_{ij}$. r_{ij} is the inscribed radius in the cross section of pore throat ij.

The algorithm for the generation of pore throats, based on the inscribed radius of pore bodies, has been explained in detail in Joekar-Niasar and Hassanizadeh (2011c).

9.2.2 Governing Equations

9.2.2.1 Assumptions

Our pore-network model is developed based on the following assumptions:

(i) The time required for filling a single pore throat compared to the time required for filling the pore bodies is negligible. So, we have not included the volume of the pore throats in the calculation of the saturation (or mass) of the wetting fluid. Indeed, experimental observations in natural porous media have shown that the volume of pore throats is almost negligible compared with the volume of pore bodies.

(ii) For a fully saturated pore body and pore throat, the hydraulic resistance to flow in pore bodies is assumed to be negligible compared with that of pore throats.

(iii) Fluids are assumed to be immiscible and incompressible and the solid matrix is assumed to be rigid.

(iv) Transient effects in flow have been neglected (low Reynolds number). This allows using the LW equation for fluid fluxes through pores.

9.2.2.2 General Pore-Scale Equations for Two-Phase Flow

Given a pore channel ij, the volumetric flux of phase α through the channel in a laminar Newtonian flow (analogous to the LW equation for circular cross sections) can be given by

$$Q_{ij}^{\alpha} = K_{ij}^{\alpha}(p_i^{\alpha} - p_j^{\alpha} + \rho^{\alpha}g(z_i - z_j)), \quad \alpha = w, n, \tag{9.2}$$

where K_{ij}^{α} is the conductivity of the pore channel as a function of geometry and fluid occupancy, ρ^{α} is the density of fluid α, g is gravity, and z_i is the height of the pore body i compared to the benchmark. The positive flow direction is from pore body i to pore body j. Assuming incompressibility of the phases, the volumetric balance equation for the pore body i reads as

$$\sum_{j \in N_i} \sum_{\alpha} Q_{ij}^{\alpha} = 0, \quad \alpha = w, n, \tag{9.3}$$

where N_i is the set of all pore throats connected to the pore body i. Equation 9.3 cannot be solved explicitly because p_i^n and p_i^w are both unknown. For a given fluid configuration, Equation 9.3 is augmented by an equation for the pore-scale capillary pressure p_i^c, which is known to be a function of the local curvature of the interface, κ_i:

$$p_i^c = p_i^n - p_i^w = f(\kappa_i) \tag{9.4}$$

For computational ease, the curvature is related to the local volume fraction of the phases (see Section 9.2.3). Thus, we can rewrite (9.4) as follows:

$$p_i^c = p_i^n - p_i^w = f(s_i^w) \tag{9.5}$$

where s_i^{α} denotes the volume fraction of phase α in the pore body i.

The equation of volume balance for phase α within a pore body may be written as

$$V_i \frac{\Delta s_i^{\alpha}}{\Delta t} = -\sum_{j \in N_i} Q_{ij}^{\alpha}, \quad \alpha = w, n, \tag{9.6}$$

where V_i is the volume of the pore body i. Note that, given the fact that $s_i^w + s_i^n = 1$, summation of Equation 9.6 for two fluids directly results in Equation 9.3.

9.2.2.3 Pressure Field Solver

To reduce the computational load, the governing equations are reformulated in terms of a total pressure $\bar{p}_i$ in a pore body, defined as the volume fraction-weighted average of fluid pressures:

$$\bar{p}_i = s_i^w p_i^w + s_i^n p_i^n \tag{9.7}$$

Equation 9.3 can be rewritten for $\bar{p}_i$:

$$\sum_{j \in N_i} (K_{ij}^w + K_{ij}^n)(\bar{p}_i - \bar{p}_j) = -\sum_{j \in N_i} [(K_{ij}^n \rho^n + K_{ij}^w \rho^w)g(z_i - z_j)]$$

$$-\sum_{j \in N_i} [(K_{ij}^n s_i^w - K_{ij}^w (1 - s_i^w))p_i^c + (K_{ij}^w (1 - s_j^w) - K_{ij}^n s_j^w)p_j^c] \tag{9.8}$$

In this equation, the right-hand side as well as the coefficients of the left-hand side depend on volume fraction only. This linear system of equations was solved for $\bar{p}_i$ by diagonally scaled biconjugate gradient method using SLATEC mathematical library (Fong et al. 1993).

9.2.2.4 Saturation Update

Commonly in the literature, Equation 9.6 has been used to calculate the new s_i^w values explicitly. This procedure, however, will result in numerical problems especially for capillary-dominated and viscous fingering flow regimes (e.g., Koplik and Lasseter 1985). Those problems have been partially resolved by some ad hoc rules (Koplik and Lasseter 1985; Dias and Payatakes 1986; Van der Marck et al. 1997; Aker et al. 1998). However, those approaches were not tested for a wide range of flow rates and viscosity ratios to illustrate the stability of the procedure.

We use a procedure analogous to fractional flow formulation in order to control the nonlinearities under such flow conditions. Details can be found in Joekar-Niasar et al. (2010). This results in a semi-implicit discretization of Equation 9.6 as follows:

$$\left(\frac{V_i}{\Delta t} - \sum_{j \in N_i} \frac{K_{ij}^n K_{ij}^w}{K_{ij}^{tot}} \frac{\partial p_{ij}^c}{\partial s_{ij}^w} \right)(s_i^w)^{k+1} + \left(\sum_{j \in N_i} \frac{K_{ij}^n K_{ij}^w}{K_{ij}^{tot}} \frac{\partial p_{ij}^c}{\partial s_{ij}^w} \right)(s_j^w)^{k+1}$$

$$= \frac{V_i}{\Delta t}(s_i^w)^k + \sum_{j \in N_i} \left(\frac{K_{ij}^n}{K_{ij}^{tot}} Q_{ij}^{tot} + (K_{ij}^n \rho^n + K_{ij}^w \rho^w)g(z_i - z_j) \right) \tag{9.9}$$

where superscript k denotes time step level and $\partial p_{ij}^c / \partial s_{ij}^w$ is calculated from the upstream pore body. One should note that since Q_{ij}^{tot} and K_{ij}^α are calculated from time step k, this scheme is not fully implicit. Here also, the diagonally scaled biconjugate gradient method from SLATEC mathematical library (Fong et al. 1993) was used to solve Equation 9.9. The stability of the model for a wide range of flow conditions (from capillary-dominated regime to viscous-fingering regime) and a wide range of viscosity ratios has been demonstrated in Joekar-Niasar et al. (2010).

9.2.2.5 Time Stepping

The time step, denoted by Δt_i, is determined based on the time of filling of pore bodies by the nonwetting phase or wetting phase, for drainage and imbibition simulations, respectively. The wetting phase volume fraction in a pore body varies between 1 and $s_{i,min}^w$ as we assume that a pore body may be drained down to a minimum volume fraction (which is controlled by global capillary pressure). On the other hand, if local imbibition occurs, the wetting phase volume fraction in a pore body can go back to 1. So, we calculate Δt_i for all pore bodies, depending on the local process, from the following formulas:

$$\Delta t_i = \begin{cases} \dfrac{V_i}{|q_i^n|}(s_i^w - s_{i,min}^w) & \text{for local drainage}, q_i^n > 0 \\[3mm] \dfrac{V_i}{|q_i^n|}(1 - s_i^w) & \text{for local imbibition}, q_i^n < 0 \end{cases} \tag{9.10}$$

where the accumulation rate of the nonwetting phase is defined as $q_i^n = \sum_{j \in N_i} Q_{ij}^n$. Then, the global time step is chosen to be the smallest Δt_i.

$$\Delta t_{global} = \min\{\Delta t_i\} \tag{9.11}$$

It should be noted that we have imposed a truncation criterion of 10^{-6} in evaluating $s_i^w - s_{i,min}^w$ and $1 - s_i^w$ in order to reduce the computational cost for solving the equations.

9.2.3 Local Rules

The Darcy-scale behavior of the two-phase flow in porous media is strongly controlled by pore-scale mechanisms. Pore invasion, snap-off, trapping, and mobilization are some of the pore-scale mechanisms that should be included in pore-network modeling.

9.2.3.1 Entry Capillary Pressure for a Pore Throat

A pore throat will be invaded by the nonwetting phase if the capillary pressure in a neighboring pore body becomes larger than the entry capillary pressure of the pore throat. For a pore throat with a square cross section, the entry capillary pressure can be calculated as follows (Ma et al. 1996):

$$p_{e,ij}^c = \frac{\sigma^{nw}}{r_{ij}} \left(\frac{\theta + \cos^2\theta - \pi/4 - \sin\theta\cos\theta}{\cos\theta - \sqrt{\pi/4 - \theta + \sin\theta\cos\theta}} \right) \tag{9.12}$$

where r_{ij} is the radius of the inscribed circle of the pore throat cross section, and θ is the contact angle.

9.2.3.2 Snap-Off

As mentioned before, fluid entrapment at pore scale and the subsequent fluid distribution at macroscale are controlled by the connectivity of the phases. Snap-off and pinch-off are important mechanisms that can lead to the discontinuity of a phase. The importance of these mechanisms on the Darcy-scale trapping depends on the macroscopic flow. For example, under drainage, since there is a continuous supply of the nonwetting phase, snap-off does not play an important role. However, under imbibition, where the nonwetting phase is the receding phase, the local pressure field can result in snap-off. If the wetting phase pressure increases under wicking, the local capillary pressure will decrease. Thus, corner flow will lead to snap-off if piston-like movement is not possible. If the nonwetting phase becomes disconnected from the outlet boundary, it will get trapped in the pore space.

Snap-off in the pore throats occurs at a critical capillary pressure based on the geometrical configuration of the capillary interface. It is the minimum capillary pressure at which the interfaces are still pinned to the solid boundaries, defined as (Vidales et al. 1998):

$$p_{ij}^c \leq \frac{\sigma^{nw}}{r_{ij}}(\cos\theta - \sin\theta) \tag{9.13}$$

We assume that as soon as snap-off occurs, the nonwetting phase retreats instantaneously into the two neighboring pore bodies, and the pore throat is filled up with the wetting phase.

9.2.3.3 Conductivities of Pore Throats and Pore Bodies

Conductivities of pore throats are determined based on their size and fluid occupancy. One of the following three cases may occur:

(a) A pore throat and the connecting pore bodies are filled by the wetting phase only. For this case, the following equation was obtained by (Azzam and Dullien 1977):

$$K_{ij}^w = \frac{\pi}{8\mu^w l_{ij}}(r_{ij}^{eff})^4, \quad r_{ij}^{eff} = \sqrt{\frac{4}{\pi}}r_{ij}$$

$$K_{ij}^n = 0,$$

(9.14)

where μ^w is the viscosity of the wetting phase and l_{ij} is the length of the pore throat.

(b) A pore throat is occupied by both phases. Then, following Ransohoff and Radke (1988), we can write

$$K_{ij}^w = \frac{4-\pi}{\epsilon\mu^w l_{ij}}(r_{ij}^c)^4$$

$$K_{ij}^n = \frac{\pi}{8\mu^n l_{ij}}(r_{ij}^{eff})^4$$

(9.15)

where $r_{ij}^c = (\sigma^{nw}/p_{ij}^c)$ and

$$r_{ij}^{eff} = \frac{1}{2}\left(\sqrt{\frac{r_{ij}^2 - (4-\pi)r_{ij}^{c2}}{\pi}} + r_{ij}\right).$$

In Equation 9.15, ϵ is a resistance factor that depends on geometry, surface roughness, crevice roundness, and other specifications of the cross section. A detailed explanation about ϵ can be found in Zhou et al. (1997). As mentioned earlier, the pore throat capillary pressure p_{ij}^c is set equal to the capillary pressure of the upstream pore body.

(c) A pore throat is filled by the wetting phase but the nonwetting phase is present in the neighboring pore bodies. It is only considered if both phases are present in a pore body. The conductivity of the nonwetting phase in a pore body is usually much larger than in a pore throat. Thus, only pore throat conductivity for the nonwetting phase is considered. Let the conductivity of the wetting phase in a pore body i next to the pore throat ij be denoted by $K_{i,ij}^w$. Then, the wetting phase conductivity along the edges in a pore body with half-corner angle of $\beta = (11/36)\pi$, the relation given by Patzek (2001) is employed.

$$K_{i,ij}^w = \frac{4r_i^{c4}}{\mu^w((\sqrt{6}/2)R_i - 2r_{ij})} \frac{\cos^4(\theta + \beta)}{\sin^4(\beta)}(a_1\Gamma^2 + a_2\Gamma + a_3)$$

$$\Gamma = \frac{\sin^2\beta}{4\cos^2(\theta + \beta)} \frac{(\theta/\beta) + -(\pi/2) + (\cos\theta\cos(\theta + \beta)/\sin\beta)}{\left[1 - (\theta + \beta - (\pi/2))(\sin\beta/\cos(\theta + \beta))\right]^2}$$

(9.16)

where $r_i^c \approx (\sigma^{nw}/p_i^c)$, $a_1 = -18.2066$, $a_2 = 5.8829$, and $a_3 = -0.35181$.

Where there is a disconnected blob of the nonwetting phase in a pore body, it is assumed that the blob will cover the opening of the pore throat that has the maximum outward wetting phase gradient. Thus, the effective length for resistance should be corrected. Then, in Equation 9.16, $(\sqrt{6}/2)R_i - 2r_{ij}$ will be replaced by the length of the blob. The length of a blob, l_b, with a given volume V_b can be geometrically approximated. It is obtained by solving the following equation for l_b:

$$4\sin\beta(l_b^3\cos^2\beta + 3l_b^2r_{ij}\cos\beta + 3l_br_{ij}^2) - 8r_i^{c2}(l_b\cos\beta - r_{ij})$$

$$\times\left[(\cos(\beta + \theta)\cos\theta/\sin\beta) - ((\pi/2) - \beta - \theta)\right] = V_b \qquad (9.17)$$

Finally, the total wetting phase conductivity of a pore throat, bordered by two pore bodies containing the nonwetting phase, can be approximated as $(1/\bar{K}_{ij}^w) = (1/K_{ij}^w) + (1/K_{i,ij}^w) + (1/K_{j,ij}^w)$, where K_{ij}^w is given by Equation 9.15. In some works, such as Zhao and Ioannidis (2011) and Raoof and Hassanizadeh (2011), the resistance of pore bodies has been considered.

It should be noted that in formulations for conductivity relations, although the wetting phase is assumed to be always connected, the hydraulic resistance can be very small. Therefore, in our simulations, a threshold for the ratio of K_{ij}^w/K_{ij}^{tot} has been defined, so that the wetting phase conductance below that threshold is set to zero. This will be in analogy to a fully dry system.

9.2.3.4 Local Capillary Pressure Curves

There are experimental observations showing that oscillation of the interface in a pore, or a back-and-forth movement within a pore, can occur depending on the pore space geometry and topology (Culligan et al. 2004; Or and Moebius 2010). When a pore gets drained, the capillary interface in another pore which is hydraulically connected might become unstable. This will lead to a simultaneous pore-scale imbibition. This is due to the fact that change of the moving interfaces (and consequent change of boundary conditions of the invading fluid) has a direct influence on the pressure field of the fluids depending on their viscosities. Thus, there might be mixed drainage and imbibition mechanisms depending on the geometry, topology, and the viscous forces.

It is required to define the local capillary pressure within a pore as a function of the curvature of fluid–fluid interface through *Young–Laplace* equation. For a given fluid–fluid interface position within a pore body, we can determine the corresponding capillary pressure and wetting phase volume fraction. Therefore, for a given pore body, a unique relationship between capillary pressure, p_i^c, and local wetting phase volume fraction, s_i^w, is obtained. The resulting local p_i^c-s_i^w curves for drainage and imbibition for given pore body geometries have been presented in Joekar-Niasar and Hassanizadeh (2011c) and they are not repeated here.

9.2.3.5 Mobilization of a Nonwetting Blob

A discontinuous phase blob will be mobilized if the local capillary pressure, in the downstream side of the blob, can overcome the local entry capillary pressure. But then, due to the continuity of the pressure field, an interface upstream will have to recede. The local capillary pressure, which is related to the curvature of the interface, is controlled by the local nonwetting and wetting phases pressure difference. It means that to include the mobilization, the pressure gradient in both (continuous and discontinuous) phases should be calculated. Details of the mobilization algorithm in the DYPOSIT model can be found in Joekar-Niasar and Hassanizadeh (2011a).

Since the discretization level in the DYPOSIT model is a pore body, there is only one pressure value for each phase in a pore body. But, for determining

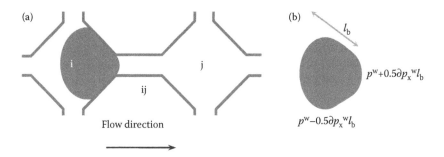

FIGURE 9.3
(a) Schematic presentation of an individual blob in a pore body on the verge of invading a pore throat. (b) Assuming a constant nonwetting phase pressure inside the blob, the capillary pressure at the tip of the blob can be estimated from the local wetting phase pressure gradient (see Equation 9.19).

whether a disconnected blob in a single pore body (Figure 9.3) will be mobilized, subscale information is required. We have estimated the wetting phase pressure drop along a single nonwetting blob with length l_b as follows:

$$\Delta p_{i-ij}^{w} = \frac{K_{i-ij}^{w\,-1}}{K_{i-ij}^{w\,-1} + K_{ij}^{w\,-1} + K_{j-ij}^{w\,-1}} \Delta p_{ij}^{w} \qquad (9.18)$$

where Δp_{i-ij}^{w} is the wetting phase pressure drop along the blob and Δp_{ij} is the pressure drop between the pores i and j. The conductivities are defined in Equations 9.15 and 9.16.

We assume that the nonwetting phase pressure inside the blob to be constant, and assign the values of p_i^w and p_i^n from global calculations to the center of l_b. Then, the capillary pressure at the tip of the blob, $p_{i_{tip}}^c$, will be estimated as follows:

$$p_{i_{tip}}^c = p_i^c + 0.5\Delta p_{i-ij}^{w} \qquad (9.19)$$

If the $p_{i_{tip}}^c$ is larger than the entry capillary pressure of the downstream pore throat given by Equation 9.12, the blob will be mobilized.

9.2.4 Boundary and Initial Conditions

The network size is chosen to be $20 \times 20 \times 80$ pore bodies. The network is oriented vertically along the gravity vector. The statistical distributions of pore bodies and pore throats of the network are shown in Figure 9.4.

Constant pressures at the top and bottom are applied. At the bottom, the porous medium is fully saturated with the wetting phase (water) pressure assigned to be zero. At the top, there is only nonwetting phase (air) and its pressure is set to be zero too. The side boundary conditions are assumed to be periodic so that the coordination number in all pores is equal to 6 and normal fluxes on the side boundaries will be the same but with opposite signs.

For the initially wet simulations, an initial saturation (and corresponding capillary pressure) is chosen at which all pores have been invaded by the nonwetting (air) phase. To determine this saturation, the characteristic P^c–S^w curve for drainage for the given network and fluids' properties is obtained first. This curve, shown in Figure 9.5, has been simulated using the quasi-static pore-network model explained in Joekar-Niasar and Hassanizadeh (2011b).

Based on the simulation results, for a residual saturation of 0.07, corresponding to the capillary pressure of 7 kPa (constant over the whole network), all pores have been drained. But two scenarios for wetting phase conductivity for this system have been considered: (i) initially hydraulically

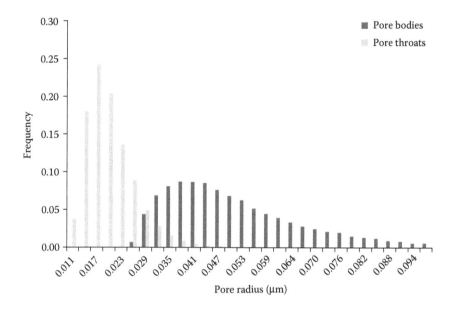

FIGURE 9.4
Statistical distribution of pore bodies and pore throats of the network.

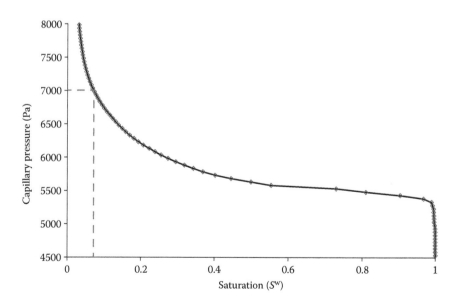

FIGURE 9.5
A characteristic capillary pressure–saturation drainage curve of the pore network. Dashed line points to the initial capillary pressure and saturation values over the network for starting the wicking (imbibition) problem.

connected corner flow scenario whereby the wetting phase is connected everywhere and its conductivity is very low but nonzero and (ii) initially hydraulically disconnected corner flow scenario, where the system will be initially nonconductive to the wetting phase and it behaves as an initially dry domain. From here onward, for brevity, these two scenarios are called "initially connected" (C) and "initially disconnected" (D).

The fluids are assumed to be air and water with densities and viscosities equal to $\mu^n = 0.000015$ Pa.s, $\mu^w = 0.001$ Pa.s, and $\rho^n = 1.225$ kg/m³, $\rho^w = 998$ kg/m³, respectively. The interfacial tension and contact angle are assumed to be 0.0725 N/m and 4°, respectively. Gravity acceleration is set to $g = 9.8$ m/s². The results are presented in the next section.

9.3 Results and Discussion

To show the effectiveness of the pore-network model in simulating the wicking process, the variation of the wetting front position versus time is shown in Figure 9.6. Wetting front position has been calculated based on the

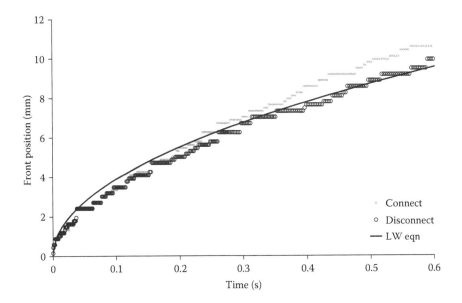

FIGURE 9.6
Variation of wetting-phase front position with time for the initially connected and initially disconnected corner flow scenarios. The data points for the initially disconnected medium are shown by open circles and the data points for the initially connected domain are represented by crosses. The solid line represents the Lucas–Washburn equation fitted to the data points.

position where saturation is equal to the mean of the base saturation (initial saturation of the domain) and peak saturation (maximum saturation of the front). The results are shown for the two scenarios of initially connected (C) and initially disconnected (D) domains. The original LW equation for a single capillary is also fitted to the front position versus time data points. In this equation, $x = a\sqrt{t}$, in which a as a fitting parameter is a measure of hydraulic conductivity of the domain. In both scenarios, the agreement between the LW curve and the data points becomes less with time (similar to the findings discussed in Section 9.1) and it is worse for the case of an initially connected porous medium.

Figure 9.6 shows that the velocity of the wetting front (slope of the curve) in an initially disconnected scenario is smaller than that of an initially connected one (in analogy with the initially dry versus the initially wet domains, respectively). The difference is visible in our simulations even for such a small physical size. Our simulations show that at such a physical scale the gravitational force is not playing a significant role compared to the capillary forces. Thus, we can find some similarities between our capillary rise simulations compared to the infiltration experiments of Bauters et al. (2000), as explained below.

9.3.1 Experiments of Bauters et al. (2000)

They investigated the infiltration of water in a sand column with dimensions 30(w) × 55(h) × 0.94(d) cm. In their experiments, gravity and capillary forces were in concert. They performed several experiments in a 20/30 sand* column with different initial water contents equal to 0, 0.005, 0.01, 0.015, 0.02, 0.03, 0.04, and 0.047 cm³/cm³. Since the porosity of this column was about 0.348, these water contents covered a range of wetting phase saturation from 0 to 0.16. A constant infiltration rate equal to 2 cm³/min through a needle was applied on the sand surface.

The macroscopic infiltration front was visualized by x-ray imaging. Figure 9.7a and b shows the position of the infiltration front at various times in an initially dry sand and an initially wet sand (water content 0.01 cm³/cm³), respectively. As it can be seen, the front velocity in the initially wet medium is higher. They found that there is a nonmonotonic dependency of the wetting front velocity on initial saturation. For the initial water content of 0.01, the maximum wetting front velocity of 1.67 mm/s (10 cm/min) was measured. As it can be seen in Figure 9.7c, this velocity is larger than the velocity of the front in an initially dry domain. In spite of their interesting findings, no fundamental reasoning for this observation was provided in Bauters et al. (2000). Furthermore, they found that the degrees of saturation behind the fronts were different for different initial conditions.

---

* Standard sand, predominantly graded to pass a 850 μm (No. 20) sieve and be trained on a 600 μm (No. 30) sieve and the 150 μm (No. 100) sieve.

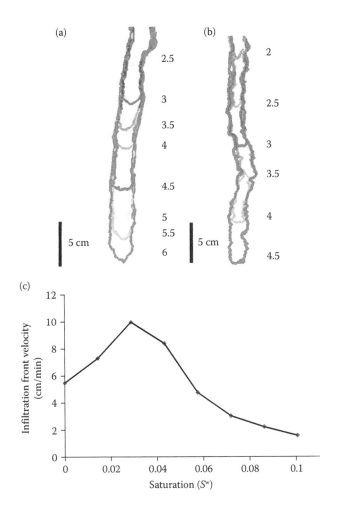

FIGURE 9.7
Experimental results of Bauters et al. (2000): Tracings of the infiltration front in (a) a totally dry sand and (b) an initially wet sand (water content 0.01 cm³/cm³). The numbers accompanying the tracings are the minutes after the infiltration started. The scale indicates 5 cm (c) infiltration front velocity as a function of the initial saturation. (Adapted from Bauters, T. W. J. et al. 2000. *Journal of Hydrology* 231–232:244–254.)

The observation on the wetting front velocity might be explained by two hypotheses: (i) There is a larger active capillary force in an initially wet medium that can suck the water faster into the porous medium. (ii) In an initially wet medium, faster macroscopic flow is not due to the higher flow rate, but due to smaller pore filling efficiency caused by snap-off. That is, the macroscopic interface velocity is only apparently larger. These two hypotheses are discussed below.

9.3.2 Hypothesis One: Larger Capillary Force Cause Larger Inflow Driving Velocity

In the capillary rise problem, capillary and gravitational forces are the main forces. However, in our simulations, due to the very small physical size of the network, the gravitational force is not important. In an initially connected domain, there exist many menisci that may become mobilized. This may lead to a higher suction rate. It means that the initially connected system has a larger driving force that can suck in more water. However, the comparison of the boundary pore velocities for these two scenarios, shown in Figure 9.8, does not show a significant difference. We remark that at the bottom boundary in both scenarios, the saturations are the same (equal to one) and cross-sectional areas are the same. The trend of velocity versus time is qualitatively in agreement with the experimental data reported in the literature (see Figure 10 in Lago and Araujo 2001). They have plotted the wetting front velocity versus time in beads packing with

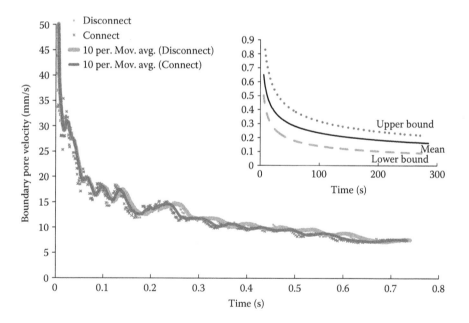

FIGURE 9.8

Temporal variation of pore flow velocity at the boundary for the initially connected domain (crosses) and initially disconnected domain (squares). Solid lines show results of moving window averaging with a window of 10 data points. The inset figure shows the wetting front velocity versus time in beads packing experiments (Lago and Araujo 2001) with characteristic grain diameters between 150 and 250 μm. The full band of the data scatter for these grain sizes has been shown by upper bound and lower bound curves. Curves are redrawn from the early stage data in Figure 10 in Lago and Araujo (2001). (Adapted from Lago, M., and M. Araujo. 2001. *Journal of Colloid and Interface Science* 234(1):35–43.)

characteristic diameters between 150 and 250 μm. The interface velocity of the earlier time data is plotted in the inset in Figure 9.8. The magnitude of the wetting front velocity is much smaller than the boundary pore velocity in our simulations. Note that the wetting front velocity has been calculated by dividing the front position by the elapsed time, but the boundary pore velocity has been calculated within the pores where porosity has not been included.

Initially, the temporal trends of the flow velocity in both systems are identical. Since the medium is filled with air with a small viscosity, the resistance of the network is initially small. However, in the later stage, the wetting fluid (whose viscosity is two orders of magnitude larger than air) fills the domain and causes much more hydraulic resistance. This resistance becomes stronger as the wetting phase is connected overall. Another interesting aspect of Figure 9.8 is the oscillatory trend of the velocity which damps out with time. However, damping of the velocity is faster in the initially connected domain compared to the initially disconnected domain.

Based on our analysis, we can conclude that the higher wetting-phase front velocity in the disconnected scenario cannot be due to the higher suction flow rate, as the suction rates in both scenarios are the same.

9.3.3 Hypothesis Two: Smaller Pore Filling Efficiency Caused by Snap-Off

Since the inflow velocities in both scenarios were almost the same, a second hypothesis should be studied. When the wetting fluid is connected everywhere, although with very low conductivity, it will bulk gradually with the invasion of the wetting phase. This growth of the wetting corner flow will increase the probability of snap-off compared to the initially disconnected system. This will lead to more trapped nonwetting (air) phase. Given the constant boundary velocities in both scenarios, a decrease in the available pore space will increase the apparent wetting-phase front velocity.

This can be illustrated by examining the saturation profiles in these two systems, as shown in Figure 9.9a and b. Two points are interesting in these saturation profiles: (a) In an initially disconnected system, the saturation profile behind the front is almost constant (at least in the early stage of wicking). But in an initially connected domain, the saturation along the network is decreasing. This is because the trapped phase cannot be swept out. In other words, given the almost same boundary inflow velocities in both systems in capillary force-controlled systems (Figure 9.8), trapping may be the major reason that causes the macroscopic wetting-phase front to move faster. In other words, the wetting efficiency in the disconnected system is higher. (b) The fingering in the initially disconnected system is stronger than in the initially connected one. This is in agreement with the experiments of Bauters et al. (2000). This can be seen by comparing the three-dimensional presentations of saturation snapshots in the networks

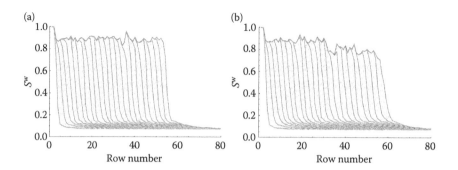

FIGURE 9.9
Saturation profiles at different time steps along the pore network for (a) initially disconnected and (b) initially connected pore networks.

as shown in Figure 9.10a and b for initially disconnected and initially connected domains, respectively.

As it can be seen for the same network saturation of 50%, the wetting-phase front in the initially connected system is closer to the top boundary, while the boundary pore velocities in both cases are the same. Significantly more trapped air is visible in the initially connected system compared to the initially disconnected one due to the snap-off in pore throats. Furthermore, the gray-scale color ahead of the wetting front shows that the fingering in the case of the initially disconnected system is stronger than the initially connected one.

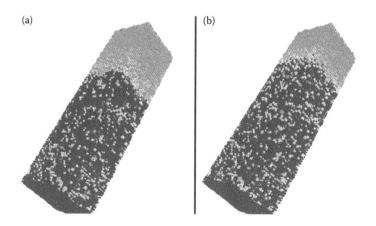

FIGURE 9.10
Three-dimensional representations of air and water distribution in the network at total wetting phase saturation equal to 0.5. (a) Initially disconnected and (b) initially connected domain. Obviously, the wetting front velocity in the initially connected domain is larger than that in the disconnected system. The spreading ahead of the wetting-phase front in the initially disconnected system (a) illustrates the more pronounced fingering compared to the initially connected one.

9.4 Conclusions and Summary

We have developed a dynamic two-phase flow pore-network model using the two-pressure algorithm. This model allows us to simulate dynamics of wicking (imbibition), including the most important pore-level invasion mechanisms such as capillary pinning, snap-off, mobilization of the nonwetting phase, pinch-off of the wetting phase, pore-level countercurrent flow, piston-like movement, corner flow, and capillary diffusion. The model is stable under a very wide range of flow rates and viscosity ratios (Joekar-Niasar and Hassanizadeh 2011c).

The wicking process for two different initial conditions has been simulated in a network representing a small domain; both systems have the same very low residual saturations. In the first one, the wetting phase in the corners is assumed not to be connected and it is referred to as an "initially disconnected system" (wetting phase permeability is set to zero). If the wetting phase becomes disconnected in the corners of pores, it makes the system to behave as a dry porous medium. In the second system, the network has the same saturation but (very low) conductivity is assigned to the wetting phase; this is referred to as an "initially connected system." It is important to note that in both systems, the initial capillary pressures that are active on the interfaces are the same. In the initially disconnected system, these capillary interfaces do not communicate with each other, and in the initially connected domain, they do.

We should note that our results have been obtained for a very small system (cm scale) with water and air as the fluids. Results show that the macroscopic interface velocity in the initially connected system is larger than that of the initially disconnected one. This is not because of the larger suction rate but because of the reduction of available void space due to the trapping of air. In an initially connected system, the probability of the trapping is larger than that of the initially disconnected one due to snap-off. It leads to a larger amount of the trapped air in the domain and consequently a reduction of the available void space for the wetting phase. Thus, the macroscopic wetting-phase front will travel faster for an initially connected domain.

Fingering in the initially disconnected system is stronger than in the initially connected one. These fingers appear due to the capillary forces (as the major driving forces) in the wicking process. Since the connectivity of the wetting phase (water) is larger in the initially connected domain, energy dissipation is stronger. Thus, the capillary fingering is damped with the increasing connectivity of the wetting phase.

We should note that although the gravitational forces have been included in the simulations, much larger domains are required to show the effect of the gravitational forces on the wicking process. However, this will require much more computational time/cost for the simulations. Nevertheless, this model alleviates the major drawbacks mentioned in Section 9.1 related to wicking in porous media, and it provides physical insights into the issues that are difficult to measure and/or observe in the laboratory.

Acknowledgments

This research was funded by a King Abdullah University of Science and Technology (KAUST) Center-in-Development Grant awarded to Utrecht University (grant number KUK-C1-017-12).

Nomenclature

a^{nw}	Specific interfacial area (m^{-1})
A_{dr}^{nw}	Interfacial area in a pore body at $s^w = s_i^{dr}$ (m^2)
A_i^{nw}	Interfacial area in a pore body (m^2)
C_a	Capillary number (–)
g	Gravity vector (m/s^2)
h	Height of the capillary rise (m)
K_{ij}^{α}	Conductance of phase α in pore throat ij (m^4/(Pa.s))
$K_{i,ij}^{w}$	Wetting phase conductivity in pore body i next to pore throat ij (m^4/(Pa.s))
l_b	Length of a blob (m)
L_i	Total length of the edges in pore body i (m)
l_{ij}	Length of pore throat ij (m)
n_{pb}	Number of pore bodies (–)
p_i^{α}	Pressure of phase α in pore body i (Pa)
$\overline{p^c}$	Average capillary pressure (Pa)
p_i^c	Capillary pressure in pore body i (Pa)
$p_{e,ij}^c$	Entry capillary pressure of pore throat ij (Pa)
p_{ij}^c	Capillary pressure of pore throat ij (Pa)
$p_{i_{tip}}^c$	Capillary pressure of the tip of blob in pore i (Pa)
$\overline{p}_i$	Total pressure in pore i (Pa)
R_i	Inscribed radius in pore body i (m)
r_{ij}	Radius of pore throat ij (m)
r_{ij}^c	Radius of capillary pressure in pore throat ij (m)
r_{ij}^{eff}	Effective hydraulic radius in pore throat ij (m)
s_i^{α}	Volume fraction of phase α in pore body i (–)
S^w	Average saturation of the wetting phase (–)
s_{ij}^w	Volume fraction of phase α in pore throat ij (–)
t	Time (s)
Q_{ij}^{α}	Flux of phase α in pore throat ij (m^3/s)
q_i^n	Flux of nonwetting phase in pore body i (m^3/s)
V_i	Volume of pore body i (m^3)
V_b	Volume of a blob (m^3)

z_i z-coordinate (height) of pore body i (m)
$\mathbb{N}_i$ Total number of pore throats connected to pore i (–)
β Half-corner angle in a pore throat (radius)
Γ A geometrical coefficient for calculating the corner conductance (–)
Δp^w_{i-ij} Wetting phase pressure drop along the blob (Pa)
Δt_i Minimum residence time in pore body i (s)
Δt_{global} Global residence time over the network (s)
κ_i Curvature of the interface in pore body i (m^{-1})
σ^{nw} Interfacial tension (n/m)
π 3.1415
θ Contact angle (radius)
$\in$ Resistance factor (–)
μ^α Viscosity of phase α (Pa · s)

References

Aker, E. K., J. Maloy, A. Hansen, and G. G. Batrouni. 1998. A two-dimensional network simulator for two-phase flow in porous media. *Transport in Porous Media* 32:163–186.

Azzam, M. I. S., and F. A. L. Dullien. 1977. Flow in tubes with periodic step changes in diameter: A numerical solution. *Chemical Engineering Science* 32:1445–1455.

Bauters, T. W. J., D. A. DiCarlo, T. S. Steenhuis, and J. Y. Parlange. 2000. Soil water content dependent wetting front characteristics in sands. *Journal of Hydrology* 231–232:244–254.

Bell, J. M., and F. K. Cameron. 1906. The flow of liquids through capillary spaces. *The Journal of Physical Chemistry* 10(8):658–674.

Chebbi, R. 2007. Dynamics of liquid penetration into capillary tubes. *Journal of Colloid and Interface Science* 315(1):255–260.

Culligan, K. A., D. Wildenschild, B. S. B. Christensen, W. Gray, M. L. Rivers, and A. F. B. Tompson. 2004. Interfacial area measurements for unsaturated flow through a porous medium. *Water Resources Research* 40:W12413.

Dias, M. M., and A. C. Payatakes. 1986. Network models for two-phase flow in porous media: Part 1. Immiscible microdisplacement of non-wetting fluids. *Journal of Fluid Mechanics* 164:305–336.

Fong, K. W., T. H. Jefferson, T. Suyehiro, and L. Walton. 1993. Guide to the SLATEC common mathematical library.

Fries, N., and M. Dreyer 2008. An analytic solution of capillary rise restrained by gravity. *Journal of Colloid and Interface Science* 320:259–263.

Gombia, M. A., V. B. Bortolotti, R. J. S. C. Brown, M. D. Camaiti, and P. A. Fantazzini. 2008. Models of water imbibition in untreated and treated porous media validated by quantitative magnetic resonance imaging. *Journal of Applied Physics* 103(9):094913.

Joekar-Niasar, V., and S. M. Hassanizadeh. 2011a. Uniqueness of specific interfacial area–capillary pressure–saturation relationship under non-equilibrium conditions in two-phase porous media flow. *Transport in Porous Media,* doi: 10.1007/s11242-012-9958-3.

Joekar-Niasar, V., and S. M. Hassanizadeh. 2011b. Specific interfacial area; The missing state variable in two-phase flow. *Water Resources Research* 47:W05513.

Joekar-Niasar, V., and S. Majid Hassanizadeh. 2011c. Effect of fluid properties on non-equilibrium capillarity effects: Dynamic pore-network modeling. *International Journal of Multiphase Flow* 37(2):198–214.

Joekar-Niasar, V., S. M. Hassanizadeh, and H. K. Dahle. 2010. Non-equilibrium effects in capillarity and interfacial area in two-phase flow: Dynamic pore-network modelling. *Journal of Fluid Mechanics* 655:38–71.

Koplik, J., and T. J. Lasseter. 1985. Two-phase flow in random network models of porous media. *Society of Petroleum Engineers Journal* 25:89–110.

Lago, M., and M. Araujo. 2001. Capillary rise in porous media. *Journal of Colloid and Interface Science* 234(1):35–43.

Lucas, R. 1918. About the time law of the capillary rise of fluids. *Kolloid Zeitschrift* 23:15–22.

Ma, S., G. Mason, and N. R. Morrow. 1996. Effect of contact angle on drainage and imbibition in regular polygonal tubes. *Colloids and Surfaces A.* 117:273–291.

Marmur, A., and R. D. Cohen. 1997. Characterization of porous media by the kinetics of liquid penetration: The vertical capillaries model. *Journal of Colloid and Interface Science* 189(2):299–304.

Nordhaug, H. F., M. Celia, and H. K. Dahle. 2003. A pore network model for calculation of interfacial velocities. *Advances in Water Resources* 26:1061–1074.

Or, D., and F. Moebius. 2010, May. Linking pore scale pressure bursts and interfacial jumps during fluid displacement with porous media pore space characteristics. European Geosciences Union (EGU) General Assembly.

Patzek, T. W. 2001. Verification of a complete pore network simulator of drainage and imbibition. *SPE Journal* 6:144–156.

Ransohoff, T. C., and C. J. Radke. 1988. Laminar flow of a wetting liquid along the corners of a predominantly gas-occupied noncircular pore. *Journal of Colloid and Interface Science* 121:392–401.

Raoof, A., and M. Hassanizadeh. 2012. A new formulation for pore-network modeling of two-phase flow. *Water Resources Research* 48:W01514.

Ridgway, Cathy J., A. C. Gane Patrick, and J. Schoelkopf. 2002. Effect of capillary element aspect ratio on the dynamic imbibition within porous networks. *Journal of Colloid and Interface Science* 252(2):373–382.

Schoelkopf, J., A. C. Patrick Gane, J. R. Cathy, and G. Peter Matthews. 2002. Practical observation of deviation from Lucas-Washburn scaling in porous media. *Colloids and Surfaces A: Physicochemical and Engineering Aspects* 206 (1–3):445–454.

Van der Marck, S. C., T. Matsuura, and J. Glas. 1997. Viscous and capillary pressures during drainage: Network simulations and experiments. *Physical Review E* 56:5675–5687.

Vidales, A. M., J. L. Riccardo, and G. Zgrabli. 1998. Pore-level modelling of wetting on correlated porous media. *Journal of Physical Distribution, Applied Physics* 31:2861–2868.

Washburn, E. W. 1921. The dynamics of capillary flow. *Physical Review* 17:273–283.

Zhao, W., and M. A. Ioannidis. 2011. Gas exsolution and flow during supersaturated water injection in porous media: I. Pore network modeling. *Advances in Water Resources* 34:2–14.

Zhou, D., M. J. Blunt, and F. M. Orr. 1997. Hydrocarbon drainage along corners of noncircular capillaries. *Journal of Colloid and Interface Science* 187:11–21.

10

A Fractal-Based Approach to Model Wicking

Jianchao Cai, Boming Yu, and Xiangyun Hu

CONTENTS

Spontaneous imbibition is a ubiquitous natural and fundamental phenomenon existing extensively in a variety of processes. Based on the fractal characteristics of pore space and tortuous streamtubes/capillaries in natural porous media, this chapter presents a study on characterizing the physical processes in wicking phenomena in air-saturated single tortuous capillary and porous media. For this purpose, the first section is devoted to briefly introducing the fractal theory, and then the fractal characteristics of porous media are given in Section 10.2. In Section 10.3, models for capillary rise in a single capillary and for spontaneous imbibition in porous media are derived, including the effect of gravity and the controversy over the validity of the Lucas–Washburn (LW) equation is also discussed in this section. The results and discussions are presented in Section 10.4, and the summary and conclusions are given in Section 10.5.

10.1 Fractal Theory

Euclidean geometry describes ordered objects such as points, curves, surfaces, and cubes using integer dimensions 0, 1, 2, and 3, respectively. Their measures are invariant with respect to the unit of measurement used. However, numerous objects found in nature (Mandelbrot 1982), such as river networks, coastlines, snowflakes, rough surfaces, islands, mountain ranges, lakes, trees, and systems of blood vessels are disordered and irregular, and these irregularly shaped objects or spatially nonuniform phenomena cannot be described by Euclidean geometry due to the scale-dependent measures of length, area, and volume. The dimensions of such objects need not necessarily be an integer and are called Hausdorff dimensions, or simply fractal dimensions and these objects are called fractals. The term "fractal" was coined by Benoit B. Mandelbrot in 1975 and was named from the Latin fractus, meaning "broken" or "fractured" (Mandelbrot 1977). Some classical geometry structures such as the Contor sets, Koch curve, Sierpinski triangle, and carpet, and Menger sponge are examples of exactly self-similar fractals or regular fractals, which exhibit self-similarity over an infinite range of length scales. Their dimensions are also called similarity dimensions (Feder 1988). In nature, however, exactly self-similar fractals in a global sense are rarely found, and many objects are not exactly self-similar such as coastlines and islands on earth. These objects are statistically self-similar and exhibit self-similarity in some average sense and over a certain local range of scales, which are called statistical fractals.

Until now, there has not been an integrated and accurate definition of fractal. As stated by Benoit B. Mandelbrot (1982), a fractal is "a rough or fragmented geometric shape that can be split into parts, each of which is (at least approximately) a reduced-size copy of the whole," a property called self-similarity. A fractal often has the following classical features (Mandelbrot 1967, 1982; Feder 1988): (a) It is too irregular to be easily described in traditional Euclidean geometry. (b) It has a fine structure and arbitrarily small scales. (c) It exhibits self-similarity (at least approximately or stochastically). (d) It generally has a simple and recursive definition. (e) It has a Hausdorff dimension which is greater than its topological dimension (although this requirement is not met by space-filling curves such as the Hilbert curve). This fact is what gives fractals many unique properties; for example, the Sierpinski triangle has an area of 0. The unique property of fractal objects is that they are independent of the unit of measurement and follow scaling law in the form of (Mandelbrot 1982):

$$M(\varepsilon) \sim \varepsilon^{D_f} \tag{10.1}$$

where M can be the length of a line, the area of a surface, or the volume of a body; D_f is the fractal dimension; and ε is the scale of measurement. Equation 10.1

indicates the self-similar property and implies that the value of fractal dimension remains constant over a range of length scales. The fractal dimension is usually estimated as the slope of a linear fit of the data on a log–log plot of a measure $M(\varepsilon)$ against the scale ε. For the strictly self-similar objects, the fractal dimension can be derived mathematically as (Mandelbrot 1982):

$$D_{\mathrm{f}} = \frac{\log N}{\log(1/r)} \tag{10.2}$$

where N represents an object of N parts scaled down by a ratio of r, and the fractal dimension D_{f} is the similarity dimension. The fractal dimension can capture what is lost in the traditional geometrical representation of form.

Since the presentation of fractal theory by Mandelbrot, this new system of geometry has had a significant impact on many fields such as physics, soil mechanics, chemistry, physiology, medicine, seismology, technical analysis, fluid mechanics, and so on. Porous media such as soil, sandstones in an oil/water reservoir, packed beds in chemical engineering, fabrics used in composite engineering, and wicks in heat pipes consist of numerous irregular pores of different sizes spanning several orders of magnitude in length scales. The pore microstructures, both the pore size and the pore interfaces, of such porous media exhibit fractal characteristics (Mandelbrot 1982; Katz and Thompson 1985; Jacquin and Adler 1987; Feder 1988; Krohn 1988a,b; Chang and Yortsos 1990; Adler and Thovert 1993; Perfect and Kay 1995; Smidt and Monro 1998; Yu et al. 2001; Perfect et al. 2004; Yu 2008; Li and Horne 2009; Kou et al. 2009; Cai and Yu 2010) and follow the fractal power law Equation 10.1, and these media are thus called fractal porous media. The fractal dimensions used in this chapter are applicable to both exactly self-similar fractals (such as the Sierpinski triangle (Figure 10.1) and Koch curve (Figure 10.2)) and statistical fractals (such as fractal/random porous media) (Feng et al. 2004; Xu et al. 2008; Yu 2008).

10.2 Fractal Characteristics of Porous Media

The disordered nature of pore microstructures suggests the existence of fractal characteristics formed by both pores and tortuousness of capillaries. Using scanning electronic microscopy (SEM) and optical data, Katz and Thompson (1985) presented experimental evidence indicating that the pore spaces of several sandstones are fractals and are self-similar over three to four orders of magnitude in length scale extending from 1 nm to 100 μm. They argued that the pore volume is a fractal with the same fractal dimension as the pore–rock interface. The fractal dimension varies from sample to sample with extreme values of 2.57–2.87. Krohn (1988a) measured the fractal properties of sandstones, shales, and carbonates using a statistical analysis of structural features on fracture surfaces and found that fractal behavior is associated with power

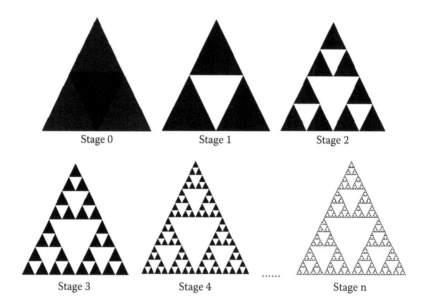

Stage 0   Stage 1   Stage 2

Stage 3   Stage 4   ......   Stage n

FIGURE 10.1
Sierpinski triangle, an exactly self-similar fractal, which is constructed by using a deterministic algorithm based on an equilateral triangle. First, using the midpoints of each side as the vertices of a new triangle, and then removing from the original, leaves three new triangles. Then, this process is iterated from each remaining triangle to remove the "middle," leaving behind three smaller triangles, each of which has dimensions one-half of those of the parent triangle. This iterate can be applied infinitely. For this fractal, $N = 3$ and $r = 1/2$, so the fractal dimension $D_f = \ln 3/\ln 2 = 1.585$.

law behavior for a number of features as a function of the feature size on the pore–rock interface. The fractal dimensions range from 2.27 to 2.89, and the long-length limits to the fractal regime range from 2 μm to over 50 μm.

A porous medium is usually defined as a structure that consists of pores with different sizes, and these pores are randomly distributed in solid space. The pores in porous media are analogous to the islands on the earth's surface and to the spots on engineering surfaces.

Based on the study of the geomorphology of the earth, Mandelbrot (1982) showed that the total number, N, of islands of areas (A) greater than a particular area, a, follows the power law

$$N(A > a) \sim a^{-D_f/2} \tag{10.3}$$

The equality in this law relation can be invoked by using the area of the largest island as (Majumdar and Bhushan 1990)

$$N(A \geq a) = \left(\frac{a_{max}}{a} \right)^{D_f/2} \tag{10.4}$$

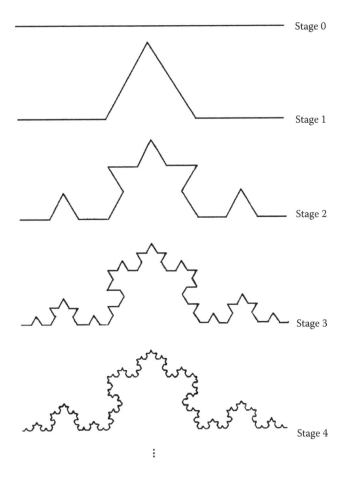

Stage 0

Stage 1

Stage 2

Stage 3

Stage 4

⋮

FIGURE 10.2
Koch curve. This curve is also generated by a recursive prescription. At first, the generator operates on a straight-line segment by dividing it into three parts, and then replacing the middle part by two sides of an equilateral triangle. Again, this generator is applied to each of the four segments of the new curve. This generator can be applied infinitely. For this curve, $N = 4$, $r = 1/3$, so the fractal dimension $D_f = \ln 4 / \ln 3 = 1.262$.

The contact between two rough surfaces can be modeled as one surface contacting a smooth plane. Thus, a large number of contact spots of different sizes coexist at the contact plane and are spread randomly over the surface. It is shown that the contact spots on the engineering surfaces also follow this power law relation. Differentiating Equation 10.4 with respect to a results in a number of islands/spots whose areas are within the range a to $a + da$

$$-dN = \frac{D_f}{2} a_{max}^{D_f/2} a^{-(D_f/2+1)} da \qquad (10.5)$$

where $a_{max} = g\lambda^2_{max}$ and $a = g\lambda^2$, with λ and g being the diameter and the geometry factor of a spot, respectively. The cumulative size distribution of pores or islands should also follow the same fractal scaling law. Thus, Equation 10.4 can be rewritten as (Yu and Cheng 2002)

$$N(\varepsilon \geq \lambda) = \left(\frac{\lambda_{max}}{\lambda}\right)^{D_f} \tag{10.6}$$

where λ is the pore size and λ_{max} is the maximum pore size of porous media. The fractal dimension is in range of $0 < D_f < 2$ and $0 < D_f < 3$ in the two- and three-dimensional spaces, respectively. Equation 10.6 holds for both exactly and statistically self-similar fractal geometries and describes the fractal scaling law relationship of the cumulative pore population. The validity of Equation 10.6 can easily be tested by examining the Sierpinski carpets (Yu and Li 2001; Yu et al. 2009). The equality of the right-hand sides of Equations 10.4 and 10.6 implies that there is only one largest island on the earth's surface or only one largest pore in a set of fractal pores.

From Equation 10.6, the number of pores of sizes lying between λ and $\lambda + d\lambda$ can be expressed as

$$-dN = D_f \lambda^{D_f}_{max} \lambda^{-(D_f+1)} d\lambda \tag{10.7}$$

Equation 10.6 indicates that the total number of pores of size greater or equal to a value λ decreases as the pore size increases. The total number of pores or islands or spots, from the smallest diameter λ_{min} to the largest diameter λ_{max}, can be obtained from Equation 10.6 as

$$N_t(\varepsilon \geq \lambda_{min}) = \left(\frac{\lambda_{max}}{\lambda_{min}}\right)^{D_f} \tag{10.8}$$

Dividing Equation 10.7 by Equation 10.8 yields

$$-\frac{dN}{N_t} = D_f \lambda^{D_f}_{min} \lambda^{-(D_f+1)} d\lambda = f(\lambda) d\lambda \tag{10.9}$$

where

$$f(\lambda) = D_f \lambda^{D_f}_{min} \lambda^{-(D_f+1)} d\lambda \tag{10.10}$$

is the probability density function for pore size distribution in fractal porous media. According to the probability theory, the probability density function should satisfy the normalization condition, which gives

$$\int_{\lambda_{min}}^{\lambda_{max}} f(\lambda)d\lambda = 1 - \left(\frac{\lambda_{min}}{\lambda_{max}}\right)^{D_f} = 1 \tag{10.11}$$

It is shown that Equation 10.11 holds if and only if (Yu and Li 2001)

$$\left(\frac{\lambda_{min}}{\lambda_{max}}\right)^{D_f} \cong 0 \tag{10.12}$$

is satisfied. Due to the fractal dimension $D_f > 0$, Equation 10.12 indicates that $\lambda_{min} \ll \lambda_{max}$ must be satisfied for Equation 10.12 to hold. Otherwise a big error may be caused in analysis, and even the porous medium may be nonfractal. For example, if $\lambda_{min} = \lambda_{max}$, Equations 10.11 and 10.12 do not hold. Therefore, caution must be taken in the fractal analysis of a porous medium, and Equation 10.12 can be considered a criterion for determining whether a porous medium can be characterized by fractal theory and technique. Several investigators (Katz and Thompson 1985; Krohn and Thompson 1986; Malcai et al. 1997; Feng et al. 2004) found that the ratio $\lambda_{min}/\lambda_{max}$ is usually in the range of 10^{-2}–10^{-4} in real porous media. In general, $\lambda_{min}/\lambda_{max} \sim 10^{-2}$ or $<10^{-2}$ in porous media; thus, Equation 10.12 holds for porous media approximately. Therefore, the fractal theory and technique can be used to analyze properties of porous media (Yu and Li 2001).

Equation 10.12 does not imply that λ_{min} must tend to zero. In general, natural porous media exhibit fractal behavior only in a limited range of length scales (Katz and Thompson 1985; Thompson et al. 1987; Krohn 1988a). The lower and upper limits of this domain of self-similarity can be identified with the corresponding limits of the pore sizes λ_{min} and λ_{max}, respectively (Yu et al. 2009; Cai et al. 2010c).

Besides, the fractal scaling law for the distribution of pore size, the convolutedness of the capillary pathways in porous media is another important property. Pores with various sizes are arbitrarily located in a natural porous medium and may be connected to form tortuous capillaries with variable cross-sectional area, with diameter and tortuous length being λ and $L_f(\lambda)$, respectively. Due to the tortuous nature of the capillaries, the tortuous length $L_f(\lambda)$, is greater than or equal to the straight length L_s, that is, $L_f(\lambda) \geq L_s$. The tortuous flow paths in which fluid flows may be similar to the exactly self-similar fractal Koch curve (with fractal dimension being 1.26 as shown in Figure 10.2) or coastlines, satisfy $L_f(\varepsilon) = \varepsilon^{1-D_T}$, where $L_f(\varepsilon)$ is the length measured by scale ε and where D_T is fractal dimension for tortuous capillaries (Mandelbrot 1982; Feder 1988). Wheatcraft and Tyler (1988) presented a fractal scaling/tortuosity relationship for flow through a heterogeneous porous medium, and the scaling relationship is given by

$$L_f(\varepsilon) = \varepsilon^{1-D_T} L_s^{D_T} \tag{10.13}$$

which indicates that the length traveled by a particle relates to the scale ε of the measuring unit (also the smallest scale of measurement, which is of the order of the pore size), and the fractal dimension D_T. Yu and Cheng (2002) argued that the diameters of capillaries are analogous to the length scale ε, which means that the smaller the diameter of a capillary, the longer the capillary. Therefore, the relationship between the diameter and length of capillaries also exhibits a similar fractal scaling law

$$L_f(\lambda) = \lambda^{1-D_T} L_s^{D_T} \tag{10.14}$$

where $1 < D_T < 2$ in a two-dimensional plane and $1 < D_T < 3$ in a three-dimensional space. The fractal dimension D_T represents the extent of convolutedness of pathways for fluid flow through a porous medium. $D_T = 1$ represents a straight capillary path, and a higher value of D_T corresponds to a highly tortuous channel and in the limiting case of $D_T = 2$ corresponds to such a highly tortuous line that it fills a plane, and $D_T = 3$ corresponds to such a highly tortuous line that it fills a three-dimensional space (Wheatcraft and Tyler 1988; Yu and Cheng 2002; Yu 2008). Equation 10.14 is one of the fractal scaling laws characterizing the fractal properties of pore media. It should be noted that Equation 10.14 is defined based on the assumption that the porous medium consists of a bundle of tortuous capillaries with smooth wall, and otherwise, Equation 10.14 and the fractal dimension D_T are unnecessary. If the roughness of capillary wall or particles is concerned, the surface fractal dimension will be used (Cai et al. 2010d).

In the above equations, the fractal dimensions D_f and D_T are the key parameters for the description of the microstructure of fractal porous media. Katz and Thompson (1985) predicted the porosity from the fractal dimension D_f and presented a correlation

$$\phi = C \left(\frac{l_{min}}{l_{max}} \right)^{3-D_f} \tag{10.15}$$

where ϕ is the porosity of porous sandstones, D_f (2–3 in three dimensions) is the fractal dimension for the pore space, C is a fitting constant for matching the experimental data and is on the order of unity, and l_{min} and l_{max} are the lower and upper limits, respectively, of self-similar regions of pores.

Yu and Li (2001) derived a relation between porosity and fractal dimension D_f based on the Sierpinski triangle and carpet models (exactly self-similar fractals). The relation is given by

$$\phi = \left(\frac{\lambda_{min}}{\lambda_{max}} \right)^{d_E - D_f} \tag{10.16}$$

$$D_f = d_E - \frac{\ln \phi}{\ln(\lambda_{min}/\lambda_{max})} \tag{10.17}$$

where d_E is the Euclidean dimension with $d_E = 2$ and 3 in the two- and three-dimensional spaces, respectively. When $d_E = 3$, Equation 10.16 is identical with Equation 10.15 (as $C = 1$). Equations 10.16 and 10.17 have been shown to be valid not only for exactly self-similar fractal geometries but also for statistically self-similar fractals such as random/disordered saturated porous media (Yu and Li 2001).

Since tortuosity depends on the scale of measurement and the fractal dimension of a streamline, it appears that the fractal dimension is more fundamental (Majumdar 1992). Thus, the relation between the tortuosity and fractal dimension is proposed as (Yu 2005)

$$\tau = \frac{L_f}{L_s} = \left(\frac{L_s}{\lambda}\right)^{D_T - 1} \tag{10.18}$$

where τ is the tortuosity (≥ 1) of flow paths followed by the transported fluids. In Equation 10.18, if $\lambda = \lambda_{min}$, which means λ_{min} is the size of measuring unit or the smallest scale of the pore size, which has a finite value, the τ has a finite value. Equation 10.18 implies that the smaller the diameter of a capillary, the longer the actual length of the capillary and the higher the tortuosity. Equation 10.18 also indicates that the statistical self-similarity exists in the range of $\lambda_{min} \sim L_s$, in which the D_T may be different for different capillaries. However, we can find the averaged D_T for a medium with given porosity (Yu 2005). By assuming that the average pore size λ_{av} corresponds to the average tortuosity in Equation 10.18, the average tortuosity τ_{av} can be expressed as

$$\tau_{av} = \left(\frac{L_s}{\lambda_{av}}\right)^{D_T - 1} \tag{10.19}$$

From Equation 10.19, the averaged fractal dimension D_T for tortuous flow paths in porous media can be obtained as (Yu 2005)

$$D_T = 1 + \frac{\ln \tau_{av}}{\ln(L_s/\lambda_{av})} \tag{10.20}$$

In the above expressions, the average pore diameter λ_{av} can be obtained based on the probability density function (Yu 2005)

$$\lambda_{av} = \int_{\lambda_{min}}^{\lambda_{max}} f(\lambda)d\lambda = \frac{D_f}{D_f - 1}\lambda_{min}\left[1 - \left(\frac{\lambda_{min}}{\lambda_{max}}\right)^{D_f - 1}\right] \tag{10.21}$$

For flow through different types of porous media, such as porous media made with spherical or cubical particles, different models for the average tortuosity τ_{av} are needed (Comiti and Renaud 1989; Yu and Li 2004).

Just as there is a difference between fractal flow distance and straight-line flow distance, there is a similar difference between fractal velocity v_f and straight-line velocity v_s for fluid flow in a single fractal capillary tube, which are denoted as (Wheatcraft and Tyler 1988)

$$v_f = dL_f / dt \tag{10.22}$$

$$v_s = dL_s / dt \tag{10.23}$$

The scaling relationship between v_f and v_s can be found by differentiating Equation 10.14 with respect to time t that results in (Wu and Yu 2007; Yu et al. 2009)

$$v_f = D_T L_s^{D_T - 1} \lambda^{1 - D_T} v_s \tag{10.24}$$

If a straight capillary ($D_T = 1$) is concerned, $v_f = v_s$. Upon integration of Equation 10.24 from the smallest pore/capillary to the largest one, the average velocity $\bar{v}_f$ in fractal porous media can be expressed as (Yu et al. 2009):

$$\bar{v}_f = \frac{D_T D_f L_s^{D_T - 1} \lambda_{min}^{1 - D_T}}{D_T + D_f - 1} v_s \tag{10.25}$$

where $\bar{v}_f = d\bar{L}_f / dt$ is defined as the actual average velocity over all tortuous capillaries. Equation 10.25 also indicates that if porosity is unity, all streamlines are straight with $D_T = 1$, and resulting in a two-dimensional pore plane with $D_f = 2$, and thus $\bar{v}_f = v_s$, and this is expected.

Equations 10.7, 10.8, 10.10, 10.12, and 10.14 form the basis for the analysis of transport properties of porous media such as permeability, flow resistance, and dispersion conductivity, which are closely related to the microstructures of the media (Yu 2008). By integrating the two scaling relationships (Equations 10.24 and 10.25) between fractal velocity and straight-line velocity, a fractal-based approach to model wicking is presented in the next section (Cai et al. 2010a,b; Cai and Yu 2010, 2011).

10.3 Fractal-Based Model for Spontaneous Imbibition

10.3.1 Introduction

There is an increasing interest in precisely evaluating the magnitude and role of the capillary effect in infiltration processing because spontaneous capillary imbibition is an important fundamental phenomenon existing extensively in many fields such as oil recovery, geology, groundwater engineering, polymer composite manufacturing, soil science, hydrology, and so on. The study of spontaneous imbibition is essential not only in evaluating the wettability of the solid–liquid systems (Ma et al. 1994) but also in predicting the production performance in these oil/gas reservoirs developed by water flooding (Handy 1960; Zhang et al. 1996; Li and Horne 2001; Morrow and Mason 2001; Alava et al. 2004; Standnes 2004, 2010; Tavassol et al. 2005; Li 2007; Hatiboglu and Babadagli 2008). The rate of imbibition is usually related to microstructures of porous media, fluid properties and fluid–solid interactions expressed through the absolute and relative permeabilities, shapes and boundary conditions of samples, viscosity, initial water saturation, interfacial tension, and wettability.

In early studies, Lucas (1918) and Washburn (1921) proposed the classical LW equation to analyze theoretically and experimentally spontaneous water imbibition and proposed a simple equal-radii cylindrical model. However, the size of pores in natural porous media varies from point to point and is even found to extend several orders of magnitude. Clearly, this model could not properly quantify the capillary imbibition process due to the complex structure of natural porous media. Some theoretical models describe the capillary imbibition by taking into account geometrically different shaped pores (Dullien et al. 1977; Hammecker et al. 1993; Leventis et al. 2000), and by introducing various corrections related to the microstructures (tortuosity and pore shape) of the rocks (Benavente et al. 2002) to modify the LW equation. These models were constructed to have a simple mathematical formulation, but model prediction results were sometimes in poor agreement with experimental data.

As explained above, the capillary imbibition models were usually based on the independent parallel capillaries and no interaction between the capillaries was considered. The wetting liquid velocity depends on the radii of cylindrical tubes, and there is also a pressure difference between the liquid in different capillaries at a fixed distance from the inlet. In order to overcome this difficulty, the extended capillary model, in which capillaries are interacting laterally and cross-flow may take place, was also proposed (Dong et al. 1998, 2005; Ruth and Bartley 2003; Unsal et al. 2007). Dong et al. (1998) developed a model which considered spontaneous imbibition through two interacting capillaries (are completely filled with the nonwetting phase) of different diameters. Ruth and Bartley (2003) presented a "perfect-cross-flow" model through improving theoretical construction that qualifies as a porous medium by attempting to describe a scenario where the resistance to flow between the tubes is zero and there is no pressure difference between the same fluid in different capillaries at the same location. Dong et al. (2005) extended their previous capillary model, by considering flow in a bundle of capillaries and presented the "interacting capillary bundle" model. The behavior of the model is closer to that of a real porous media than the model with independent tubes.

Although the "perfect-cross-flow" model and the "interacting capillary bundle" model predict a behavior that differs from that of the classic "independent cylindrical tube bundle" model, the equations do not have analytical solutions, and the model is partway between network models, which are complex and only give simulation solutions, and the independent tube model which gives analytic solutions (Unsal et al. 2007). Ruth and Bartley (2003) argued that the "perfect-cross-flow" model and the "independent cylindrical tubes bundle" model bound the behavior of real porous media. In present analysis, we assume that porous media are comprised of a bundle of independent and differently sized capillaries parallel to each other.

On the other hand, Handy (1960) presented an analytical model to characterize the process of spontaneous water imbibition into gas-saturated rocks (gas–liquid–rock systems) by neglecting gravity and assuming piston-like displacement of the imbibition process and infinite mobility of the gas phase; the model can be described as:

$$N_{wt}^2 = 2A^2 \frac{P_c K \phi S_{wf}}{\mu} t \tag{10.26}$$

where N_{wt} is the volume of wetting fluid imbibed, A is the cross-section area of the core, ϕ is porosity, μ is the viscosity, and t is the time; K, P_c, and S_{wf} are respectively the effective permeability, the capillary pressure, and the wetting phase saturation. However, according to Equation 10.26, the amount of water imbibed into porous media is infinite when the imbibition time approaches infinity, and this is physically impossible for all systems, including the vertical imbibition systems.

In several cases, gravity force should not be neglected as this depends on the ratio of gravity to the capillary pressure gradient. By analyzing the disadvantages of using the Handy model to characterize spontaneous imbibition in porous media, and then taking into account the gravity based on the hypothesis that the spontaneous imbibition of water vertically rising in a core sample follows Darcy's law, Li and Horne (2001) derived a linear relationship between the spontaneous imbibition rate and the reciprocal of the recovery by water imbibition. From Li and Horne's imbibition model, the infinite value from Equation 10.26 would not appear. In addition, permeability and capillary pressure can be calculated separately from the test data by using their proposed correlations for water imbibition into porous media.

Although many analytical and experimental studies of spontaneous imbibition in porous media were reported in the literature, study of the spontaneous imbibition process based on the fractal characteristics of porous media are limited. Cai et al. (2010a,b, 2011) studied the capillary rise in a single tortuous capillary and spontaneous imbibition in porous media based on fractal theory. Li and Zhao (2012) proposed a fractal model to predict a power law relationship between spontaneous imbibition rate and time. It has been shown that the pore space of natural porous media is fractal, and tortuous streamtubes/capillaries in natural porous media also exhibit fractal behavior. In the following section, we attempt to apply fractal geometry to derive analytical expressions for characterizing the physical process of capillary rise in a single tortuous capillary and spontaneous capillary imbibition into air-saturated porous media. While analyzing the physical process, the viscosity and density factors of gas are neglected, and the cocurrent imbibition (i.e., the flowing direction of the wetting phase is the same with that of the nonwetting phase) is only considered, the analogy may be extended to the countercurrent imbibition process.

10.3.2 Capillary Rise in a Single Fractal Capillary

Since flow paths in natural porous media are usually tortuous (Bear 1972) and exhibit fractal behavior (Wheatcraft and Tyler 1988; Majumdar 1992), the first aim of the present section is to characterize the capillary rise in a single fractal capillary.

When a tortuous capillary of any shape (diameter λ) is brought into contact with a wetting liquid, the liquid rises up within the capillary due to capillary pressure ($P_c = 4\sigma \cos \theta/\lambda$, with θ being the contact angle). Assuming a quasi-steady state, the fully developed laminar flow of an incompressible Newtonian liquid subject to the hydrostatic pressure P_h ($P_h = \rho g L_s$, where g is the gravitational acceleration) and capillary pressure P_c, the differential height of liquid within the capillary with time may be expressed by the LW equation (Washburn 1921):

$$\frac{dL_f}{dt} = \frac{\lambda^2}{32\mu L_f}\left(\frac{4\sigma \cos\theta}{\lambda} - \rho g L_s\right) \tag{10.27}$$

where σ is surface tension and ρ is density of wetting liquid, $L_f \equiv L_f(t)$ is the length of the liquid column at time t, and the capillary is assumed to have uniform internal circular cross section. Initially, the liquid rises rapidly and then slowly approaches the equilibrium height L_e, where the capillary pressure is balanced by the hydrostatic pressure. L_e is calculated as:

$$L_e = \frac{4\sigma \cos\theta}{\lambda \rho g} \tag{10.28}$$

Equation 10.28 indicates the equilibrium height to be independent of the shape of the capillary.

Inserting Equations 10.14 and 10.24 into Equation 10.27 yields

$$\frac{dL_s}{dt} = \frac{\lambda^{2D_T-1}\sigma\cos\theta}{8\mu D_T L_s^{2D_T-1}} - \frac{\lambda^{2D_T}\rho g}{32\mu D_T L_s^{2D_T-2}} \tag{10.29}$$

Based on Equation 10.29, once the tortuosity fractal dimension is determined, the relation between vertical height and capillary rise time can be found by numerical integration. Equation 10.29 also implies that the equilibrium height, as predicted by the fractal theory, is independent of the shape of the capillary.

For a short period of capillary rise time, the second term in the bracket on the right-hand side of Equation 10.27 can be neglected, and then Equation 10.29 can be deduced using the initial condition $L_s(0) = 0$ as follows:

$$L_s = \left(\frac{\sigma \cos \theta}{4\mu\lambda^{1-2D_T}} \right)^{1/2D_T} t^{1/2D_T} \tag{10.30}$$

From Equation 10.30, for the tortuous capillary described by fractal geometry, the height of capillary rise against time follows $L_s \sim t^{1/2D_T}$, for straight capillary $(D_T = 1)$, $L_s \sim t^{1/2}$.

The accumulated weight of liquid imbibed into a single tortuous capillary is

$$w = \frac{1}{4}\rho\pi\lambda^2 L_f \tag{10.31}$$

Based on Equation 10.29, the differential expression for liquid weight increase in a capillary rising process based on fractal geometry can be calculated as

$$\frac{dw}{dt} = \frac{\pi\rho\sigma\lambda^{2+D_T} \cos\theta}{32\mu L_s^{D_T}} - \frac{\pi\rho^2 g\lambda^{3+D_T}}{128\mu L_s^{D_T-1}} \tag{10.32}$$

Note that there are two dependent variables, w and L_s, in this ordinary differential equation. So we need another equation involving w and L_s in order to solve it. By employing Equations 10.14, 10.18, and 10.29, the following relation can be obtained:

$$\frac{dw}{dt} = \frac{\pi^2\rho^2\sigma\lambda^5 \cos\theta}{128\mu} \frac{1}{w} - \frac{\pi\rho^2 g}{128\mu} \frac{\lambda^4}{\tau} \tag{10.33}$$

From Equation 10.33, once the fractal dimension is determined, the relation between weight increase and the time of rise can be found by numerical integration. The total imbibed weight can be obtained from Equation 10.33 for $dw/dt = 0$

$$w = \frac{\pi\lambda\tau\sigma \cos\theta}{g} \tag{10.34}$$

Again, for a short period of capillary rise time, Equation 10.33 can be simplified to yield

$$w = \frac{\rho \pi \lambda^2}{8} \sqrt{\frac{\lambda \sigma \cos \theta}{\mu}} \sqrt{t}$$ (10.35)

From Equation 10.35, in the early rising stage, the accumulated weight of liquid imbibed into a single tortuous capillary is independent of the shape of capillary based on fractal geometry. The same conclusion can also be obtained when the fractal character of capillary is not considered.

Equations 10.29 through 10.35 are the basic fractal description of capillary rise in a single capillary. By analyzing these equations, it is found that the shape of the capillary may not affect the equilibrium height but can influence the rise velocity. The effect of gravity on the capillary rise can be neglected in the very early capillary rise stage, especially for the larger capillaries. In the early rise stage, the accumulated weight of liquid imbibed into a single tortuous capillary is independent of the shape of the capillary.

10.3.3 Fractal Model for Spontaneous Imbibitions in Porous Media

Spontaneous imbibition experiments were usually performed by bringing the bottom face of a core sample into contact with wetting liquid. When the bottom surface of the sample (saturated with air) touches the liquid, the liquid is spontaneously imbibed into the sample.

Based on Hagen–Poiseuille equation, the flow rate for imbibed liquid through a single tortuous capillary with hydraulic diameter λ is

$$q(\lambda) = \frac{\pi}{128} \frac{\lambda^4}{L_f(\lambda)\mu} \left(\frac{4\sigma \cos \theta}{\lambda} - \rho g L_s \right)$$ (10.36)

The total imbibition rate Q can be obtained by an integral of all flow rates from the minimum- to maximum-diameter capillaries as

$$Q = - \int_{\lambda_{min}}^{\lambda_{max}} q(\lambda)dN = \frac{\sigma \cos \theta}{32\mu} \frac{\lambda_{max}^{2+D_T}}{L_s^{D_T}} \frac{\pi D_f (1 - \beta^{2+D_T-D_f})}{2 + D_T - D_f}$$

$$- \frac{\rho g L_s}{128\mu} \frac{\lambda_{max}^{3+D_T}}{L_s^{D_T}} \frac{\pi D_f}{3 + D_T - D_f}$$ (10.37)

where dN can be found from Equation 10.7. For simplicity, the ratio $\beta = \lambda_{min}/\lambda_{min}$ is defined. Since $0 < D_f < 2$ and $1 < D_T < 2$ in the two-dimensional space, the exponent $3 + D_T - D_f > 2$, in general $\beta < 10^{-2}$, so the $\beta^{3+D_T-D_f} \ll 1$, and it is used in the second term on the right-hand side of Equation 10.37. It can be seen from Equation 10.37 that the total spontaneous imbibition rate depends on imbibition height and on the structure parameters (D_f, D_T, λ_{max}, and β) of porous media, fluid properties (μ, ρ, and σ), as well as the contact angle between the extraneous liquid and solid surface.

Using the following equation for the total pore area (A_p) in fractal porous media (Xu and Yu 2008)

$$A_p = \frac{\pi \lambda_{max}^2 D_f}{4(2 - D_f)}(1 - \phi)$$

(10.38)

The actual average velocity ($\bar{v}_f = d\bar{L}_f/dt$) over all tortuous capillaries can be obtained by using relation $\bar{v}_f = Q/A_p$, and then combining it with Equation 10.25, the average velocity ($v_s = dL_s/dt$) of the rising front of the liquid in porous media can be obtained:

$$\frac{dL_s}{dt} = \frac{(D_T + D_f - 1)(2 - D_f)}{D_T D_f (2 + D_T - D_f)} \frac{\sigma \cos\theta}{8\mu} \frac{\lambda_{max}^{D_T}}{L_s^{2D_T-1}\lambda_{min}^{1-D_T}} \frac{1 - \beta^{2+D_T-D_f}}{1 - \phi}$$
$$- \frac{(D_T + D_f - 1)(2 - D_f)}{D_T D_f (3 + D_T - D_f)} \frac{\rho g}{32\mu} \frac{\lambda_{max}^{1+D_T}}{L_s^{2D_T-2}\lambda_{min}^{1-D_T}} \frac{1}{1 - \phi}$$

(10.39)

For straight capillaries ($D_T = 1.0$), Equation 10.39 can be simplified as:

$$\frac{dL_s}{dt} = \frac{2 - D_f}{3 - D_f} \frac{\sigma \cos\theta}{8\mu} \frac{\lambda_{max}}{L_s} \frac{1 - \beta^{3-D_f}}{1 - \phi} - \frac{2 - D_f}{4 - D_f} \frac{\rho g \lambda_{max}^2}{32\mu} \frac{1}{1 - \phi}$$

(10.40)

Equations 10.39 and 10.40 are the differential expressions for the liquid infiltration height. They show a linear relationship between the straight-line imbibition velocity (dL_s/dt) and the reciprocal of the imbibition height ($1/L_s$). From Equation 10.39, the imbibition height L_s of wetting liquid as a function of time can be obtained by numerical integration. It is also seen from Equation 10.39 that the maximum imbibition height (L_m) can be found if we set $dL_s/dt = 0$, which yields

$$L_m = \frac{3 + D_T - D_f}{2 + D_T - D_f} \frac{4\sigma\cos\theta}{\rho g \lambda_{max}}(1 - \beta^{2+D_T-D_f})$$

(10.41)

It is noticed that the maximum imbibition height obtained from Equation 10.39 is different from that obtained from Equation 10.40 because the fractal character of tortuosity is considered in Equation 10.39. Equation 10.41 indicates that the equilibrium height is not related to fluid viscosity.

For an initially short period of infiltration time, the second term on the right-hand side of Equation 10.39 can be neglected; thus, after integral, we have:

$$L_s^{2D_T} = \frac{(D_T + D_f - 1)(2 - D_f)}{D_f(2 + D_T - D_f)} \frac{\sigma \cos\theta}{4\mu} \frac{\lambda_{max}^{D_T}}{\lambda_{min}^{1-D_T}} \frac{(1 - \beta^{2+D_T-D_f})}{1 - \phi} t \qquad (10.42)$$

Equation 10.42 shows that the imbibition height follows a law of $L_s \sim t^{1/2D_T}$, which is different from the LW behavior $L_s \sim t^{1/2}$. This is attributed to the fact that in the present model the fractal character of tortuous capillaries is considered. Li and Zhao (2012) also found that the imbibition height behavior does not follow the square root of imbibition time based on the fractal geometry. In fact, early in 1961, Laughlin and Davis (1961) found that the LW equation does not hold in general for the wicking behavior of wool felt and cotton fabrics in a light grade of lubricating oil. They modified the LW equation as $L_s(t) \sim t^k$, where k is the time exponent. Cai and Yu (2011) discussed the effect of tortuosity on the capillary imbibition in wetting porous media and proposed the analytical expression $k = 1/2\,D_T$; the predictions by the present model can reproduce approximately the global trend of the variation of the time exponent with porosity changing.

The imbibition weight of wetting liquid can be found from

$$W = A\phi\rho \int_{\lambda_{min}}^{\lambda_{max}} L_f(\lambda)f(\lambda)d\lambda = A\phi\rho \frac{D_f L_s^{D_T}\lambda_{min}^{1-D_T}}{D_T + D_f - 1} \qquad (10.43)$$

Differentiating Equation 10.43 yields

$$\frac{dW}{dt} = A\phi\rho \frac{D_T D_f L_s^{D_T-1}\lambda_{min}^{1-D_T}}{D_T + D_f - 1} \frac{dL_s}{dt} \qquad (10.44)$$

Inserting Equation 10.39 into Equation 10.44 and employing Equation 10.43 yields

$$\frac{dW}{dt} = \frac{(A\phi\rho)^2 D_f(2 - D_f)}{(2 + D_T - D_f)(D_T + D_f - 1)} \frac{\sigma \cos\theta}{8\mu} \frac{\lambda_{max}^{D_T}}{\lambda_{min}^{D_T-1}} \frac{1 - \beta^{2+D_T-D_f}}{1 - \phi} \frac{1}{W}$$

$$- \frac{A\phi\rho D_f(2 - D_f)}{\tau(D_T + D_f - 1)(3 + D_T - D_f)} \frac{\rho g \lambda_{max}}{32\mu} \frac{\lambda_{max}^{D_T}}{\lambda_{min}^{D_T-1}} \frac{1}{1 - \phi} \qquad (10.45)$$

Equation 10.45 is a differential expression for the weight increase due to fluid imbibition. The imbibition weight of wetting liquid as a function of time can be obtained when the imbibition height is calculated from Equation 10.45 by numerical integration.

For a bundle of straight capillaries $(D_T = 1.0)$, Equation 10.45 can be expressed as

$$\frac{dW}{dt} = (A\phi\rho)^2 \frac{2 - D_f}{3 - D_f} \frac{\lambda_{max}\sigma\cos\theta}{8\mu} \frac{1 - \beta^{3-D_f}}{1 - \phi} \frac{1}{W}$$

$$- A\phi\rho \frac{2 - D_f}{4 - D_f} \frac{\rho g\lambda_{max}^2}{32\tau\mu} \frac{1}{1 - \phi} \qquad (10.46)$$

Equations 10.45 and 10.46 show a linear relationship between dW/dt and $1/W$ with the straight line not being through the origin, and this is expected because the effect of gravitational force is considered in the present model.

Similarly, in an initially short period of imbibition time, the second term on the right-hand side of Equation 10.45 can be neglected which then yields:

$$W = A\phi\rho \left[\frac{D_f(2 - D_f)}{(2 + D_T - D_f)(D_T + D_f - 1)} \frac{\sigma\cos\theta}{4\mu} \frac{\lambda_{max}^{D_T}}{\lambda_{min}^{D_T-1}} \frac{1 - \beta^{2+D_T-D_f}}{1 - \phi} \right]^{1/2} t^{1/2} \quad (10.47)$$

Equation 10.47 shows that the accumulated imbibed weight in porous media is proportional to $t^{1/2}$ in initial imbibition stage with the gravity factor being neglected. This is consistent with the LW equation. From Equation 10.47, it can be seen that the factors affecting the capillary imbibition are the structure parameters (D_f, D_T, λ_{min}, λ_{max}, β, ϕ, and A) of porous media, fluid properties (ρ, σ, μ) and the fluid–solid interaction (θ). However, the models by conventional methods cannot reveal such mechanisms, and therefore the present fractal approach may provide a better understanding of the mechanisms about imbibitions. An analytical fractal approach to model wicking properties of saturated porous media is thus proposed. It is expected that the proposed approach may have potential in the analysis of oil–water systems, in which the fracture factor can further be considered, as long as the fluid properties and microstructural parameters of a porous medium have been determined.

10.4 Experimental Validation

To test the proposed model for characterizing the spontaneous water imbibition into air-saturated single fractal capillary and fractal porous media, the parameters of the proposed imbibition model are given as follows.

Water is considered as imbibition liquid, and the relevant parameters $\sigma = 0.0727$ N/m, $\rho = 1.0$ g/cm^3, and $\mu = 0.001$ Pa s are applied. The tortuosity for the fractal dimension for a given capillary can be determined by the box-counting method (Yu and Cheng 2002) or calculated by (Yu 2005):

$$D_T = 1 + \frac{\ln \tau}{\ln L_e / \lambda} \qquad (10.48)$$

where L_e is the capillary rise equilibrium height corresponding to diameter λ, and can be calculated from Equation 10.28.

Figure 10.3 shows the capillary rise height with time at different values of the fractal dimension for tortuosity based on Equation 10.29 through numerical integration. As expected, the capillary rise increases rapidly in the initial period of time and then increases slowly and approaches equilibrium. The rapid increase of capillary rise before the rising front reaches a certain height and is attributed to the fact that this process may be a capillary pressure-dominated process as we assumed in the model derivation. Then the increase becomes small when approaching the equilibrium, and this may be the effect of gravity on the capillary rise process. Note that an increase in fractal dimension is responsible for a decrease in capillary rise rate and capillary rise takes more time to approach equilibrium. This is attributed to increased flow resistance due to the convolution present in highly tortuous capillaries. It is worth pointing out that the three curves presented in Figure 10.3 eventually reach the same equilibrium height. This is expected because capillaries with the same diameter, as stated above (Equation 10.28), have the same equilibrium height and are independent of the shape of the capillary.

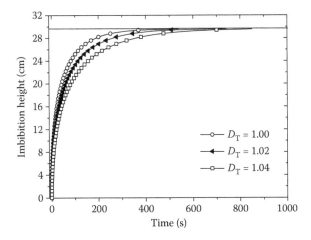

FIGURE 10.3
The height of capillary rises as a function of time (Equation 10.29) at different values of tortuosity fractal dimension at $\lambda = 0.01$ cm, $\cos \theta = 1$.

For spontaneous imbibition in fractal porous media, the pore fractal dimension D_f can be calculated from Equation 10.17, where $d = 2$ is used. Since it is very costly and time consuming to measure the maximum pore size experimentally, the available theoretical-prediction models that can relate the maximum pore size to porosity and other parameters of porous media are quite limited (Yu 2008; Cai and Yu 2010). The maximum pore diameter may be approximately obtained as a compromise between two arrangements of circular particles (Yun et al. 2009):

$$\lambda_{max} = (\lambda_{max1} + \lambda_{max2})/2 \tag{10.49}$$

Here, λ_{max1} is calculated based on the model for equilateral-triangle arrangement (Yu and Cheng 2002)

$$\lambda_{max1} = \frac{D_s}{2}\sqrt{\frac{2\phi}{1-\phi}} \tag{10.50}$$

and λ_{max2} is calculated based on the model for square arrangement (Wu and Yu 2007):

$$\lambda_{max2} = \frac{D_s}{2}\left[\sqrt{\frac{\phi}{1-\phi}} + \sqrt{\frac{\pi}{4(1-\phi)}} - 1\right] \tag{10.51}$$

In reality, solid particles may be arranged in *neither* equilateral-triangle arrangement *nor* square arrangement in porous media, and the maximum pore size may change with different rock types and sandstones. Therefore, a more realistic model for maximum pore size in porous media is desirable. Recently, Cai and Yu (2010) presented a preliminary analysis on the maximum pore size based on the spontaneous imbibition effect and fractal geometry.

In Equations 10.50 and 10.51, D_s is the mean diameter of particles which can be predicted based on the rearranged Kozeny–Carman equation (Carman 1937):

$$D_s = \frac{4(1-\phi)}{\phi}\sqrt{\frac{K_{kc}}{\phi}} \tag{10.52}$$

where K is permeability, and Kozeny constant k_c is equal to 4.8 ± 0.3. The D_s can also be determined by experiments.

The tortuousity fractal dimension D_T is determined by Equation 10.19, where parameters τ_{av} and L_s/λ_{av} can be found by (Yu and Li 2004; Xu and Yu 2008)

$$\tau_{av} = \frac{1}{2}\left[1 + \frac{1}{2}\sqrt{1-\phi} + \frac{\sqrt{(\sqrt{1-\phi} - 1)^2 + (1-\phi)/4}}{1 - \sqrt{1-\phi}}\right] \tag{10.53}$$

$$\frac{L_s}{\lambda_{av}} = \frac{D_f - 1}{2\beta} \sqrt{\frac{1 - \phi}{\phi} \frac{\pi}{D_f(2 - D_f)}} \tag{10.54}$$

It was generally recognized that the minimum pore size is smaller by two orders of magnitude compared to the maximum pore size in porous media. The $\beta = \lambda_{min}/\lambda_{max} = 0.01$ (Yu and Li 2001) is, therefore, taken here. Thus, when K, k_c, and ϕ are given, D_f, D_T, and λ_{max} are determined, and the spontaneous imbibition height and weight in porous media can be calculated by the proposed expressions. The imbibition height versus structure parameters is shown in Figures 10.4 through 10.6, in which the relevant parameters $\theta = 30°$ and $D_s = 0.02$ cm are used, and water is used as imbibed liquid. Figure 10.4 shows the imbibition height versus imbibition time at different porosities through numerical integration of Equation 10.39. The imbibition height increases rapidly in the initial period of imbibition time and then increases more slowly and approaches the equilibrium, which takes a long time for wetting liquid. From Figure 10.4, it is seen that the higher equilibrium height corresponds to lower porosity. Also note that a decrease in porosity is responsible for a decrease in the capillary imbibition rate and the imbibition takes more time to approach the equilibrium. This phenomenon may be explained in more detail as follows. Lower porosity and smaller pore throats lead to lower permeability, and this implies that liquid flow in porous medium becomes difficult, and which in turn means that the capillary imbibition rate becomes smaller. However, smaller throat size also implies a higher capillary pressure, leading to larger pull on the liquid column and this means that the capillary imbibition rate becomes larger. So there is interplay between these two competing phenomena.

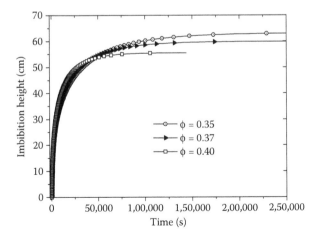

FIGURE 10.4
Plot of imbibition height versus the time of imbibition at different porosities through numerical integration of Equation 10.39.

Figure 10.5 shows a little difference between the predictions by Equation 10.39 (the gravity is included) and by Equation 10.42 (the gravity is neglected) at early imbibition time. This verifies that the gravity factor can be neglected in the early period of imbibition time. However, the gravity factor may have a significant influence on the imbibition height after about 2500 s, at which the imbibition height predicted by Equation 10.39 is much higher than that by Equation 10.36 and quickly tends to infinity. Therefore, the imbibition height predicted by Equation 10.36 may be more reasonable than Equation 10.39 and the gravity factor should be taken into account.

Figure 10.6 shows that the equilibrium height changes with the fractal dimensions for pores (D_f) and for tortuosity (D_T) based on Equation 10.38. It is found from Figure 10.6a that the pore fractal dimension has a significant effect on the equilibrium height. This can be explained by the fact that the equilibrium height is significantly dependent on porosity, while the pore fractal dimension is also significantly dependent on porosity. However, the tortuosity fractal dimension has a slight influence on the equilibrium height as shown in Figure 10.6b. The fractal dimension for tortuosity of tortuous capillaries represents the heterogeneity of real porous media, and the larger fractal dimension corresponds to a more heterogeneous medium, which has lower capillary imbibition equilibrium height.

In Figures 10.3 through 10.6, water is chosen as the imbibed liquid, whose parameters are given in Figure 10.3, and $\theta = 30°$ and $D_s = 0.02$ cm are given in Figure 10.6. Other parameters such as D_f, D_T, and λ_{max} are calculated from Equations 10.17, 10.20, and 10.49, respectively.

To verify the validity of the proposed model, the theoretical predictions are tested by comparison with the available experimental data (Schembre et al. 1998; Li and Horne 2001; Olafuyi et al. 2007). The comparisons of the model predictions for imbibition weight with the experimental results are shown in Figures 10.7 through 10.9, in which water was used as the imbibed liquid. An unconsolidated porous medium (glass-bead pack) was used as the core sample in Li and Horne's measurements, and the consolidated porous media (Berea sandstone and Bentheim) were used in Schembre et al.'s and Olafuyi et al.'s experiments. In all cases, the experiments were conducted under isothermal condition. The detailed sample characteristics are listed in Table 10.1.

The Kozeny constants $k_c = 4.6$, 5.1, and 5.5 were respectively applied to fit the experimental data in Figure 10.7 for a glass-bead pack sample, in Figure 10.8 for Berea sandstone sample, and in Figure 10.9 for Bentheim core. In all these comparisons, the Kozeny constants are comparable to the value of 4.8 ± 0.3 as suggested by Carman (1937).

Before the imbibition front reaches a certain height above the level of the water surface of a container, the imbibition process is a capillary pressure-dominated process as assumed in the model derivation. In Figures 10.7 through 10.9, it can indeed be seen that the experimental results clearly show a straight-line relation at the initial period of imbibed time. It is worth

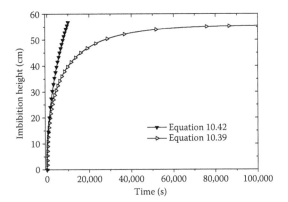

FIGURE 10.5
A comparison for the effect of gravity factor on the imbibition height based on Equation 10.39 and Equation 10.42 at $\phi = 0.4$.

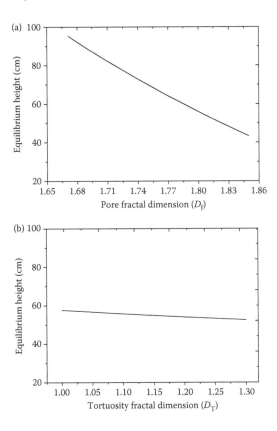

FIGURE 10.6
Plot of the equilibrium height versus the fractal dimensions (a) for pores and (b) for tortuosity based on Equation 10.41.

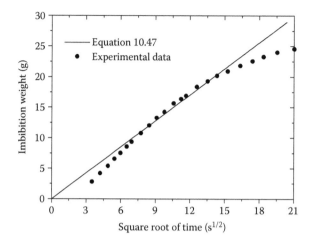

FIGURE 10.7
A comparison of the weight of water imbibed versus the square root of imbibition time by the present model predictions (based on Equation 10.47) with the experiment data. (Data from Li, K. W. and R. N. Horne. 2001. *SPE J.* 6: 375–384.)

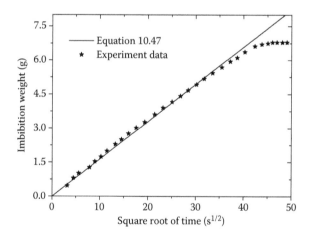

FIGURE 10.8
A comparison of the model predictions (based on Equation 10.47) of the weight water imbibed versus the square root of imbibition time with the experiment data at different porosities. (Data from Schembre, J. M. et al. 1998. *SPE 46211, Presented at the 1998 Western Regional Meeting,* Bakerfield, California, May 10–13.)

pointing out that the initial period of imbibition time with straight linear relation is different in each case. For example, the time period is about 400 s in Figure 10.7, about 1600 s for Berea sandstone sample as shown in Figure 10.8, and about 30 s for Bentheim core in Figure 10.9. The reason may be different pore structures and permeabilities as well as the wettability of rock samples.

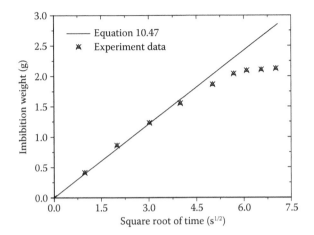

FIGURE 10.9
A comparison of the spontaneous imbibition weight of water versus the square root of imbibition time by the model predictions (based on Equation 10.47) with the experiment data at porosity 0.23. (Data from Olafuyi, O. A. et al. 2007. *SPE 109724, Presented at Asia Pacific Oil and Gas Conference and Exhibition*, Jakarta, Indonesia, 30 October–1 November.)

TABLE 10.1

Rock Sample Characteristics and the Contact Angles with Water on Surfaces

Rock Type	Shape	Length (cm)	Diameter (cm)	Porosity (%)	Permeability (Darcy)	Contact Angle (°)
Glass-bead pack (Li and Horne 2001)	Cylindrical	29.5	3.40	38.6	25.7	50
Berea (Schembre et al. 1998)	Cylindrical	11.0	2.00	15.0	0.5	26
Bentheim (Olafuyi et al. 2007)	Cylindrical	2.56	2.56	23.0	2.89	0

In Figure 10.7, the imbibed volume/weight is linear up until 400 s before constantly dropping with $t^{1/2}$. The model overestimates the imbibition rate at the start and end of the test. Equation 10.47 fits well with experimental data (see Figures 10.8 and 10.9) in a short period of imbibition time, where the volume changes linearly with $t^{1/2}$. This verifies the validity of the theoretical predictions by Equation 10.47. It is seen that the amount of imbibed water depends linearly on the square root of time in some period of imbibition time. In fact, the effect of gravity on the imbibition process increases with the increase of height of the imbibition front (Li and Horne 2001). Thus, after a certain period of imbibition time, the influence of gravity becomes increasingly important. In addition to increasing gravity, the driving capillary pressure must pull the water from the basin further up as the front goes upwards, that is, more and more of the capillary pressure is consumed in transporting the water upwards. Thus, the imbibition rate slows down with time and causes deviation of the measured points from the straight line (see Figures 10.7 through 10.9).

It should also be noted that in Figure 10.7 the experiment points in the initial period of imbibition time do not go through the origin. Li and Horne (2001) explained that this phenomenon might be caused by gravity and several other reasons: (a) due to the influence of the buoyancy caused by inserting the core sample into the water in a container, the data for early spontaneous imbibition into porous media is usually difficult to record accurately; (b) the core sample may not be fully submerged in water and thus causes the water not to imbibe immediately. In addition, the imbibition rate/length at the immediate start of the test does not necessarily have to be linear with $t^{1/2}$ (Cai et al. 2010c), as some countercurrent imbibition occurs at the start, and the imbibition may switch to co-current after some period of imbibition time (Bourblaux and Kalaydjian 1990).

10.5 Summary and Conclusions

In summary, analytical expressions for characterizing capillary rise in a single tortuous capillary and spontaneous imbibition process of wetting fluid into gas-saturated porous media are proposed according to the fractal theory. The following conclusions are drawn:

1. The height and mass of imbibed liquid, in which the gravitational factor is included, is expressed as a function of the fractal dimensions for pores and for tortuous capillaries, the minimum and maximum hydraulic diameter of pores, and the ratio for minimum to maximum hydraulic diameters, porosity, fluid properties as well as the fluid-solid interaction. The imbibed weight predicted by the present model is in good agreement with the experimental data.

2. The imbibition rate decreases with time primarily because the gravity increases with the imbibition front height. For a whole imbibition period, the gravity factor should be included.

3. It is found that the average imbibition height versus time follows $t^{1/2D_T}$ law, and the LW law is the special case of the proposed model.

4. A reasonable explanation from the effect of the convolutedness property of tortuous capillaries on the spontaneous imbibition in

porous media is given. The time exponent of capillary imbibition is a function of fractal dimension for tortuosity of tortuous capillaries/streamtubes in porous media.

5. It is expected that the proposed model in this work may have potential in analysis of multiphase systems as long as the properties of fluids and microstructural parameters of a porous medium are determined.

6. Since fractures of natural porous media (e.g., low permeability reservoirs) also show fractal character, combing the fractal behavior of pore space and tortuous streamtubes/capillaries, the wicking phenomena in porous media with dual porosity may also be analyzed by fractal theory.

Acknowledgments

This work was supported by the National Natural Science Foundation of China through Grant Number 10932010 and 41102080, the China Postdoctoral Science Foundation through Grant Number 2011M500123, and the Fundamental Research Funds for the Central Universities, China University of Geosciences (Wuhan) through Grant Number CUGL100204.

Nomenclature

A	Area or total cross-sectional area
a	Area, as defined by Equation 10.3
A_p	Total pore cross-sectional area
C	Constant, as defined by Equation 10.15
D	Fractal dimension
d_E	Euclidean dimension
D_f	Pore fractal dimension
D_T	Tortuosity fractal dimension
g	Geometry factor
K	Effective permeability
k	Time exponent
k_c	Kozeny constant
l	Self-similar as defined by Equation 10.15
L_e	Equilibrium height in a single capillary
L_f	Actual length that the flow travels
L_m	Maximum imbibition height in porous media

L_s Straight-line length
M Length or area or volume or mass of a body
N Number of pores or capillaries
N_{wt} Imbibed volume in porous media
P_c Capillary pressure (Pa)
P_h Hydrostatic pressure
Q Total flow rate
q Flow rate
S_{wf} Water saturation behind imbibition front (fraction)
t Imbibition time (s)
v Velocity
w Imbibed weight in a single capillary (g)
W Imbibed weight in porous media (g)

Greek Letters

β Ratio as defined as $\lambda_{min}/\lambda_{max}$
σ Surface tension (N/m)
μ Viscosity (Pa s)
ε Scale of measurement
λ Pore diameter (cm)
ϕ Porosity
θ Contact angle
ρ Liquid density (g/cm^3)
τ Tortuosity

Subindexes

av   Average value
max   Maximum value
min   Minimum value
p   pore
s   straight

References

Adler, P. M. and J. F. Thovert. 1993. Fractal porous media. *Transp. Porous Media* 13: 41–78.

Alava, M., M. Dubé, and M. Rost. 2004. Imbibition in disordered media. *Adv. Phys.* 53: 83–175.

Bear, J. 1972. *Dynamics of Fluids in Porous Media*. New York: Elsevier.

Benavente, D., P. Lock, M. Á. G. Del Cura, and S. Ordóñez. 2002. Predicting the capillary imbibition of porous rocks from microstructure. *Transp. Porous Media* 49: 59–76.

Bourblaux, B. J. and F. J. Kalaydjian. 1990. Experimental study of concurrent and countercurrent flows in natural porous media. *SPE Res. Eng.* 5: 361–368.

Cai, J. C. and B. M. Yu. 2010. Prediction of maximum pore size of porous media based on fractal geometry. *Fractals* 18: 417–423.

Cai, J. C. and B. M. Yu. 2011. A discussion of the effect of tortuosity on the capillary imbibition in porous media. *Transp. Porous Media* 89: 251–263.

Cai, J. C., B. M. Yu, M. F. Mei, and L. Luo. 2010a. Capillary rise in a single tortuous capillary. *Chin. Phys. Lett.* 27: 054701.

Cai, J. C., B. M. Yu, M. Q. Zou, and L. Luo. 2010b. Fractal characterization of spontaneous co-current imbibition in porous media. *Energy Fuels* 24: 1860–1867.

Cai, J. C., B. M. Yu, M. Q. Zou, and M. F. Mei. 2010c. Fractal analysis of invasion depth of extraneous fluids in porous media. *Chem. Eng. Sci.* 65: 5178–5186.

Cai, J. C., B. M. Yu, M. Q. Zou, and M. F. Mei. 2010d. Fractal analysis of surface roughness of particles in porous media. *Chin. Phys. Lett.* 27: 024705.

Carman, P. C. 1937. Fluid flow through granular beds. *Trans. Inst. Chem. Eng.* 15: 150–167.

Chang, J. and Y. C. Yortsos. 1990. Pressure transient analysis of fractal reservoir. *SPE Form. Eval.* 5: 31–38.

Comiti, J. and M. Renaud. 1989. A new model for determining mean structure parameters of fixed beds from pressure measurements: Application to beds packed with parallelepipedal particles. *Chem. Eng. Sci.* 44: 1539–1545.

Dong, M. Z., F. A. L. Dullien, L. M. Dai, and D. M. Li. 2005. Immiscible displacement in the interacting capillary bundle model part I. Development of interacting capillary bundle model. *Transp. Porous Media* 59: 1–18.

Dong, M. Z., F. A. L. Dullien, and J. Zhou. 1998. Characterization of waterflood saturation profile histories by the "Complete" capillary number. *Transp. Porous Media* 31: 213–237.

Dullien, F. A. L., M. S. El-Sayed, and V. K. Batra. 1977. Rate of capillary rise in porous media with nonuniform pores. *J. Colloid Interface Sci.* 60: 497–506.

Feder, J. 1988. *Fractals.* New York: Plenum.

Feng, Y. J., B. M. Yu, M. Q. Zou and D. M. Zhang. 2004. A generalized fractal geometry model for the effective thermal conductivity of porous media base on self-similarity. *J. Phys. D. Appl. Phys.* 37: 3030–3040.

Hammecker, C., J.-D. Mertz, C. Fischer and D. Jeannette. 1993. A geometrical model for numerical simulation of capillary imbibition in sedimentary rocks. *Transp. Porous Media* 12: 125–141.

Handy, L. L. 1960. Determination of effective capillary pressures for porous media from imbibition data. *Pet. Trans. AIME.* 219: 75–80.

Hatiboglu, C. U. and T. Babadagli. 2008. Pore-scale studies of spontaneous imbibition into oil-saturated porous media. *Phys. Rev. E* 77: 066311.

Jacquin, C. G. and P. M. Adler. 1987. Fractal porous media II: Geometry of porous geological structures. *Transp. Porous Media* 2: 571–596.

Katz, A. J. and A. H. Thompson. 1985. Fractal sandstone pores: Implications for conductivity and formation. *Phys. Rev. Lett.* 54: 1325–1328.

Kou, J. L., Y. Liu, F. M. Wu, et al. 2009. Fractal analysis of effective thermal conductivity for three-phase (unsaturated) porous media. *J. Appl. Phys.* 106: 054905.

Krohn, C. E. 1988a. Fractal measurements of sandstones, shales, and carbonates. *J. Geophys. Res.* 93: 3297–3305.

Krohn, C. E. 1988b. Sandstone fractal and Euclidean pore volume distributions. *J. Geophys. Res.* 93: 3286–3296.

Krohn, C. E. and A. H. Thompson. 1986. Fractal sandstone pores: automated measurements using scanning-electron-microscope images. *Phys. Rev. B.* 33: 6366–6374.

Laughlin, R. D. and J. E. Davis. 1961. Some aspects of capillary absorption in fibrous textile wicking. *Textile Res. J.* 31: 904–910.

Leventis, A., D. A. Verganelakis, M. R. Halse, J. B. Webber, and J. H. Strange. 2000. Capillary imbibition and pore characterisation in cement pastes. *Transp. Porous Media* 39: 143–157.

Li, K. W. 2007. Scaling of spontaneous imbibition data with wettability included. *J. Contam. Hydrol.* 89: 218–230.

Li, K. W. and R. N. Horne. 2001. Characterization of spontaneous water imbibition into gas-saturated rocks. *SPE J.* 6: 375–384.

Li, K. W. and R. N. Horne. 2009. Experimental study and fractal analysis of heterogeneity in naturally fractured rocks. *Transp. Porous Media* 78: 217–231.

Li, K. W. and H. Zhao. 2012. Fractal prediction model of spontaneous imbibition rate. *Transp. Porous Media* 91: 363–376.

Lucas, R. 1918. Rate of capillary ascension of liquids. *Kolloid-Zeitschrift* 23: 15–22.

Ma, S., N. R. Morrow, X. Zhou, and X. Zhang. 1994. Characterization of wettability from spontaneous imbibition measurements. *Paper CIM 94-47, Presented at the 45th Annual Technical Meeting of the Petroleum Society of CIM held in Calgary, Canada,* June 12–15.

Majumdar, A. 1992. Role of fractal geometry in the study of thermal phenomena. *Annu. Rev. Heat Transfer.* 4: 51–110.

Majumdar, A. and B. Bhushan. 1990. Role of fractal geometry in roughness characterization and contact mechanics of surfaces. *J. Tribol.* 112: 205–216.

Malcai, O., D. A. Lidar, O. Biham, and D. Avnir. 1997. Scaling range and cutoffs in empirical fractals. *Phys. Rev. E.* 56: 2817–2828.

Mandelbrot, B. B. 1967. How long is the coast of Britain? Statistical self-similarity and fractional dimension. *Science* 156: 636–638.

Mandelbrot, B. B. 1977. *Fractals: Form, Chance and Dimension.* San Francisco: W. H. Freeman

Mandelbrot, B. B. 1982. *The Fractal Geometry of Nature.* New York: W. H. Freeman.

Morrow, N. R. and G. Mason. 2001. Recovery of oil by spontaneous imbibition. *Curr. Opin. Colloid Interface Sci.* 6: 321–337.

Olafuyi, O. A., Y. Cinar, M. A. Knackstedt, and W. V. Pinczewski. 2007. Spontaneous imbibition in small cores. *SPE 109724, Presented at Asia Pacific Oil and Gas Conference and Exhibition,* Jakarta, Indonesia, 30 October–1 November.

Perfect, E. and B. D. Kay. 1995. Applications of fractals in soil and tillage research: A review. *Soil Till. Res.* 36: 1–20.

Perfect, E., A. B. Kenst, M. Díaz-Zorita, and J. H. Grove. 2004. Fractal analysis of soil water desorption data collected on disturbed samples with water activity meters. *Soil Sci. Soc. Am. J.* 68: 1177–1184.

Ruth, D. and J. Bartley. 2003. A perfect-cross-flow model for two phase flow in porous media. In: *Proceedings of the 2002 International Symposium of the Society of Core Analysts,* Monterey, California, SCA2002-05: 1–12.

Schembre, J. M., S. Akin, L. M. Castanier, and A. R. Kovscek. 1998. Spontaneous water imbibition into diatomite. *SPE 46211, Presented at the 1998 Western Regional Meeting*, Bakerfield, California, May 10–13.

Smidt, J. M. and D. M. Monro. 1998. Fractal modeling applied to reservoir character-ization and flow simulation. *Fractals* 6: 401–408.

Standnes, D. C. 2004. Experimental study of the impact of boundary conditions on oil recovery by co-current and counter-current spontaneous imbibition. *Energy Fuels* 18: 271–282.

Standnes, D. C. 2010. Scaling group for spontaneous imbibition including gravity. *Energy Fuels* 24: 2980–2984.

Tavassol, Z., R. W. Zimmerman, M. J. Blunt, D. B. Das, and S. M. Hassanizadeh. 2005. Analytic analysis for oil recovery during counter-current imbibition in strongly water-wet systems. *Trans. Porous Media* 58: 173–189.

Thompson, A. H., A. J. Katz, and C. E. Krohn. 1987. The microgeometry and transport properties of sedimentary rock. *Adv. Phys.* 36: 625–694.

Unsal, E., G. Mason, D. W. Ruth, and N. R. Morrow. 2007. Co- and counter-current spontaneous imbibition into groups of capillary tubes with lateral connections permitting cross-flow. *J. Colloid Interface Sci.* 315: 200–209.

Washburn, E. W. 1921. Dynamics of capillary flow. *Phys. Rev.* 17: 273–283.

Wheatcraft, S. W. and S. W. Tyler. 1988. An explanation of scale-dependent dispersiv-ity in heterogeneous aquifers using concepts of fractal geometry. *Water Resour. Res.* 24: 566–578.

Wu, J. S. and B. M. Yu. 2007. A fractal resistance model for flow through porous media. *Int. J. Heat Mass Transfer* 50: 3925–3932.

Xu, P. and B. M. Yu. 2008. Developing a new form of permeability and Kozeny-Carman constant for homogeneous porous media by means of fractal geometry. *Adv. Water Resour.* 31: 74–81.

Xu, P., A. S. Mujumdar and B. M. Yu. 2008. Fractal theory on drying: A review. *Drying Technol.* 26: 640–650.

Yu, B. M. 2005. Fractal character for tortuous streamtubes in porous media. *Chin. Phys. Lett.* 22: 158–160.

Yu, B. M. 2008. Analysis of flow in fractal porous media. *Appl. Mech. Rev.* 61: 050801.

Yu, B. M. and P. Cheng. 2002. A fractal permeability model for bi-dispersed porous media. *Int. J. Heat Mass Transfer* 45: 2983–2993.

Yu, B. M. and J. H. Li. 2001. Some fractal characters of porous media. *Fractals* 9: 365–372.

Yu, B. M. and J. H. Li. 2004. A geometry model for tortuosity of flow path in porous media. *Chin. Phys. Lett.* 21: 1569–1571.

Yu, B. M., J. C. Cai, and M. Q. Zou. 2009. On the physical properties of apparent two-phase fractal porous media. *Vadose Zone J.* 8: 177–186.

Yu, B. M., L. J. Lee, and H. Q. Cao. 2001. Fractal characters of pore microstructures of textile fabrics. *Fractals* 9: 155–163.

Yun, M. J., B. M. Yu, and J. C. Cai. 2009. Analysis of seepage characters in fractal porous media. *Int. J. Heat Mass Transfer* 52: 3272–3278.

Zhang, X., N. R. Morrow, and S. Ma. 1996. Experimental verification of a modified scaling group for spontaneous imbibition. *SPE Res. Eng.* 11: 280–285.

11

Modeling Wicking in Deformable Porous Media Using Mixture Theory

Daniel M. Anderson and Javed I. Siddique

CONTENTS

In this chapter, we present a model for capillary wicking of a liquid into a deformable porous material. The model is derived using a mixture theory approach in a general setting for solid and liquid constituents. We explore the predictions of the model for a one-dimensional capillary rise and deformation configuration. This problem involves two free boundaries—an upper boundary associated with the capillary rise height of the fluid into the dry porous material and a lower boundary associated with the deforming solid interface in contact with a fluid bath. We outline results for both the zero gravity capillary rise/deformation case where similarity solutions can be obtained as well as the non-zero gravity capillary rise/deformation case where we identify time-dependent and steady-state solutions. Different permeability and solid stress models are investigated. We also present a comparison of the model predictions with experimental results available on capillary rise of water into deformable sponges.

11.1 Introduction

In this chapter, we examine the phenomena of capillary wicking with a focus on mixture theory models of deformable porous materials. Wicking, or the imbibition of a liquid into a porous material through capillary suction, occurs in a wide variety of physical processes. These include familiar ones such as the clean up of liquid spills using paper towels or sponges and the design of fabrics ranging from yarn (Pezron et al. 1995; Monaenkova and Kornev 2010) to high-tech fabrics for the optimization of wicking and cooling/drying properties in sports performance clothing (socks, shirts, caps, etc.). They also occur in environmental phenomena ranging from flows through sand or soils (e.g., Bear 1972) to the capillary rise of water into snow (Coléou et al. 1999; Jordan et al. 1999) and industrial processes such as coating and imbibition of iron ore pellets to optimize their performance (Pavlovets 2010). These examples include scenarios where the penetration of the fluid into a porous material is accompanied by the deformation of the porous material. A fascinating historical perspective on the foundations of porous media, including mixture theory and early competing ideas on the description of deformable porous solids, is given in the book by de Boer (2000).

The natural environment has long been a source of problems involving flows in porous materials. Some of the earliest work on flows in deformable porous materials were driven by an interest in the consolidation of soils under different types of loading (Biot 1941a,b, 1955, 1956; Biot and Clingan 1941). The description of flows and deformation in clays and soils still motivates research (e.g., Achanta et al. 1994; Murad et al. 1995; Bennethum 2007). There has also been much interest in fluid flows and deformable porous materials driven by biological applications related to the study of soft tissue, such as articular cartilage, skin, arterial walls, and the cornea. Numerous authors have contributed to various aspects of this work using mixture theory-based models (e.g., Lai and Mow 1980; Holmes 1983, 1984, 1985, 1986; Mow et al. 1984; Hou et al. 1989; Holmes and Mow 1990; Kwan et al. 1990; Lai et al. 1991; Gu et al. 1993; Barry and Aldis 1990 1991, 1992, 1993; Barry et al. 1991; Myers et al. 1995; Ateshian et al. 2004). Similar models have been used to study the motion of blood cells through small capillaries with deformable porous walls (Wang and Parker 1995; Damiano et al. 1996). Humphrey (2003) reviewed the biomechanics of these soft tissues. Industrial processes such as manufacturing of composite materials involve injection of a fluid into a deformable porous "preform" (e.g., Sommer and Mortensen 1996; Preziosi et al. 1996; Michaud et al. 1999; Ambrosi and Preziosi 2000; Billi and Farina 2000). The mixture theory models used to study these infiltration problems involve an external pressure-driven flow with an advancing fluid front moving through the porous material. Other applications include those in the paper, printing, and inkjet industry (Chen and Scriven 1990;

Selim et al. 1997; Clarke et al. 2002; Holman et al. 2002; Fitt et al. 2002; Anderson 2005). Related models to the one we examine here arise in the study of suspensions and gels (Manley et al. 2005; Kim et al. 2007). Diersch et al. (2010a,b) study unsaturated flows in absorbent swelling porous materials with application toward absorbent hygiene products. They model a three-phase system of solid, liquid, and gas, including phenomena such as large solid deformation and chemical reactions and present solutions for two- and three-dimensional swelling in absorbent gelling materials such as diapers.

Capillary wicking is an important phenomena in the dynamics of droplets on porous substrates. These problems can involve a competition between spreading, contact line dynamics, and capillary wicking as well as gravitational drainage into the porous substrate (e.g., Davis and Hocking 1999, 2000). The dynamics, infiltration times, and imprint shapes of sessile droplets spreading on porous substrates has been explored computationally and experimentally in a series of recent papers using capillary network models (Markicevic et al. 2009, 2010; Markicevic and Navaz 2009, 2010) and continuum models (Navaz et al. 2008; D'Onofrio et al. 2010). Hilpert and Ben-David (2009) have recently made comparisons with the spreading and imbibition experiments on sessile droplets by Clarke et al. (2002). Anderson (2005) explored some of these issues theoretically for spreading and imbibition of droplets on deformable porous substrates.

To set the stage for the mixture theory-based models for capillary flows in deformable porous materials, we briefly review some important aspects of wicking in related contexts such as rigid porous materials. We begin with the classical description of capillary rise into a porous material by Washburn (1921). In this model, it was assumed that the porous material was a collection of capillary tubes, and wicking through these tubes was characterized as Poiseuille flow driven by the pressure difference acting to force the liquid in the tubes. In the absence of gravitational and inertial effects, the volume of liquid that penetrates into the porous material in time t is proportional to $\sqrt{t}$.

The classical Washburn theory has been the framework into which new phenomena incorporated and the standard against which other models measured. Szekely et al. (1971), for example, included the gravitational and inertial effects into the classical Washburn model. Schuchardt and Berg (1990) modified the Washburn equation by allowing the vertical capillaries to have a radius that decreases linearly in time (i.e., the solid swells). Masoodi et al. (2007) compared the Washburn model to two other models based on a Darcy formulation and an energy balance argument. All three share the $\sqrt{t}$ dependence for the zero gravity case but differ in the proportionality constant. Masoodi and Pillai (2010) describe a new Darcy-based model for wicking in a porous material that swells. Here, the mass conservation equation is augmented with source and sink terms that model the liquid absorption into the solid matrix and the rate of increase of the solid volume. This equation is then coupled to the standard Darcy equation. The model is compared to the swelling-modified Washburn model of Schuchardt and Berg (1990) as well as to experiments. Further extensions of these models include the effects of an externally imposed pressure (Masoodi et al. 2010) and three-dimensional effects (Masoodi et al. 2011). Hilpert (2009a) extended the Washburn model by incorporating the effects of a dynamic contact angle on wicking in a horizontal capillary tube. Hilpert (2009b) then extended this to account for nonzero gravity in inclined capillary tubes. Fries and Dreyer (2008) identify an analytical solution in the form $h(t)$ for the Lucas–Washburn model of capillary rise including gravity. Unlike the classical Lucas–Washburn solution given implicitly by $t(h)$, this solution $h(t)$ is expressed explicitly in terms of the Lambert W function. Hilpert (2010a) has recently used the Lambert W function to identify explicit solutions for the infiltration of capillary tubes for the case of a dynamic contact angle between the liquid and the capillary wall. Hilpert (2010b,c) has also identified a variety of explicit and semianalytical solutions for liquid withdrawal from the capillary tube with dynamic and constant contact angles.

One of the predictions of the classical Washburn model including gravity effects is that the initial capillary rise dynamics following the $\sqrt{t}$ behavior makes a transition to a steady-state capillary rise height. Delker et al. (1996) performed experiments with capillary rise of water through glass beads and observed that the initial $t^{1/2}$ regime was followed by a new, nonsteady regime characterized by a capillary rise height following the power law $t^{1/4}$. Lago and Araujo (2001) reported a similar power law scaling close to $t^{1/4}$ that follows the initial $t^{1/2}$ dynamics. Lockington and Parlange (2004) have presented a theoretical model that incorporates saturation gradients at the wetting front. Their results effectively provide a Washburn-like model for a rigid porous material that incorporates saturation gradients and captures the general trends of these capillary rise experiments.

Siddique et al. (2009) have recently observed results analogous to the experiments of Delker et al. (1996) and Lago and Araujo (2001) in one-dimensional experiments on the capillary rise of water into deformable sponges. That is, the capillary rise height and the solid deformation follow early time $t^{1/2}$ scalings but then follow a long time regime with a power law of approximately $t^{0.2}$. Their theoretical predictions, like the classical Washburn predictions for rigid porous materials, failed to predict this long-time scaling regime. We discuss this set of experiments as well as their general mixture theory-based modeling in more detail in the next sections.

In the next section, we use the mixture theory to derive governing equations for the flow of a Newtonian fluid through a deformable porous medium. We then explore solutions to the model for a geometry and boundary conditions appropriate for one-dimensional capillary wicking of a liquid into a deformable porous material. We compare predictions of the model to experimental measurements of capillary wicking of water into a deformable sponge.

11.2 Mixture Theory Framework for a Solid/Liquid System

Mixture theory can be used to describe a mixture of any number of phases, or constituents, and is formulated in a thermodynamic framework involving mass, momentum, and energy balances for each constituent along with the corresponding entropy production inequalities (e.g., Green and Naghdi 1967; Bowen and Wiese 1969; Atkin and Craine 1976; Bowen 1980, 1982). Central to the formulation is the idea that each point in space can be thought of as an infinitesimally-small control volume occupied by the various constituents in proportions that can vary from control volume to control volume. That is, as in standard continuum mechanics formulations, each control volume encompasses *enough* material so that *many* particles of each constituent are contained within it. These general theories have been applied to mixtures of two ideal fluids (e.g., Green and Naghdi 1967), mixtures of two ideal gases (e.g., Atkin and Craine 1976), mixtures of elastic materials (Bowen and Wiese 1969), and porous material mixtures of solid and viscous liquid (e.g., Bowen 1980, 1982; Mow et al. 1980; Holmes 1986; Barry and Aldis 1992, 1993; Preziosi et al. 1996). The model that we present below for flow through a deformable porous material follows closely to that of Barry and Aldis (1992, 1993), Preziosi et al. (1996), Ambrosi and Preziosi (2000), and Billi and Farina (2000).

We begin with the assumption that at each point in the mixture, represented by an infinitesimally small control volume δV, we can establish well-defined quantities such as solid volume fraction and liquid volume fraction. We proceed by defining δV_s and δV_ℓ as the actual volumes of solid and liquid contained within the control volume (so $\delta V_s + \delta V_\ell = \delta V$). We then define solid and liquid volume fractions

$$\phi = \frac{\delta V_s}{\delta V}, \quad \chi = \frac{\delta V_\ell}{\delta V},$$ (11.1)

where we note that $\phi + \chi = 1$. In a similar fashion, quantities such as apparent liquid and solid densities ρ_s and ρ_ℓ at a given point in space and time are defined by

$$\rho_s = \frac{\delta m_s}{\delta V}, \quad \rho_\ell = \frac{\delta m_\ell}{\delta V},$$ (11.2)

where δm_s and δm_ℓ are the actual solid and liquid masses contained in the control volume. A mixture density ρ_m can be defined as

$$\rho_m = \rho_s + \rho_\ell = \phi \rho_s^T + \chi \rho_\ell^T,$$ (11.3)

with $\rho_s = \phi \rho_s^T$ and $\rho_\ell = \chi \rho_\ell^T$, where ρ_s^T and ρ_ℓ^T are the true densities of the two material phases involved. In addition to volume fractions and densities, quantities such as liquid and solid velocities, liquid pressure, and solid stress can also be defined at each point in space.

Governing equations for the dependent variables can be obtained by appealing to global mass and momentum balance laws applied in the mixture. We outline these below for our system of interest but note that more general treatments can be found elsewhere (e.g., Atkin and Craine 1976; Bowen 1980).

11.2.1 Mass Conservation

A general mass conservation statement for the solid phase for an arbitrary fixed control volume V with boundary ∂V and outward unit normal $\hat{n}$ is given by

$$\frac{d}{dt}\int_V \rho_s dV + \int_{\partial V} \rho_s v_s \cdot \hat{n} dA = 0, \tag{11.4}$$

where v_s is the velocity vector associated with the solid. More general mixture theory models include an integral on the right-hand side of this equation accounting for the solid mass production generated from the other constituent (e.g., see Atkin and Craine 1976, Equation 2.10) as might occur if solid–liquid phase transformation was present. We do not include these effects here since our liquid and porous solid under consideration do not exchange mass. Similarly, for the liquid phase, we have

$$\frac{d}{dt}\int_V \rho_\ell dV + \int_{\partial V} \rho_\ell v_\ell \cdot \hat{n} dA = 0, \tag{11.5}$$

where v_ℓ is the velocity vector associated with the liquid. Application of the divergence theorem leads to the local statement of conservation of solid

$$\frac{\partial \rho_s}{\partial t} + \nabla \cdot (\rho_s v_s) = 0, \tag{11.6}$$

and conservation of liquid

$$\frac{\partial \rho_\ell}{\partial t} + \nabla \cdot (\rho_\ell v_\ell) = 0. \tag{11.7}$$

Further, if one defines a mixture velocity v_m through the relation $\rho_m v_m = \rho_s v_s + \rho_\ell v_\ell$, then adding these two mass balances leads to

$$\frac{\partial \rho_m}{\partial t} + \nabla \cdot (\rho_m v_m) = 0. \tag{11.8}$$

11.2.2 Momentum Conservation

Again, following Atkin and Craine (1976), a general linear momentum conservation statement for the solid phase is given by

$$\frac{d}{dt}\int_V \rho_s v_s \, dV + \int_{\partial V} \rho_s v_s (v_s \cdot \hat{n}) dA = \int_V (\rho_s b_s + \pi_s) dV + \int_{\partial V} T_s \cdot \hat{n} \, dA, \quad (11.9)$$

where b_s is a body force (e.g., $b_s = g$ where g is gravitational acceleration), π_s is the solid momentum transfer associated with interactions with other constituents (e.g., in our case a drag force accounting for friction between solid and liquid phases), and T_s is the partial stress tensor of the solid phase. The partial stress tensor of a constituent p in a mixture is the stress tensor of that constituent when the other constituent (i.e., solid or liquid) is also present and in general is different from the stress tensor associated with constituent p in the absence of other constituents (for further discussion, see Billi and Farina 2000). The partial stress tensor cannot be measured directly, but, as pointed out below, a related stress tensor for the mixture as a whole can be (Preziosi et al. 1996). A similar momentum conservation statement for the liquid phase is given by

$$\frac{d}{dt}\int_V \rho_\ell v_\ell \, dV + \int_{\partial V} \rho_\ell v_\ell (v_\ell \cdot \hat{n}) dA = \int_V (\rho_\ell b_\ell + \pi_\ell) dV + \int_{\partial V} T_\ell \cdot \hat{n} \, dA, \quad (11.10)$$

where b_ℓ is a body force (e.g., $b_\ell = g$), π_ℓ is the liquid momentum transfer associated with interactions with other constituents, and T_ℓ is a partial stress tensor for the liquid. Application of the divergence theorem and local mass conservation Equations 11.6 and 11.7 leads to the local statements of conservation of linear momentum for solid and liquid phases

$$\rho_s \left(\frac{\partial v_s}{\partial t} + (v_s \cdot \nabla)v_s \right) = \nabla \cdot T_s + \rho_s b_s + \pi_s, \quad (11.11)$$

$$\rho_\ell \left(\frac{\partial v_\ell}{\partial t} + (v_\ell \cdot \nabla)v_\ell \right) = \nabla \cdot T_\ell + \rho_\ell b_\ell + \pi_\ell. \quad (11.12)$$

Newton's third law requires that interaction forces between the solid and liquid phases are equal and opposite, so that $\pi_s + \pi_\ell = 0$. Adding the two momentum balance equations yields, after some manipulation

$$\rho_m \left(\frac{\partial v_m}{\partial t} + (v_m \cdot \nabla)v_m \right) = \nabla \cdot T_m + \rho_s b_s + \rho_\ell b_\ell, \quad (11.13)$$

where the stress tensor for the mixture is

$$\mathbf{T}_m = \mathbf{T}_s + \mathbf{T}_\ell - \rho_s(v_s - v_m) \otimes (v_s - v_m) - \rho_\ell(v_\ell - v_m) \otimes (v_\ell - v_m), \quad (11.14)$$

where $\otimes$ denotes the vector outer, or dyadic, product (see also Billi and Farina 2000, Equations 11 through 13). From this point onward, we shall assume that $\mathbf{b}_s = \mathbf{b}_\ell = \mathbf{g}$. As pointed out by Atkin and Craine (1976), the absence of the interaction force terms π_s and π_ℓ in Equation 11.13 indicates that these forces are internal effects that do not enter the momentum balance for the mixture as a whole. As shown by Atkin and Craine (1976), a balance of angular momentum leads to the condition that the sum of the antisymmetric part of the partial stresses $\mathbf{T}_s + \mathbf{T}_\ell$ vanishes. This implies that $\mathbf{T}_m$ is symmetric.

The above balances provide the basic governing equations necessary for our present model. What is still lacking is a prescription for the stresses and the friction forces. We note that thermodynamically consistent forms for these quantities can be found by appealing to the second law of thermodynamics requiring nonnegative entropy production in the mixture. A general discussion of the underlying principles can be found in de Groot and Mazur (1984). Coleman and Noll (1963) also present a related calculation that includes elastic and viscous effects. The detailed mixture theory derivations given by Atkin and Craine (1976) and Bowen (1980) follow the general procedure and write down energy balances for the solid and liquid phases and postulate an energy balance for the mixture as a whole. They introduce an entropy inequality expressing the requirement for the second law of thermodynamics that the entropy production of the mixture is nonnegative. This entropy inequality involves quantities such as the stress tensors and friction forces as well as quantities such as the heat flux in nonisothermal systems. Rather than reproducing these detailed calculations here, we simply present the proposed forms of these stresses and friction forces motivated by this procedure and used in previous work on deformable porous materials. With this in mind, we follow other authors (e.g., see Holmes 1986; Barry and Aldis 1992, 1993) and introduce stress tensors

$$\mathbf{T}_s = -\phi p\mathbf{I} + \sigma_s, \quad \mathbf{T}_\ell = -\chi p\mathbf{I} + \sigma_\ell, \quad (11.15)$$

where p is the liquid pressure and σ_s and σ_ℓ are solid and liquid stress tensors that still need to be specified. It follows that the stress tensor for the mixture can be expressed in terms of isotropic and nonisotropic parts

$$\mathbf{T}_m = -p\mathbf{I} + \sigma_m, \quad (11.16)$$

where the excess stress tensor σ_m (Billi and Farina 2000) is given by

$$\sigma_m = \sigma_s + \sigma_\ell - \rho_s(v_s - v_m) \otimes (v_s - v_m) - \rho_\ell(v_\ell - v_m) \otimes (v_\ell - v_m). \quad (11.17)$$

Mixture theory models for deformable porous materials begin to differ at the point when σ_m is specified for systems of interest. Barry and Aldis (1992, 1993), for example, address one-dimensional and radial flows in soft biological tissues in which the viscous stress σ_ℓ and nonlinear fluid terms in (11.17) are effectively absent and take $\sigma_m = \sigma_s$ from linear elasticity (e.g., see Chau and Pagano 1967). Holmes (1984) also models the solid as a linear elastic material. Preziosi et al. (1996), for application of fluid injection in porous preforms, discuss the merits of modeling the wet sponge as an elastic material, or, in order to address viscoelastic effects, as a Voigt–Kelvin solid or as an anelastic solid. They argue that a constitutive equation for the excess stress tensor σ_m, as it corresponds to the mixture rather than the individual constituents present in the mixture, could in principle be obtained by conducting experiments on the wet porous material mixture.

In addition to the specification of the stress, the governing equations also require specification of the friction force terms $\pi_\ell = -\pi_s$. As noted above, thermodynamically consistent forms for these terms can be sought through examination of an entropy inequality. In specifying these for the model we outline here, we follow Barry and Aldis (1992, 1993) but also note Holmes (1985) and Damiano et al. (1996), for example, and take

$$\pi_\ell = -\pi_s = K(v_s - v_\ell) - p\nabla\phi, \tag{11.18}$$

where p is the liquid pressure and K is a drag coefficient related to the permeability k and fluid viscosity μ by $K = \mu\,(1 - \phi)^2/k$. We assume that the permeability is in general a function of the solid fraction, so $k = k(\phi)$. This assumption is appropriate for the isotropic case but would need to be revisited for anisotropic materials. A rigid porous material with uniform solid fraction has $\pi_\ell = -Kv_\ell$. We also point out here that a modified friction force term of the form

$$\pi_\ell = K\,|\,v_s - v_\ell\,|^{n-1}\,(v_s - v_\ell) - p\nabla\phi, \tag{11.19}$$

where n is a power law index ($n < 1$ for a shear thinning fluid and $n > 1$ for a shear thickening fluid) has recently been proposed by Siddique and Anderson (2011) for application to a non-Newtonian power law fluid.

In the present case, we shall make the following assumptions to simplify the model equations. The true densities of the solid and liquid phases, ρ_s^T and ρ_ℓ^T, are constant. We shall neglect the inertial terms in the solid and liquid phase and neglect the effects of liquid stress σ_ℓ. Then the liquid momentum balance (11.12) becomes

$$v_\ell - v_s = \frac{k}{\mu(1 - \phi)}\left[-\nabla p + \rho_\ell^T \mathbf{g}\right]. \tag{11.20}$$

and the mixture momentum balance (11.13) becomes

$$0 = -\nabla p + \nabla \cdot \sigma_s + \rho_m \mathbf{g}. \tag{11.21}$$

These equations, along with mass conservation equations

$$\frac{\partial \phi}{\partial t} + \nabla \cdot (\phi v_s) = 0, \quad \frac{\partial \phi}{\partial t} - \nabla \cdot ((1 - \phi) v_\ell) = 0, \tag{11.22}$$

where $\chi = 1 - \phi$, make up the governing equations for our system.

In our one-dimensional model, we shall take $\sigma_s = \sigma(\phi)\mathbf{I}$ where $\sigma(\phi)$ is a scalar function. This form, or an equivalent, has also been used for one-dimensional problems by Preziosi et al. (1996), Barry and Aldis (1992), and Fitt et al. (2002). Siddique et al. (2009) used a simple linear model for $\sigma(\phi)$ so that $d\sigma/d\phi$ (henceforth denoted as $\sigma'(\phi)$) was constant. A number of other authors report a nonlinear relationship for polyurethane sponges (Beavers et al. 1975, 1981a, 1981b; Parker et al. 1987; Lanir et al. 1990; Sommer and Mortensen 1996; Preziosi et al. 1996). Mathematical expressions for this have been obtained in the form of a piecewise-defined function (Beavers et al. 1975) and a degree seven polynomial (Sommer and Mortensen 1996) for polyurethane sponges. As explained in more detail in the next section, we explore both a linear and a nonlinear function for $\sigma(\phi)$. We also examine the results for a collection of permeability functions $k(\phi)$.

11.3 One-Dimensional Capillary Rise of Liquid into a Deformable Porous Material

The configuration under consideration involves a one-dimensional deformable sponge-like material. The upper end of the deformable porous material is held fixed throughout the experiment. We assume that the moment when the infiltration front first contacts the sponge is represented by $t = 0$, the capillary rise of liquid starts at position $z = 0$ between the deformable sponge and an infinite bath of fluid open to atmospheric pressure (i.e., $p = p_A$ at $z = 0$ for all time). For $t > 0$, the imbibition of liquid starts in an initially dry porous material due to capillary suction in the pore space of the porous material assuming the capillary pressure $p_c < 0$, which results in the deformation of the porous material. In Figure 11.1, $z = h_\ell(t)$ represents the upper interface of the wet porous material and $z = h_s(t)$ represents the lower interface formed after the deformation of the material (note that h_s will be negative for the case of swelling). We also assume that the pressure in the fluid bath is hydrostatic. This implies that $p = p_A - \rho_\ell^T g h_s$ at $z = h_s(t)$. The dry porous material is rigid and has uniform solid fraction ϕ_0.

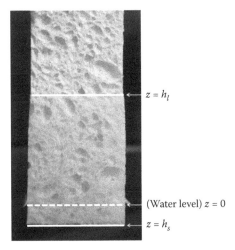

FIGURE 11.1
Experimental set up used by Siddique et al. (2009). (From Siddique, J.I., Anderson, D.M., and Bondarev, A. 2009. *Phys. Fluids* 21: 013106.)

Below, we present the dimensionless model after defining the following scaling parameters:

$$\bar{z} = \frac{z}{L}, \quad \bar{t} = \frac{t}{T}, \quad \bar{w}_s = \frac{w_s}{(L/T)}, \quad \bar{w}_\ell = \frac{w_\ell}{(L/T)}, \quad \bar{p} = \frac{p}{\Sigma_0}$$

$$\bar{h}_s = \frac{h_s}{L}, \quad \bar{h}_\ell = \frac{h_\ell}{L}, \quad \bar{\sigma}(\phi) = \frac{\sigma(\phi)}{\Sigma_0}, \quad \bar{k}(\phi) = \frac{k(\phi)}{K_0}, \tag{11.23}$$

where the length and time scales $L = \Sigma_0/\rho_\ell^T g$ and $T = L^2\mu/K_0\Sigma_0$ are suggested by balancing the terms in the two momentum equations. Here, Σ_0 and K_0 are representative scales for the solid stress and the permeability, respectively. Precise values of these two scales are not specified at this stage but will be part of a fitting procedure later when the theoretical predictions are compared to experimental measurements. After incorporating the above assumptions and dimensionless parameters, the remaining unknowns can be divided into two categories. The first one is the set of variables in the wet material region such as the solid volume fraction ϕ, the vertical component of liquid phase velocity $\bar{w}_\ell$, the vertical component of solid phase velocity $\bar{w}_s$, the liquid pressure $\bar{p}$, the solid stress function $\bar{\sigma}(\phi)$, and the permeability $\bar{k}(\phi)$. The second category contains the boundary positions $\bar{h}_s$ and $\bar{h}_\ell$. The dimensionless set of equations for the one-dimensional material deformation after dropping the bars can be written as

$$\frac{\partial \phi}{\partial t} + \frac{\partial}{\partial z}(\phi w_s) = 0, \tag{11.24}$$

$$\frac{\partial \phi}{\partial t} - \frac{\partial}{\partial z}[(1 - \phi)w_\ell] = 0, \tag{11.25}$$

$$w_\ell - w_s = -\frac{k(\phi)}{(1 - \phi)}\left(\frac{\partial p}{\partial z} + 1\right), \tag{11.26}$$

$$0 = -\frac{\partial p}{\partial z} + \frac{\partial \sigma}{\partial z} - (\rho\phi + 1), \tag{11.27}$$

where $\rho = \rho_s^T / \rho_\ell^T - 1$. Also note that the gravity vector $\mathbf{g} = -g\hat{\mathbf{k}}$, where $\hat{\mathbf{k}}$ is a unit vector in the positive z direction. Equations 11.24 through 11.27 are consistent with the work of previous authors (Preziosi et al. 1996; Anderson 2005; Siddique et al. 2009).

Equations 11.24 through 11.27 after some mathematical manipulation reduce to a single partial differential equation for solid volume fraction ϕ. After subtracting (11.25) from (11.24) and integrating, we find $c(t) = \phi w_s + (1 - \phi) w_\ell$. Here, $c(t)$ is a constant of integration function and can be determined from the boundary conditions as described below. Combining (11.26) and the formula for $c(t)$ allows us to write the formula for solid w_s and liquid w_ℓ velocities

$$w_s = c(t) + k(\phi)\left(\frac{\partial p}{\partial z} + 1\right). \tag{11.28}$$

$$w_\ell = c(t) - \frac{\phi k(\phi)}{(1 - \phi)}\left(\frac{\partial p}{\partial z} + 1\right). \tag{11.29}$$

The assumption that stress is a function of the solid volume fraction $\sigma = \sigma(\phi)$ gives

$$\frac{\partial p}{\partial z} = \sigma'(\phi)\frac{\partial \phi}{\partial z} - (\rho\phi + 1). \tag{11.30}$$

Now, combining Equations 11.24, 11.28, and 11.30 yields a partial differential equation for solid volume fraction in the wet sponge region $h_s(t) \leq z \leq h_\ell(t)$

$$\frac{\partial \phi}{\partial t} + c(t)\frac{\partial \phi}{\partial z} = -\frac{\partial}{\partial z}\left[\phi k(\phi)\left\{\sigma'(\phi)\frac{\partial \phi}{\partial z} - \rho\phi\right\}\right]. \tag{11.31}$$

The boundary conditions applied at the liquid–wet material interface $z = h_s(t)$ are

$$w_s(h_s^+, t) = \frac{dh_s}{dt}, \quad p(h_s^+, t) = -h_s(t), \quad \sigma(h_s^+, t) = 0. \tag{11.32}$$

Equation 11.32 is the kinematic condition, hydrostatic pressure assumption in the liquid bath, and zero stress condition, respectively.

The boundary conditions applied at the wet material–dry material interface $z = h_\ell(t)$ are

$$w_\ell(h_\ell^-, t) = \frac{dh_\ell}{dt}, \quad p(h_\ell^-, t) = p_c/\Sigma_0, \tag{11.33}$$

where p_c/Σ_0 is a dimensionless constant capillary pressure. Here, Equation 11.33 are the kinematic and capillary pressure conditions, respectively. In Equations 11.32 and (11.33)$_2$ we take $p_A = 0$ for simplicity.

The expression for $c(t)$ can be derived following a procedure described by Preziosi et al. (1996), Anderson (2005), and Siddique et al. (2009). In particular, evaluating $c(t) = \phi w_s + (1 - \phi)w_\ell$ on both sides of the wet/dry interface, using $w_s(h_\ell^+) = 0$, $\phi(h_\ell^+) = \phi_0$, and assuming $w_\ell(h_\ell^+) = w_\ell(h_\ell^-)$ yields $w_s(h_\ell^-) = c(t)(\phi(h_\ell^-) - \phi_0)/(\phi(h_\ell^-)(1 - \phi_0))$. Combining this with Equations 11.28 and 11.30 evaluated at $z = h_\ell^-$ gives

$$c(t) = -\frac{(1 - \phi_0)}{\phi_0} \left[\frac{\phi k(\phi)}{(1 - \phi)} \left(\sigma'(\phi)\frac{\partial\phi}{\partial z} - \rho\phi \right) \right]_{h_\ell^-}. \tag{11.34}$$

The boundary conditions (11.32)$_1$ and (11.33)$_1$ along with Equations 11.28 through 11.30 yield equations for interface positions

$$\frac{dh_s}{dt} = c(t) + k(\phi)\left(\sigma'(\phi)\frac{\partial\phi}{\partial z} - \rho\phi \right)\bigg|_{h_s^+}, \tag{11.35}$$

$$\frac{dh_\ell}{dt} = c(t) - \frac{\phi k(\phi)}{(1 - \phi)}\left(\sigma'(\phi)\frac{\partial\phi}{\partial z} - \rho\phi \right)\bigg|_{h_\ell^-}. \tag{11.36}$$

With this, we have the set of equations for all the variables we need.

In the present study, we shall consider two possible forms for $\sigma(\phi)$. Figure 11.2 shows $\sigma(\phi)$ versus ϕ for the linear case $\sigma(\phi) = \phi_r - \phi$ considered in Siddique et al. (2009) (dashed curve) and for the degree seven polynomial used by Sommer and Mortensen (1996) (solid curve). Both cases use $\phi_r = \phi_r^{SM} \equiv 0.135179$, which we obtained numerically by solving $\sigma(\phi_r) = 0$ from the Sommer and Mortensen data. The choice for $\Sigma_0 = 5 \times 10^5$ Pa was chosen to have the linear fit visually match the overall trend of the Sommer and Mortensen data. Other options such as a least squares fit of their data would also be possible; however, that in general would lead to a different value of ϕ_r for the linear fit. We found it instructive in comparing the linear and nonlinear stress functions to maintain the same value of ϕ_r between the two cases.

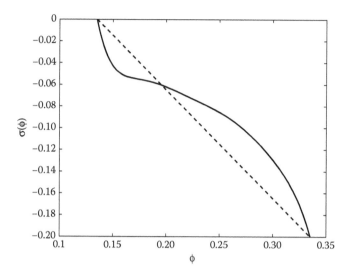

FIGURE 11.2
Plot showing $\sigma(\phi)$ versus ϕ for the linear case $\sigma(\phi) = \phi_r - \phi$ considered in Siddique et al. (2009) (dashed curve) and for the degree seven polynomial used by Sommer and Mortensen (1996) (solid curve). Both cases use $\phi_r = \phi_r^{SM}$. The coefficients listed in Sommer and Mortensen have been made dimensionless using the pressure scale $\Sigma_0 = 5 \times 10^5$ Pa. (Data from Siddique, J.I., Anderson, D.M., and Bondarev, A. 2009. *Phys. Fluids* 21: 013106; Sommer, J.L. and Mortensen, A. 1996. *J. Fluid Mech.* 311: 193–217.)

In this study, we shall also investigate a variety of permeability functions $k(\phi)$. Most of the functions we examine here have also been examined in the work of Masoodi and Pillai (2010) but we also include a permeability function of the form examined by Anderson (2005) and Siddique et al. (2009) also for comparison. Other choices for the permeability function include the exponential form used by Barry and Aldis (1992, 1993) although we shall only consider those listed in Table 11.1. The fourth and fifth columns of Table 11.1 correspond to the results from the zero gravity similarity solution described in more detail in the next section.

11.3.1 Zero Gravity Effects Solution

The first fundamental case we consider is the one where the effects of gravity are absent. The case of zero gravity in our present model can be formally obtained by setting $\rho = 0$ in Equations 11.31, 11.34 through 11.36. In other words, one can recover the zero gravity case in the present model by omitting terms that are multiplied by the density ratio parameter ρ. This leads to

$$\frac{\partial \phi}{\partial t} + c(t)\frac{\partial \phi}{\partial z} = -\frac{\partial}{\partial z}\left[\phi k(\phi)\sigma'(\phi)\frac{\partial \phi}{\partial z}\right], \tag{11.37}$$

TABLE 11.1

Different Choices for Permeability Function $k(\phi)$

Number	References	$k(\phi)$	λ_s	λ_ℓ
1	Blake (1922), Kozeny (1927) Carman (1937)	$\left(\dfrac{1-\phi}{1-\phi_0}\right)^3 \left(\dfrac{\phi_0}{\phi}\right)$	−0.20031	0.20147
2	Zunker (1920) as reported by Rumpf and Gupte (1971)	$\dfrac{\phi_0(1-\phi)}{\phi(1-\phi_0)}$	−0.15999	0.16251
3	Terzaghi (1925)	$\left(\dfrac{0.87-\phi}{0.87-\phi_0}\right)^2 \left(\dfrac{\phi_0}{\phi}\right)^{0.3}$	−0.14582	0.14969
4	Anderson (2005), Siddique et al. (2009)	$\dfrac{\phi_0}{\phi}$	−0.14304	0.14601
5	Rumpf and Gupte (1971)	$\left(\dfrac{1-\phi}{1-\phi_0}\right)^{5.5} \left(\dfrac{\phi}{\phi_0}\right)$	−0.12925	0.13489
6	Rose (1945)	$\left(\dfrac{1-\phi}{1-\phi_0}\right)^{4.1} \left(\dfrac{\phi}{\phi_0}\right)$	−0.11070	0.11633
7	Fehling (1939) as reported by Rumpf and Gupte (1971)	$\left(\dfrac{1-\phi}{1-\phi_0}\right)^{4} \left(\dfrac{\phi}{\phi_0}\right)$	−0.10948	0.11510

Sources: Masoodi, R. and Pillai, K.M. 2010. *AIChE J.* 56: 2257–2267. Dullien F.A.L. 1992. *Porous Media: Fluid Transport and Pore Structure.* San Diego: Academic Press; Rumpf, H. and Gupte, A.R. 1971. *Chemie Ingenieur Technik,* 43: 367–375.

Note: The fourth and fifth columns show the computed values of λ_s and λ_ℓ corresponding to the permeability function for the particular case with $\phi_r = \phi_r^{SM}$, $\phi_0 = 0.33$, $\phi_\ell^* = 0.2$, and $\sigma(\phi) = \phi_r - \phi$.

where

$$c(t) = -\frac{(1-\phi_0)}{\phi_0}\left[\frac{\phi k(\phi)\sigma'(\phi)}{(1-\phi)}\frac{\partial\phi}{\partial z}\bigg|_{h_\ell^-}\right], \tag{11.38}$$

along with the boundary conditions

$$\frac{dh_s}{dt} = c(t) + k(\phi)\sigma'(\phi)\frac{\partial\phi}{\partial z}\bigg|_{h_s^+}, \quad \frac{dh_\ell}{dt} = c(t) - \frac{\phi k(\phi)\sigma'(\phi)}{(1-\phi)}\frac{\partial\phi}{\partial z}\bigg|_{h_\ell^-}. \tag{11.39}$$

Equation 11.37 along with the boundary condition admits a solution in terms of the similarity variable $\eta = z/(2\sqrt{t})$. Similarly, the interface positions can be expressed as

$$h_s(t) = 2\lambda_s\sqrt{t}, \quad h_\ell(t) = 2\lambda_\ell\sqrt{t}, \tag{11.40}$$

where λ_s and λ_ℓ are constants to be determined. Substituting the similarity variable η into (11.37) results in an ordinary differential equation

$$2\eta\frac{d\phi}{d\eta} + \frac{1-\phi_0}{\phi_0}\left[\frac{\phi k(\phi)\sigma'(\phi)}{(1-\phi)}\frac{d\phi}{d\eta}\right]_{\lambda_{\bar{\ell}}}\frac{d\phi}{d\eta} = \frac{d}{d\eta}\left[\phi k(\phi)\sigma'(\phi)\frac{d\phi}{d\eta}\right]. \quad (11.41)$$

Note that the coefficient of the second term on the left-hand side is constant. This equation is subject to boundary conditions $\phi(\lambda_s) = \phi_r$ and $\phi(\lambda_\ell) = \phi_\ell^*$, where $\phi_\ell^* = \phi_r - p_c/\Sigma_0$ as in Siddique et al. (2009). On substituting Equation 11.40 in Equation 11.39 yields the following equations for interface positions:

$$\lambda_s = \frac{1}{2}\left\{-\frac{1-\phi_0}{\phi_0}\left[\frac{\phi k(\phi)\sigma'(\phi)}{(1-\phi)}\frac{d\phi}{d\eta}\right]_{\lambda_{\bar{\ell}}} + \left[k(\phi)\sigma'(\phi)\frac{d\phi}{d\eta}\right]_{\lambda_s^+}\right\}, \quad (11.42)$$

$$\lambda_\ell = -\frac{1}{2\phi_0}\left[\frac{\phi k(\phi)\sigma'(\phi)}{1-\phi}\frac{d\phi}{d\eta}\right]_{\lambda_{\bar{\ell}}}. \quad (11.43)$$

This zero gravity solution was investigated for special choices of $k(\phi)$ and $\sigma(\phi)$ that admitted error function solutions for $\phi(\eta)$ (Anderson 2005; Siddique et al. 2009). We explore more general solutions here.

Figure 11.3 shows the computed similarity solution profile $\phi(\eta)$ for two different cases corresponding to the linear case $\sigma(\phi) = \phi_r - \phi$ (dashed curve) and the degree seven polynomial in Sommer and Mortensen (solid curve). For the linear case, the endpoints are $(\lambda_s, \phi_r) = (-0.14304, \phi_r^{SM})$ and $(\lambda_\ell, \phi_\ell^*) = (0.14601, 0.2)$. For the Sommer and Mortensen case, the endpoints are $(\lambda_s, \phi_r) = (-0.16191, \phi_r^{SM})$ and $(\lambda_\ell, \phi_\ell^*) = (0.13760, 0.20654)$. One interpretation of these curves is that they represent the early time solution ϕ of the nonzero gravity capillary rise problem. We note that the nonlinearity of the function $\sigma(\phi)$ in the Sommer and Mortensen model is apparent in the solution ϕ. The similarity solution is shown for only one choice of permeability function $k(\phi) = \phi_0/\phi$. We have observed that the similarity solutions for the other permeability functions under consideration show qualitatively similar $\phi(\eta)$ profiles. In Table 11.1, we have included two columns showing the values of λ_s and λ_ℓ for a variety of permeability functions using the stress function $\sigma(\phi) = \phi_r - \phi$. In this calculation, we have chosen $\phi_r = \phi_r^{SM}$, $\phi_0 = 0.33$, and $\phi_\ell^* = 0.2$. These results show that the wicking dynamics vary with the permeability function. As we discuss in more detail in the next section, steady-state solutions with nonzero gravity are independent of the permeability function.

The work of Anderson (2005) showed that a variety of regimes are possible for capillary rise and solid deformation in the zero gravity case. In that paper, the specific case with $k(\phi) \sim 1/\phi$ and $\sigma(\phi) = \phi_r - \phi$ was examined. Their

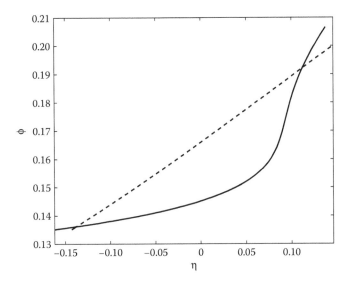

FIGURE 11.3
Computed similarity solution profile $\phi(\eta)$ for two different cases corresponding to the linear case $\sigma(\phi) = \phi_r - \phi$ (dashed curve) and the degree seven polynomial in Sommer and Mortensen (solid curve). We have used $k(\phi) = \phi_0/\phi$ with $\phi_0 = 0.33$, $\phi_\ell^* = 0.2$, and $\phi_r = \phi_r^{SM}$. (Data from Sommer, J.L. and Mortensen, A. 1996. *J. Fluid Mech.* 311: 193–217.)

configuration was associated with a fluid droplet on top of a porous substrate and so the positions of the fluid and solid are reversed relative to the capillary rise configuration considered here. With the appropriate reinterpretation of their configuration, we can summarize one of their key findings as follows. It was shown that for the zero gravity similarity solution as the fluid penetrated into the sponge, the sponge could swell, shrink, or remain undeformed. Exactly which of these cases was observed depended on ϕ_0, ϕ_r, and ϕ_ℓ^* (e.g., see Figures 2 and 3 in Anderson (2005)). We note that for the shrinkage scenario in the capillary rise configuration, it would have to be assumed that the shrinking sponge still maintains contact with the water supply (i.e., the bath).

11.3.2 Steady-State Solution

The second fundamental case we consider is the equilibrium solution in the presence of gravity effects. We seek an equilibrium solid fraction profile $\phi = \phi(z)$, equilibrium interface positions h_s^∞ and h_ℓ^∞, including the value of $\phi(h_\ell^\infty) = \phi_\ell^\infty$. Note that $\phi(h_s^\infty) = \phi_r$ is specified.

The equation governing the steady-state solution for solid volume fraction ϕ, which follows from Equation 11.27 after using Equation 11.26, can be written as

$$\sigma'(\phi)\frac{d\phi}{dz} = \rho\phi. \tag{11.44}$$

This can be written in integral form as

$$\int_{\phi_r}^{\phi} \frac{\sigma'(\phi)}{\phi} d\phi = \rho(z - h_s^{\infty}), \tag{11.45}$$

which can be viewed as implicitly determining the steady profile $\phi(z)$. Evaluating Equation 11.45 at $z = h_\ell^{\infty}$ and $\phi = \phi_\ell^{\infty}$ (ϕ_ℓ^{∞} still to be determined) leads to the result

$$h_s^{\infty} = h_\ell^{\infty} - \frac{1}{\rho} \int_{\phi_r}^{\phi_\ell^{\infty}} \frac{\sigma'(\phi)}{\phi} d\phi. \tag{11.46}$$

The steady-state liquid interface position can be computed from Equation 11.26 (i.e., $dp/dz = -1$) after using boundary conditions $(11.32)_2$ and $(11.33)_2$

$$h_\ell^{\infty} = -\frac{p_c}{\Sigma_0} = \phi_\ell^* - \phi_r. \tag{11.47}$$

Finally, we apply a global mass conservation argument stated as follows. The solid mass before the imbibition of liquid into the material is equal to the solid mass after the liquid is imbibed into the porous material. Note that the solid mass that occupies the region that will eventually be wet can be represented at time zero by $\rho_s^T \phi_0 h_\ell^{\infty}$ since the dry solid fraction is assumed to be the constant ϕ_0. Once the fluid has been imbibed into the solid and equilibrium has been achieved, this same solid mass can be represented by $\rho_s^T \int_{h_s^{\infty}}^{h_\ell^{\infty}} \phi(z) dz$. Equating these two quantities implies that

$$h_\ell^{\infty} \phi_0 = \int_{h_s^{\infty}}^{h_\ell^{\infty}} \phi(z) dz. \tag{11.48}$$

Then, from Equations 11.48 and 11.44, we find that

$$h_\ell^{\infty} \phi_0 = \int_{h_s^{\infty}}^{h_\ell^{\infty}} \phi(z) dz = \frac{1}{\rho} \int_{\phi_r}^{\phi_\ell^{\infty}} \sigma'(\phi) d\phi = \frac{1}{\rho} [\sigma(\phi_\ell^{\infty}) - \sigma(\phi_r)] = \frac{\sigma(\phi_\ell^{\infty})}{\rho}, \tag{11.49}$$

which can be used to determine the value of ϕ_ℓ^{∞}. Note that as in Siddique et al. (2009), the case $\rho = 0$ requires $\phi_\ell^{\infty} = \phi_r$. Observe that these equilibrium equations are independent of the permeability function $k(\phi)$.

Figure 11.4 shows the equilibrium solution $\phi(z)$ for two different forms for $\sigma(\phi)$ corresponding to the linear case $\sigma(\phi) = \phi_r - \phi$ (dashed curve) and the degree seven polynomial in Sommer and Mortensen (solid curve). We

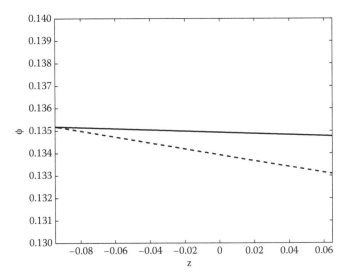

FIGURE 11.4
Long-time equilibrium solution $\phi(z)$ for two different cases corresponding to the linear case $\sigma(\phi) = \phi_r - \phi$ (dashed curve) and the degree seven polynomial in Sommer and Mortensen (solid curve). We have used $k(\phi) = \phi_0/\phi$ with $\phi_0 = 0.33$, $\phi_\ell^* = 0.2$, $\phi_r = \phi_r^{SM}$, and $\rho = 0.1$. (Data from Sommer, J.L. and Mortensen, A. 1996. *J. Fluid Mech.* 311: 193–217.)

use $k(\phi) = \phi_0/\phi$. These are the long-time analogs of the similarity solutions shown in Figure 11.3. The equilibrium interface positions are $h_s^\infty = -0.094592$ and $h_\ell^\infty = 0.064776$ with $\phi_\ell^\infty = 0.13304$ for the linear case (dashed curve) and $h_s^\infty = -0.093761$, $h_\ell^\infty = 0.064801$ with $\phi_\ell^\infty = 0.13476$ for the Sommer and Mortensen case (solid curve). The value of ϕ at $z = h_s^\infty$ in each case is $\phi_r = \phi_r^{SM}$. Notice that while the similarity solution ϕ increases from the bottom of the sponge to the wet–dry interface, the equilibrium solution has the reverse trend. Siddique et al. (2009) showed that the long-time equilibrium ϕ profile has a derivative whose sign is controlled by the sign of ρ. This is also apparent by examining the equilibrium Equation 11.44. That is, upon noting that $\sigma'(\phi) < 0$, Equation 11.44 shows that $\rho > 0$ (solid more dense than liquid) corresponds to ϕ decreasing with height (sponge relatively overcompressed at the bottom and undercompressed at the top). The case shown in Figure 11.4 has $\rho = 0.1$ which means the solid is more dense than the liquid. This leads to a relative compression of the sponge (higher solid fraction values) near the bottom of the wet sponge compared to the top of the wet sponge. The degree of this variation in ϕ throughout the sponge is influenced by the magnitude of $\sigma'(\phi)$ as can be seen in Equation 11.44. For the case shown in Figure 11.4, we note that ϕ is near ϕ_r where $\sigma'(\phi)$ is much larger for the Sommer and Mortensen case (see Figure 11.2). Correspondingly, the variation of ϕ across the wet region of the sponge is less than that for the linear case $\sigma(\phi) = \phi_r - \phi$. Finally, we note that when $\rho = 0$ (solid and liquid have matched densities) the equilibrium solid fraction profile is uniform with $\phi(z) = \phi_r$ throughout the wet sponge region.

11.3.3 Nonzero Gravity Effects Solution

We now address the time-dependent nonzero gravity case. Here, we solve the partial differential Equation 11.31 along with Equations 11.35 and 11.36 numerically. We use a method of lines approach in which we rescale the spatial variable onto a fixed domain (via $\bar{z} = (z - h_s)/(h_\ell - h_s)$) and approximate the spatial derivatives using second-order accurate finite difference discretizations. This reduced the PDE (partial differential equation) to a system of ODEs (ordinary differential equations) that were also coupled to the ODEs for the motion of the boundaries h_s and h_ℓ. This system of ODEs was solved with MATLAB® using ode23s. This approach was also used by Siddique et al. (2009) and Siddique and Anderson (2011). This approach uses the similarity solution as an initial condition for the nonzero gravity PDE solution.

Figure 11.5 shows the evolution of the interface positions h_s and h_ℓ as a function of time. Here, we show the dependence of wicking dynamics and solid deformation on the choice of permeability functions used as ordered

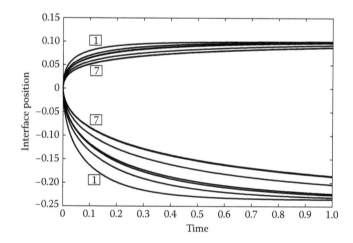

FIGURE 11.5
Evolution of interface positions for different permeability functions. The curves for different permeability functions used in Masoodi and Pillai (2010) and Siddique et al. (2009) are ordered as (1) Blake (1922), Kozeny (1927), and Carman (1937), (2) 'Zunker (1920) as reported by Rumpf and Gupte (1971), (3) Terzaghi (1925), (4) Anderson (2005), Siddique et al. (2009), (5) Rumpf and Gupte (1971), (6) Rose (1945), and (7) Fehling (1939) as reported by Rumpf and Gupte (1971). Note that curves 3 and 4 as well as curves 6 and 7 are nearly indistinguishable on this plot. We have used $\sigma(\phi) = \phi_r - \phi$ with $\phi_0 = 0.33$ $\phi_r = 0.10$, $\phi_\ell^* = 0.20$, and $\rho = 0.1$. The curves are shown in dimensionless form but we note that the length and time scales L and T from the experiments reported in Siddique et al. (2009) are 70 cm and 540 s, respectively. Therefore, the examples in this figure show capillary rise and deformation corresponding to a wet sponge region of approximately 20 cm in roughly 9 min.

in Table 11.1. We have used the linear choice of $\sigma(\phi) = \phi_r - \phi$ (Anderson 2005; Siddique et al. 2009). It is worth mentioning that for all choices of permeability functions, the early time dynamics of solid and liquid interfaces follow the nonzero gravity solution and evolve later to steady-state values of h_s^∞ and h_ℓ^∞. The results for permeability functions 3 and 4 listed in Table 11.1 are graphically very close and indicate a crossover in dynamics from the early time similarity solution to the later time evolution. As explored in detail in Siddique et al. (2009), we also observe that the steady-state positions strongly depend on the choice of parameter values used for ϕ_0, ϕ_r, ϕ_ℓ^*, and ρ.

Figure 11.6 shows the dynamic behavior of $h_s(t)$ and $h_\ell(t)$ for two different cases corresponding to the linear case $\sigma(\phi) = \phi_r - \phi$ (dashed curve) and the degree seven polynomial in Sommer and Mortensen (solid curve) with permeability function given by $k(\phi) = \phi_0/\phi$. The early and long time ϕ profiles for these dynamics correspond to those shown in Figures 11.3 and 11.4. We observe that the dynamics for these two cases are fairly similar and the interface positions approach equilibrium values h_s^∞ and h_ℓ^∞ that are also closely matched between the two cases. These equilibrium positions can vary strongly with the parameter values ϕ_0, ϕ_r, ϕ_ℓ^*, and ρ as shown by Siddique et al. (2009) for the case of linear $\sigma(\phi)$. The dynamical results show that a simple linear approximation for $\sigma(\phi)$ captures the same general trends observed in a nonlinear case.

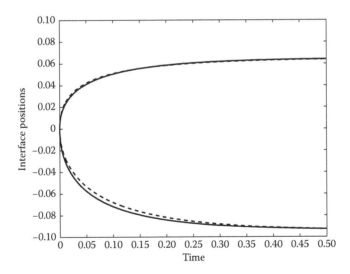

FIGURE 11.6
Dynamic behavior of $h_s(t)$ and $h_l(t)$ for two different cases corresponding to the linear case $\sigma(\phi) = \phi_r - \phi$ (dashed curve) and the degree seven polynomial in Sommer and Mortensen (solid curve). We have used $k(\phi) = \phi_0/\phi$ with $\phi_0 = 0.33$, $\phi_\ell^* = 0.2$, $\phi_r = \phi_r^{SM}$, and $\rho = 0.1$. Here, we again note for reference that the length and time scales in the experiments of Siddique et al. (2009) are $L = 70$ cm and $T = 540$ s. (Data from Siddique, J.I., Anderson, D.M., and Bondarev, A. 2009. *Phys. Fluids* 21: 013106; Sommer, J.L. and Mortensen, A. 1996. *J. Fluid Mech.* 311: 193–217.)

11.3.4 Experimental Work and Comparisons to Mixture Theory Results

The mixture theory model predictions have been compared to experimental data on capillary wicking of water into deformable sponges by Siddique et al. (2009). Their experiments reveal novel dynamic regimes for both capillary rise and solid deformation. Their theory captures the early time dynamics of capillary wicking and solid deformation but fail to capture the long-time dynamics observed experimentally. We review these experiments, make some additional comparisons with the mixture theory model, and suggest some directions for future work.

In the experiments of Siddique et al. (2009), they positioned a dry kitchen sponge above a water bath and raised the water level until the water made contact with the sponge. As the water began rising into the sponge, the wet portion of the sponge began to deform (always swelling in their experiments) so that the bottom of the sponge was below the water level. Figure 11.1 shows an image of one such sponge during imbibition and deformation. Note that the bottom of the sponge was initially at the water level—the depth of the wet sponge below the water level indicates the amount of solid deformation that had occurred up to that particular point in the experiment. For this particular configuration, the primary sponge deformation was one-dimensional. The water bath itself was large enough that there was a negligible change in the bath height during the course of the experiments due to water uptake into the sponge, sponge deformation into the bath, or other effects such as evaporation.

These revealed two dynamic capillary rise regimes. There was an early time regime ($t < 10$ s) in which the capillary rise height and solid deformation both followed a $t^{1/2}$ power law. This early time regime was followed by a long time regime ($t > 30$ min) in which the capillary rise was reported to follow a $t^{0.22}$ power law while the solid deformation continued following a $t^{0.2}$ power law. We have shown these data in Figure 11.7 for the capillary rise height and in Figure 11.8 for the solid deformation. No equilibrium height was reached for these sponges, which were all approximately 9.1 cm in length. These experimental observations are noteworthy when compared with capillary rise experiments conducted by Delker et al. (1996) as well as others by Lago and Araujo (2001) that showed an initial dynamic regime where the capillary rise height matching a $t^{1/2}$ power law was followed by a second regime in which the power law maintained the approximate form $t^{1/4}$. In particular, in both rigid porous materials and deformable porous materials, no equilibrium rise height was observed.

Figure 11.7 shows the experimental data from Siddique et al. (2009) for the wet–dry sponge interface $h_l(t)$ along with theoretical predictions from the mixture theory model using the stress function $\sigma(\phi) = \phi_r - \phi$ and three different permeability functions. The highest, middle, and lowest dashed–dotted curves h_ℓ and $-h_s$ correspond to permeability functions 1, 3, and 7 in Table 11.1, respectively. The other permeability functions examined earlier,

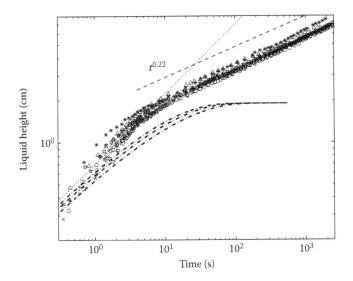

FIGURE 11.7
Experimental data from Siddique et al. (2009) for the wet–dry sponge interface $h_i(t)$ using the various symbols that were also used in that reference (see their Figure 13). Additionally, the dotted line shows a zero gravity similarity solution with $h_\ell \sim t^{1/2}$, the dashed line indicates the slope of a power law $t^{0.22}$, and the three dash–dotted lines correspond to new theoretical predictions using $\sigma(\phi) = \phi_r - \phi$ and three different permeability functions $k(\phi)$ as explained in the text. (Data from Siddique, J.I., Anderson, D.M., and Bondarev, A. 2009. *Phys. Fluids* 21: 013106.)

including the $k(\phi) = \phi_0/\phi$ as in Siddique et al., fall in the same range as the others and have not been included in the graph for clarity. For comparison, we have used their reported parameter values of $\phi_r = 0.073$, $\phi_\ell^* = \phi_0 = 0.1$, and $\rho = 0$ and have followed the same fitting procedure used there to obtain estimates for $D \equiv K_0 \Sigma_0/\mu$ and L (note that D is related to the time scale by $T = L^2/D$). In particular, their value of ϕ_r was found by comparing the volume of water displaced by a submerged wet sponge in its relaxed state with no trapped air (i.e., the total amount of solid volume of the sponge) to the volume occupied by the wet sponge (parallelepiped) in its relaxed state. Other direct measurements on wet and dry sponges were used to estimate the value of ϕ_0. Siddique et al. (2009) also made the assumption that ϕ_ℓ^* was equal to ϕ_0. They observed that $\rho > 0$ (the solid was more dense than water) but also that a nonzero value of this parameter did not significantly influence the predictions and therefore used $\rho = 0$ for simplicity. Also, D was set by matching the zero gravity similarity solution $h_\ell = 2\lambda_\ell \sqrt{Dt}$ to the experimental data $h_\ell = q_\ell \sqrt{t}$ at early times where q_ℓ was estimated at 0.7 cm^2 s^{-1} and λ_ℓ was computed based on the similarity solution for $k(\phi) = \phi_0/\phi$. We chose this value once based on the permeability function $k(\phi) = \phi_0/\phi$ and fixed it rather than changing it for each different permeability function used.

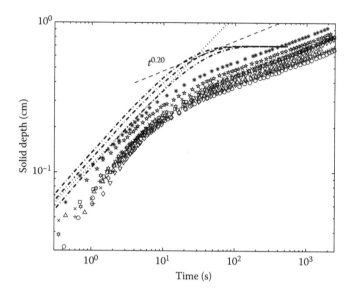

FIGURE 11.8
Experimental data from Siddique et al. (2009) for the deformed solid depth $-h_s(t)$ under the water level using the various symbols that were also used in that reference (see their Figure 14). The dotted line shows a zero gravity similarity solution with $-h_s \sim t^{1/2}$, the dashed line indicates the slope of a power law $t^{0.20}$, and the three dash–dotted lines correspond to new theoretical predictions using $\sigma(\phi) = \phi_r - \phi$ and three different permeability functions $k(\phi)$ as explained in the text. (Data from Siddique, J.I., Anderson, D.M., and Bondarev, A. 2009. *Phys. Fluids* 21: 013106.)

The length scale L ($= 70$ cm) was chosen so that the equilibrium height prediction approximately matched the height of the transition between the two dynamic regimes in the experimental data.

Once the parameters were set based on observations of the capillary rise height, the predictions for solid deformation could be made without any further parameter fitting. The experimental results for the solid deformation of Siddique et al. (2009) and the corresponding theoretical predictions are shown in Figure 11.8. As observed by Siddique et al. the early time comparison between theory and experiment should be interpreted with the knowledge that no additional parameters were fit for the solid deformation. The long time behavior observed in the experiments is not captured by the model.

The calculations shown here indicate that while the chosen form for the permeability function $k(\phi)$ can influence details of the dynamics, the main features such as the initial power law $t^{1/2}$ are preserved. We have already observed that both a linear and a nonlinear choice for $\sigma(\phi)$ lead to qualitatively similar predictions on the evolution. The predicted equilibrium positions are influenced by the choice of $\sigma(\phi)$ but are independent of permeability function $k(\phi)$. All of these observations point to the need for additional mechanisms to describe the observed long-time power law behavior observed in the experiments.

In the case of a rigid porous material, Lockington and Parlange (2004) have proposed a modification of the classical Washburn model that takes into account saturation gradients at the advancing capillary rise front. Their work showed that the inclusion of those effects could adequately capture the long-time regime power law. The work of Lockington and Parlange on rigid porous materials suggests that the inclusion of saturation gradients in a mixture theory model for capillary rise and material deformation could potentially allow one to capture the long-time dynamics observed in the experiments of Siddique et al. (2009).

11.3.5 Further Discussion of Capillary Wicking Results

Capillary wicking and solid deformation with finite volumes has been examined by Anderson (2005) and Siddique et al. (2009). Anderson (2005) showed that for one-dimensional imbibition of a finite volume of fluid into a deformable porous solid, the initial wicking and solid deformation stage was followed by a second "relaxation" stage in which the entire volume of fluid had been completely taken up by the solid but that additional solid deformation could occur. Four notable regimes were identified: (1) solid swelling occurred during the wicking stage and the final relaxed state also corresponded to an expanded solid, (2) solid swelling occurred during the wicking stage but the final relaxed state had no net swelling of the solid (i.e., the solid relaxed back to an undeformed configuration), (3) solid swelling occurred during the wicking stage but the final relaxed state corresponded to a net compression (i.e., the wet solid shrunk), and (4) solid shrinking occurred during the wicking stage and the final relaxed state of the solid also corresponded to a net compression relative to the initial state. See Figure 4 in Anderson (2005) for additional details.

Siddique et al. (2009) also considered the capillary wicking and gravitational drainage of a finite volume of fluid into a deformable porous material. Their main results involved a prediction for how quickly the entire volume of fluid could be wicked into the solid material. Their observations indicated that the time for complete uptake of the fluid into the deformable solid was reduced when gravity acted together with capillary suction. A small influence on the solid–liquid density ratio was also reported (see Siddique et al., Figure 10).

Siddique and Anderson (2011) have recently extended the analysis of Siddique et al. (2009) to address the influence of a non-Newtonian (power law) fluid on the capillary rise and solid deformation problem. A key observation for the non-Newtonian wicking problem was that the equilibrium solutions were the same as for Newtonian fluids. The dynamics for the non-Newtonian case, however, could vary dramatically from the Newtonian case. One of the most directly quantifiable differences between the Newtonian and non-Newtonian (power law) was that for the case of wicking and deformation in the absence of gravity the interface positions

h_s and h_ℓ follow a $t^{1/2}$ power law for the Newtonian case while they follow a modified power law $t^{n/(n+1)}$ where n is the power law exponent characterizing shear thinning fluids ($n < 1$) and shear thickening fluids ($n > 1$). Preliminary experiments reported in their work indicate a clear dependence of the time scale on the fluid rheology. See Siddique and Anderson (2011) for further details.

11.4 Summary

We have presented a mixture theory-based model of capillary wicking and the resultant solid deformation in porous materials. Theoretical results highlight the dependence of the wicking and deformation dynamics on models for the solid stress and permeability as well as explore special solutions related to early time similarity behavior and steady states. We have reviewed a recent experimental work on one-dimensional capillary wicking and deformation in sponges, which help to reveal both successes and shortcomings of the theoretical predictions. Through further experiments and theoretical investigations of both simple and complex models, new phenomena related to mixture theory-based wicking models in porous materials can be explored.

Acknowledgments

This material is based upon work supported by the National Science Foundation under Grant No. 0709095 (DMA and JIS) and Grant No. 1107848 (DMA). Dr. Siddique would like to acknowledge the support from Penn State York advisory board grant. The authors would also like to thank Andrei Bondarev and John Pelesko who helped provide experimental data that inspired the modeling work presented here.

Nomenclature

$\mathbf{b}_\ell$	Liquid body force
$\mathbf{b}_s$	Solid body force
$\mathbf{g}$	$= -g\hat{\mathbf{k}}$, gravitational acceleration vector
g	Gravitational acceleration scalar

h_ℓ Vertical position of the wet sponge/dry sponge interface
h_s Vertical position of the wet sponge/fluid bath interface
h_ℓ^∞ Vertical position of the wet sponge/dry sponge interface in equilibrium
h_s^∞ Vertical position of the wet sponge/fluid bath interface in equilibrium
K $= \mu\,(1 - \phi)^2/k$, drag coefficient
$\hat{\mathbf{k}}$ Unit vector in vertical direction
$k(\phi)$ Permeability of the solid matrix
K_0 Representative permeability scale
L $= \Sigma_0/(\rho_\ell^T g)$, length scale
p Pressure
p_A Atmospheric pressure
p_c Capillary pressure
t Time
T $= L^2\mu/(K_0\Sigma_0)$, time scale
$\mathbf{T}_\ell$ Partial stress tensor of the liquid phase
$\mathbf{T}_m$ Stress tensor of the mixture
$\mathbf{T}_s$ Partial stress tensor of the solid phase
v_ℓ Liquid velocity
v_m Mixture velocity
v_s Solid velocity
w_ℓ Vertical velocity component of the liquid
w_s Vertical velocity component of the solid
z Vertical coordinate

Greek Letters

η $= z/(2\sqrt{t})$, similarity variable
λ_ℓ Wet sponge/dry sponge interface position for similarity solution in Equation 11.40
λ_s Wet sponge/fluid bath interface position for similarity solution in Equation 11.40
μ Dynamic viscosity of the liquid
π_ℓ Liquid friction/interaction force
π_s Solid friction/interaction force
ρ $= \rho_s^T/\rho_\ell^T - 1$, density ratio parameter
ρ_ℓ Apparent liquid density in the mixture
ρ_m Density of the mixture
ρ_s Apparent solid density in the mixture
ρ_ℓ^T True density of bulk liquid
ρ_s^T True density of bulk solid
$\sigma(\phi)$ Scalar solid stress function
σ_ℓ Deviatoric part of the liquid stress tensor
σ_m Excess stress tensor of the mixture

σ_s Deviatoric part of the solid stress tensor
Σ_0 Representative solid stress and pressure scale
ϕ Solid fraction of the mixture
ϕ_0 Initial solid fraction
π_ℓ Solid fraction at $z = h_\ell$
ϕ_r Solid fraction for relaxed, zero stress configuration
ϕ_ℓ^* $\equiv \phi_r - p_c/\Sigma_0$, solid fraction at $z = h_\ell$ for similarity solution
ϕ_ℓ^∞ Solid fraction at $z = h_\ell$ in equilibrium
ϕ_r^{SM} Solid fraction for relaxed, zero stress configuration: Sommer and Mortensen data
$\mathcal{X}$ Liquid fraction of the mixture

References

Achanta, S., Cushman, J.H., and Okos, M.R. 1994. On multicomponent, multiphase thermomechanics with interfaces. *Int. J. Eng. Sci.* 32: 1717–1738.

Ambrosi, D. and Preziosi, L. 2000. Modeling injection molding processes with deformable porous preforms. *SIAM J. Appl. Math.* 61: 22–42.

Anderson, D.M. 2005. Imbibition of a liquid droplet on a deformable porous substrate. *Phys. Fluids* 17: 087104.

Ateshian, G.A., Chahine, N.O., Basalo, I.M., and Hung, C.T. 2004. The correspondence between equilibrium biphasic and triphasic material properties in mixture models of articular cartilage. *J. Biomech.* 37: 391–400.

Atkin, R.J. and Craine, R.E. 1976. Continuum theories of mixture: Basic theory and historical development. *Quart. J. Mech. Appl. Math.* 29: 209–244.

Barry, S.I. and Aldis, G.K. 1990. Comparison of models for flow induced deformation of soft biological tissue. *J. Biomech.* 23: 647–654.

Barry, S.I. and Aldis, G.K. 1991. Unsteady flow induced deformation of porous materials. *Int. J. Non-Linear Mech.* 26: 687–699.

Barry, S.I. and Aldis, G.K. 1992. Flow induced deformation from pressurized cavities in absorbing porous tissues. *Bull. Math. Biol.* 54: 977–997.

Barry, S.I. and Aldis, G.K. 1993. Radial flow through deformable porous shells. *J. Austral. Math. Soc. Ser. B* 34: 333–354.

Barry, S.I., Parker, K.H., and Aldis, G.K. 1991. Fluid flow over a thin deformable porous layer. *J. Appl. Math. Phys. (ZAMP)* 42: 633–648.

Bear, J. 1972. *Dynamics of Fluids in Porous Media.* New York: Dover.

Beavers, G.S., Hajji, A., and Sparrow, E.M. 1981a. Fluid flow through a class of highly-deformable porous media Part I: Experiments with air. *J. Fluids Eng.* 103: 432–439.

Beavers, G.S., Wilson, T.A., and Masha, B.A. 1975. Flow through a deformable porous material. *J. Appl. Mech.* 97: 598–602.

Beavers, G.S., Wittenberg, K. and Sparrow, E.M. 1981b. Fluid flow through a class of highly-deformable porous media: Part II: Experiments with water. *J. Fluids Eng.* 103: 440–444.

Bennethum, L.S. 2007. Theory of flow and deformation of swelling porous materials at the macroscale. *Comput. Geotech.* 34: 267–278.

Billi, L. and Farina, A. 2000. Unidirectional infiltration in deformable porous media: Mathematical modeling and self-similar solution. *Q. Appl. Math.* 58: 85–101.

Biot, M.A. 1941a. General theory of three-dimensional consolidation. *J. Appl. Phys.* 12: 155–164.

Biot, M.A. 1941b. Consolidation settlement under a rectangular load distribution. *J. Appl. Phys.* 12: 426–430.

Biot, M.A. 1955. Theory of elasticity and consolidation for a porous anisotropic solid. *J. Appl. Phys.* 26: 182–185.

Biot, M.A. 1956. Theory of deformation of a porous viscoelastic anisotropic solid. *J. Appl. Phys.* 27: 459–467.

Biot, M.A. and Clingan, F.M. 1941. Consolidation settlement of a soil with an impervious top surface. *J. Appl. Phys.* 12: 578–581.

Blake, F.C. 1922. The resistance of packing to fluid flow. *Transactions American Institute of Chemical Engineers*, Vol. 14, pp. 415–422.

Bowen, R.M. 1980. Incompressible porous media models by use of the theory of mixtures. *Int. J. Eng. Sci.* 18: 1129–1148.

Bowen, R.M. 1982. Compressible porous media models by use of the theory of mixtures. *Int. J. Eng. Sci.* 20: 697–735.

Bowen, R.M. and Wiese, J.C. 1969. Diffusion in mixtures of elastic materials. *Int. J. Eng. Sci.* 7: 689–722.

Carmen, P.C. 1937. Fluid flow through granular beds. *Transactions of the Institute of Chemical Engineers*, Vol. 15, pp. 150–166.

Chau, P.C. and Pagano, N.J. 1967 *Elasticity: Tensor, Dyadic and Engineering Approaches.* New York: Dover.

Chen, K.S.A. and Scriven, L.E. 1990. Liquid penetration into a deformable porous substrate. *J. Tech. Assoc. Pulp Paper Ind.* 73: 151–161.

Clarke, A., Blake, T.D., Carruthers, K., and Woodward, A. 2002. Spreading and imbibition of liquid droplets on porous surfaces. *Langmuir* 18: 2980–2984.

Coleman, B.D. and Noll, W. 1963. The thermodynamics of elastic materials with heat conduction and viscosity. *Arch. Rat. Mech. Anal.* 13: 167–178.

Coléou, C., Xu, K., Lesaffre, B., and Brzoska, J.-B. 1999. Capillary rise in snow. *Hydrol. Process.* 13: 1721–1732.

Damiano, E.R., Duling, B.R., Ley, K., and Skalak, T.C. 1996. Axisymmetric pressure-driven flow of rigid pellets through a cylindrical tube lined with a deformable porous wall layer. *J. Fluid Mech.* 314: 163–189.

Davis, S.H. and Hocking, L.M. 1999. Spreading and imbibition of viscous liquid on a porous base. *Phys. Fluids* 11: 48–57.

Davis, S.H. and Hocking, L.M. 2000. Spreading and imbibition of viscous liquid on a porous base. II. *Phys. Fluids* 12: 1646–1655.

de Boer, R. 2000. *Theory of Porous Media: Highlights in the Historical Development and Current State.* New York: Springer.

de Groot, S.R. and Mazur, P. 1984. *Non-Equilibrium Thermodynamics.* New York: Dover.

Delker, T., Pengra, D.B. and Wong, P. 1996. Interface pinning and the dynamics of capillary rise in porous media. *Phys. Rev. Lett.* 76: 2902–2905.

Diersch, H.-J. G., Clausnitzer, V., Myrnyy, V., Rosati, R., Schmidt, M., Beruda, H., Ehrnsperger, B.J., and Virgilio, R. 2010a. Modeling unsaturated flow in absorbent swelling porous media: Part 1. Theory. *Transp. Porous Media* 83: 437–464.

Diersch, H.-J. G., Clausnitzer, V., Myrnyy, V., Rosati, R., Schmidt, M., Beruda, H., Ehrnsperger, B.J., and Virgilio, R. 2010b. Modeling unsaturated flow in absorbent swelling porous media: Part 2. Numerical simulation. *Trans. Porous Media* DOI 10.1007/s11242-010-9650-4.

D'Onofrio, T.G., Navaz, H.K. and Markicevic, B. 2010. Experimental and numerical study of spread and sorption of VX sessile droplets into medium grain-size sand. *Langmuir* 26: 3317–3322.

Dullien, F.A.L. 1992. *Porous Media: Fluid Transport and Pore Structure*. San Diego: Academic Press.

Fitt, A.D., Howell, P.D., King, J.R., Please, C.P., and Schwendeman, D.W. 2002. Multiphase flow in a roll press nip. *Eur. J. Appl. Math.* 13: 225–259.

Fries, N. and Dreyer, M. 2008. An analytic solution of capillary rise restrained by gravity. *J. Coll. Int. Sci.* 320: 259–263.

Green, A.E. and Naghdi, P.M. 1967. A theory of mixtures. *Arch. Rat. Mech. Anal.* 24: 243–263.

Gu, W.Y, Lai, W.M., and Mow, V.C. 1993. Transport of fluid and ions through a porous-permeable charged-hydrated tissue, and streaming potential data on normal bovine articular cartilage. *J. Biomechanics* 26: 709–723.

Hilpert, M. 2009a. Effects of dynamic contact angle on liquid infiltration into horizontal capillary tubes: (Semi)-analytical solutions. *J. Coll. Int. Sci.* 337: 131–137.

Hilpert, M. 2009b. Effects of dynamic contact angle on liquid infiltration into inclined capillary tubes: (Semi)-analytical solutions. *J. Coll. Int. Sci.* 337: 138–144.

Hilpert, M. 2010a. Explicit analytical solutions for liquid infiltration into capillary tubes: Dynamic and constant contact angle. *J. Coll. Int. Sci.* 344: 198–208.

Hilpert, M. 2010b. Effects of dynamic contact angle on liquid withdrawal from capillary tubes: (Semi)-analytical solutions. *J. Coll. Int. Sci.* 347: 315–323.

Hilpert, M. 2010c. Liquid withdrawal from capillary tubes: Explicit and implicit analytical solution for constant and dynamic contact angle. *J. Coll. Int. Sci.* 351: 267–276.

Hilpert, M. and Ben-David, A. 2009. Infiltration of liquid droplets into porous media: Effects of dynamic contact angle and contact angle hysteresis. *Int. J. Multiphase Flow* 35: 205–218.

Holman, R.K., Cima, M.J., Uhland, S.A., and Sachs, E. 2002. Spreading and infiltration of inkjet-printed polymer solution droplets on a porous substrate. *J. Coll. Int. Sci.* 249: 432–440.

Holmes, M.H. 1983. A nonlinear diffusion equation arising in the study of soft tissue. *Quart. App. Math.* 41: 209–220.

Holmes, M.H. 1984. Comparison theorems and similarity solution approximations for a nonlinear diffusion equation arising in the study of soft tissue. *SIAM J. Appl. Math.* 44: 545–556.

Holmes, M.H. 1985. A theoretical analysis for determining the nonlinear hydraulic permeability of a soft tissue from a permeation experiment. *Bull. Math. Biol.* 47: 669–683.

Holmes, M.H. 1986. Finite deformation of soft tissue: analysis of a mixture model in uni-axial compression. *J. Biomech. Eng.* 108: 372–381.

Holmes, M.H. and Mow, V.C. 1990. The nonlinear characteristics of soft gels and hydrated connective tissue in ultrafiltration. *J. Biomech.* 23: 1145–1156.

Hou, J.S., Holmes, M.H., Lai, W.M., and Mow, V.C. 1989. Boundary conditions at the cartilage-synovial fluid interface for joint lubrication and theoretical verifications. *J. Biomech. Eng.* 111: 78–87.

Humphrey, J.D. 2003. Continuum biomechanics of soft biological tissues. *Proc. R. Soc. Lond. A* 459: 3–46.

Jordan, R.E., Hardy, J.P., Perron, Jr., F.E., and Fisk, D.J. 1999. Air permeability and capillary rise as measures of the pore structure of snow: An experimental and theoretical study. *Hydrol. Process.* 13: 1733–1753.

Kim, C., Liu, Y., Kühnle, A., Hess, S., Viereck, S., Danner, T., Mahadevan, L., and Weitz, D.A. 2007. Gravational stability of suspensions of attractive colloidal particles. *Phys. Rev. Lett.* 99: 028303.

Kozeny, J. 1927. Capillary motion of water in soils. Sitzungsberichte der Akademie der Wissenschaften Wien, Vol. 136, pp. 271–306.

Kwan, M.K., Lai, W.M., and Mow, V.C. 1990. A finite deformation theory for cartilage and other soft hydrated connective tissue–I. Equilibrium results. *J. Biomechanics* 23: 145–155.

Lago, M. and Araujo, M. Capillary rise in porous media. *J. Coll. Int. Sci.* 234: 35–43. [Same article also published in 2001 *Physica A* 289: 1–17.]

Lai, W.M., Hou, J.S., and Mow, V.C. 1991. A triphasic theory for the swelling and deformation behaviors of articular cartilage. *J. Biomech. Eng.* 113: 245–258.

Lai, W.M. and Mow, V.C. 1980. Drag-induced compression of articular cartilage during a permeation experiment. *Biorheology* 17: 111–123.

Lanir, Y., Sauob, S. and Maretsky, P. 1990. Nonlinear finite deformation response of open cell polyurethane sponge to fluid filtration. *J. Appl. Mech.* 57: 449–454.

Lockington, D.A. and Parlange, J.-Y. 2004. A new equation for macroscopic description of capillary rise in porous media. *J. Coll. Int. Sci.* 278: 404–409.

Manley, S., Skotheim, J.M., Mahadevan, L., and Weitz, D.A. 2005. Gravational collapse of colloidal gels. *Phys. Rev. Lett.* 94: 218302.

Markicevic, B. and Navaz, H.K. 2009. Numerical solution of wetting fluid spread into porous media. *Int. J. Num. Methods Heat Fluid Flow* 19: 521–534.

Markicevic, B. and Navaz, H.K. 2010. Primary and Secondary Infiltration of Wetting Liquid Sessile Droplet into Porous Medium. *Transp. Porous Media* 85: 953–974.

Markicevic, B., D'Onofrio, T.G., and Navaz, H.K. 2010. On spread extent of sessile droplet into porous medium: Numerical solution and comparisons with experiments. *Phys. Fluids* 22: 012103.

Markicevic, B., Li, H., Sikorski, Y., Zand, A.R., Sanders, M., and Navaz, H.K. 2009. Infiltration time and imprint shape of a sessile droplet imbibing porous medium. *J. Coll. Int. Sci.* 336: 698–706.

Masoodi, R. and Pillai, K.M. 2010. Darcy's law-based model for wicking in paper-like swelling porous media. *AIChE J.* 56: 2257–2267.

Masoodi, R., Pillai, K.M., and Varanasi, P.P. 2007. Darcy's law-based models for liquid absorption in polymer wicks. *AIChE J.* 53: 2769–2782.

Masoodi, R., Pillai, K.M., and Varanasi, P.P. 2010. Effect of externally applied liquid pressure on wicking in paper wipes. *J. Engineered Fibers Fabrics* 5: 49–66.

Masoodi, R., Tan, H., and Pillai, K.M. 2011. Darcy's law-based numerical simulation for modeling 3D liquid absorption into porous wicks. *AIChE J.* 57(5): 1132–1143.

Michaud, V.J., Sommer, J.L., and Mortensen, A. 1999. Infiltration of fibrous preforms by a pure metal: Part V. Influence of preform compressibility. *Metall. Mat. Transp. A* 30: 471–482.

Monaenkova, D. and Kornev, K.G. 2010. Elastocapillarity: Stress transfer through fibrous probes in wicking experiments. *J. Coll. Int. Sci.* 348: 240–249.

Mow, V.C., Holmes, M.H., and Lai, W.M. 1984. Fluid transport and mechanical properties of articular cartilage: A review. *J. Biomechanics* 17: 377–394.

Mow, V.C., Kuei, S.C., Lai, W.M., and Armstrong, C.G. 1980. Biphasic creep and stress relaxation of articular cartilage: Theory and experiments. *J. Biomech. Engng. Trans. ASME* 102: 73–84.

Murad, M.A., Bennethum, L.S., and Cushman, J.H. 1995. A multi-scale theory of swelling porous media: I. Application to one-dimensional consolidation. *Transp. Porous Media* 19: 93–122.

Myers, T.G., Aldis, G.K., and Naili, S. 1995. Ion induced deformation of soft tissue. *Bull. Math. Biol.* 57: 77–98.

Navaz, H.K., Markicevic, B., Zand, A.R., Sikorski, Y., Chan, E., Sanders, M., and D'Onofrio, T.G. 2008. Sessile droplet spread into porous substrates— determination of capillary pressure using a continuum approach. *J. Coll. Int. Sci.* 325: 440–446.

Parker, K.H., Mehta, R.V., and Caro, C.G. 1987. Steady flow in porous, elastically deformable materials. *J. Appl. Mech.* 54: 794–800.

Pavlovets, V.M. 2010. Capillary wicking of samples obtained by spraying wet iron ore to the pelletizer coating. *Steel Translation* 40: 699–702.

Pezron, I., Bourgain, G., and Quéré, D. 1995. Imbition of a Fabric. *J. Coll. Int. Sci.* 173: 319–327.

Preziosi, L., Joseph, D.D. and Beavers, G.S. 1996. Infiltration of initially dry, deformable porous media. *Int. J. Multiphase Flow* 22: 1205–1222.

Rose, H.E. 1945. An investigation into the laws of flow of fluids through beds of granular materials. In *Proceedings of the Institute of Mechanical Engineers*, Vol. 153. pp. 141–148.

Rumpf, H. and Gupte, A.R. 1971. Influence of porosity and particle size distribution in resistance law of porous flow. *Chem. Ing. Tech.* 43: 367–375.

Sami Selim, M., Tesavage, V.F., Chebbi, R., and Sung, S.H. 1997. Drying of Water-Based Inks on Plain Paper. *J. Imaging Science and Tech.* 41: 152–158.

Schuchardt, D.R. and Berg, J.C. 1990. Liquid transport in composite cellulose-superabsorbent fiber network. *Wood Fiber Sci.* 23: 342–357.

Siddique, J.I. and Anderson, D.M. 2011. Capillary rise of a non-Newtonian liquid into a deformable porous material. *J. Porous Media* 14: 1087–1102.

Siddique, J.I., Anderson, D.M. and Bondarev, A. 2009. Capillary rise of a liquid into a deformable porous material. *Phys. Fluids* 21: 013106.

Sommer, J.L. and Mortensen, A. 1996. Forced unidirectional infiltration of deformable porous media. *J. Fluid Mech.* 311: 193–217.

Szekely, J., Neumann, A.W. and Chuang, Y.K. 1971. The rate of capillary penetration and the applicability of the Washburn equation. *J. Coll. Int. Sci.* 35: 273–278.

Terzaghi, C. 1925. Principles of soil mechanics: III--Determination of permeability of clay. *Engineering News-Record* 95: 832–836.

Wang, W. and Parker, K.H. 1995. The effect of deformable porous surface layers on the motion of a sphere in a narrow cylindrical tube. *J. Fluid Mech.* 283: 287–305.

Washburn, E.W. 1921. The dynamics of capillary flow. *Phys. Rev.* 17: 273–283.

12

Simulating Fluid Wicking into Porous Media with the Lattice Boltzmann Method

Marcel G. Schaap and Mark L. Porter

CONTENTS

Lattice Boltzmann (LB) methods are relatively new, yet simple methods for simulating single and multiphase fluid behavior and are particularly suitable for porous media with irregular geometries. In this chapter, we briefly discuss the fundamentals of the single-phase LB method and its extensions to multiphase fluid systems. In particular, we discuss the merits of an earlier Shan–Chen-type model that was successfully used by us to simulate liquid water retention and wicking in a glass bead system. Although this model can reach excellent results, a ghost phase must be used in this model to generate phase separation. Additional drawbacks are that many simplifying assumptions must be made with regard to the domination of capillary forces over inertial, viscous, and gravitational forces and that the model type does not have an explicitly defined temperature. To bring more physical realism into the LB modeling of multiphase system we use work by Yuan and Schaefer who introduced van der Waals-type equations of state into the Shan–Chen model. The great advantage of such models is that temperature is now explicitly defined. To make this method applicable to porous systems we introduce a fluid–solid interaction term which allows us to control for contact angle. We show that a wide range of conditions can be simulated and demonstrate that liquid and vapor densities and surface tension qualitatively (but not quantitatively) agree with published experimental data for the water–vapor system. Future modifications to the parameterization of the equation of state should lead to a better quantitative match for the water–vapor and other substances of environmental interest. Finally, we simulate wicking into a two-dimensional synthetic porous medium under different gravitational conditions and show that the saturation dynamics qualitatively agree with published analytical solutions of macroscopic infiltration into porous media.

12.1 Introduction

Over the past two decades, LB methods have been used to simulate fluid dynamics in a wide variety of applications, such as laminar and turbulent flow in simple and complex geometries, multiphase separation and flow, as well as simulations of advection and dispersion (Succi, 2001). LB simulations have also successfully been applied to porous media, such as reported in Ferréol and Rothman (1994), Heijs and Lowe (1995), O'Connor and Fredrich (1999), Zhang et al. (2000), and Pan et al. (2001), among others. The LB method essentially incorporates microscopic processes that recover Navier–Stokes fluid behavior at a larger scale. While more conventional finite difference, or finite element schemes exist for solving the Navier–Stokes equation directly (e.g., Quiblier, 1984; Adler, 1990; Masad et al., 2000) a clear advantage of the LB method is its simplicity and its ability to deal with arbitrary geometries. In addition, LB methods can be coded very efficiently and run on parallel computers, reducing cost and computation time.

One of the most powerful advantages of the LB method is that it can be extended to multiphase and multicomponent fluid systems (Gunstensen and Rothman, 1991; Shan and Chen, 1993, 1994; Swift et al., 1995, 1996; Luo, 1998; He et al., 1998; Zhang et al., 2004; Zhang and Kwok, 2009). On the one hand, the LB method is especially attractive for these types of flows because the particle kinetics allows for a relatively easy way to deal with the moving and deforming interfaces. Other approaches such as traditional computational fluid dynamics and level set methods require special treatment at the interfaces. On the other, one of the major drawbacks to the LB method for these types of flows is that they have largely been limited to relatively small density and viscosity ratios. Thus, either special treatment with complex models is required to address these issues (e.g., Inamuro et al., 2004a,b; Zheng et al., 2006) or a range of assumptions is needed to represent real-world physical systems (Schaap et al., 2007; Porter et al., 2009)—which typically limits the applicability of the LB method. In an effort to preserve the simplicity of the LB method for multiphase flows, Yuan and Schaefer (2006) investigated the use of van der Waals-type equations of state (EOS). They found that density ratios greater than 1000 were easily achieved with these types of EOSs. Thus, after a brief review of the LB method and our earlier work on multiphase interactions in porous media we will focus on LB simulations for wicking in porous media using a van der Waals-type EOS.

12.2 LB Simulations

12.2.1 A Brief History

The LB method is a systematic simplification of the kinetic theory forwarded by Ludwig Boltzmann in 1872. Being part of statistical mechanics, kinetic theory deals with discrete particle systems and how these evolve toward thermodynamic equilibrium (Succi, 2001). Although mathematically exact, kinetic theory is normally computationally intractable for any macroscopically relevant size. The desire for computational effectiveness first led to Lattice Gas Automata (LGA) in which particle collections are constrained to regular grids with prescribed directions and particle–particle interactions. LGA methods were a major step forward in the sense that it became computationally feasible to macroscopically recover the Navier–Stokes equation from microscopic particle interactions. The possibility to carry out LGA simulations in media with arbitrary geometries, such as porous materials, was a particularly attractive aspect of this method. Another advantage is that LGA simulators can be programmed using only integer operations, making the method relatively fast. One of the problems with LGA, however, were the "noisy" solutions resulting from the fact that the method works with discrete particles moving over a regular lattice grid. To reduce the statistical noise it is common to average several simulations for the same media.

Further theoretical work led to the lattice Boltzmann method (LBM) in which the interaction of discrete particles is replaced with interactions of continuous particle distributions. Although LBM is computationally much more demanding than LGA, it does not suffer from statistical noise effects and does not require averaging over many replications. Developments in the early 1990s led to the lattice Bhatnagar–Gross–Krook (LBGK) models with single relaxation times that increased the simplicity even further while retaining the ability to closely approximate the Navier–Stokes equation in one to four dimensions (e.g., Qian et al., 1992). For a much more complete review of the history and theory backgrounds of LGA and LBM and its various implementations and applications the reader is referred to Chen and Doolen (1998) and Succi (2001).

12.2.2 The Single-Phase BGK LB Model

LB models are generally identified by the number of dimensions and the number of discrete directions in which neighboring lattice nodes exchange mass. Two-, three-, or four-dimensional models include D2Q9, D3Q15, D3Q19, and D4Q25 (Qian et al., 1992), all having sufficiently symmetrical geometries (cf. Succi, 2001). For example, D3Q19 is a three-dimensional model with mass (or particle) distribution in 19 directions, or links (Figure 12.1), which are classified by three types. There is one type I link that does not exchange mass with neighbors, six Type II links that exchange mass with the nearest neighbors and 12 Type III links that interact with the next-nearest neighbors. Unlike "real world" dynamical systems with particle collections that can have arbitrary locations, speeds and direction of travel, mass residing on a LBM grid is highly constrained with respect to position (only at lattice

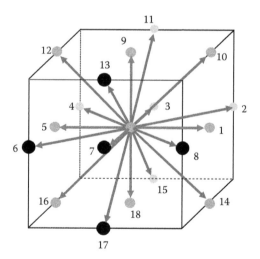

FIGURE 12.1
The D3Q19 coordination scheme indicating the interactions of the central node with 18 of its neighbors. Foreground nodes are black, whereas nodes further in the background are depicted in lighter shades of gray. Two types of links are visible in this graph: six orthogonal arrows (Type II link) point to the nearest neighbors, while 12 diagonal arrows (Type III link) point to the next-nearest neigbors. The Type I link (link 0) is not visible as it is constrained to the central node itself and does not interact with any neighbors. Note that the D3Q19 scheme does not permit interaction with the eight next–next-nearest neighbors located at the corners of the cube. The 2D D2Q9 scheme simply consists of nodes 0 through 8, that is, the midplane of the D3Q19 configuration.

nodes), direction of travel (19 directions in D3Q19) and speeds (0 for Type I, 1 for Type II and √2 for type III links). However, the amount of mass associated with a particular link as well as the macroscopic velocity associated with a node is *not* discrete, but continuous.

To evolve a LB system in time, three operations are applied repeatedly:

- Movement of particles to adjacent nodes, also called the streaming operator. This operator embodies the time step of the LB scheme.
- Calculation of new macroscopic mass and velocities for each node, using the newly acquired particle distribution.
- Reordering of the particle distribution for each node with the collision operator, using the new mass and velocity values.

The streaming operator takes care of the transport of mass from neighboring nodes, as shown in Figure 12.2 for the two-dimensional D2Q9 model. After streaming the particle distribution that was present at the center node (left) is now located at the eight surrounding nodes (right). Conversely, the members of the particle distribution of the surrounding nodes that pointed in the direction of the center node before the time step (left) are now located at the center node. Note that only one member of the particle distribution is shown for the surrounding nodes, in reality they go through the exact same redistribution process with their respective neighbors.

The streaming operator thus causes the particle distributions of each node to be replaced with particle distributions from surrounding nodes. In mathematical terms this is defined as

$$f_i(\mathbf{x} + \mathbf{e}_i, t + 1) = f_i(\mathbf{x}, t) \tag{12.1}$$

where f_i is the particle distribution with i as an index for the 19 directions; $\mathbf{x}$ is the location vector of each node (i.e., x, y, z coordinates), and t is the discrete

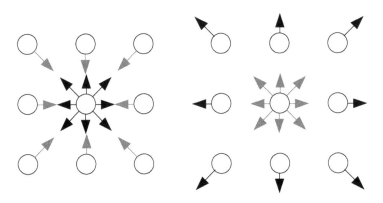

FIGURE 12.2
Streaming (mass transport) for the D2Q9 model. See text for explanation.

TABLE 12.1

Coordinate System and Link Classification for the D2Q9 (Gray) and D3Q19 Models. The Vector $\mathbf{e}_i$ Is Derived from the x, y, and z Rows for Each Link i.

Link	0	1	2	3	4	5	6	7	8	9	10	11	12	13	14	15	16	17	18
type	I	II	III	II	III	II	III	II	III	II	III	III	III	III	III	III	III	III	II
x	0	1	1	0	−1	−1	−1	0	1	0	1	0	−1	0	1	0	−1	0	0
y	0	0	1	1	1	0	−1	−1	−1	0	0	1	0	−1	0	1	0	−1	0
z	0	0	0	0	0	0	0	0	0	1	1	1	1	1	−1	−1	−1	−1	−1

time. The vector $\mathbf{e}_i$ is a coordinate system to indicate the relative positions of the neighboring nodes as defined by the x, y, and z rows in Table 12.1.

After the streaming operation is performed for all nodes in the simulation domain the macroscopic quantities density, $\rho(\mathbf{x})$, and velocity, $\mathbf{u}(\mathbf{x})$, at location $\mathbf{x}$ and time t are computed from the particle density function as

$$\rho(\mathbf{x},t) = \sum_i f_i(\mathbf{x},t) \tag{12.2}$$

and

$$\mathbf{u}(\mathbf{x},t) = \frac{1}{\rho(\mathbf{x},t)} \sum_i f_i(\mathbf{x},t)\mathbf{e}_i \tag{12.3}$$

The macroscopic quantities are subsequently used in the collision operator where they are needed to compute the equilibrium distribution, f_i^{eq}. The equilibrium distribution represents the particle distribution that should be present at a given node given $\rho(\mathbf{x})$ and $\mathbf{u}(\mathbf{x})$. The change in the *actual* particle distribution is defined with

$$\Delta f_i(\mathbf{x},t) = -\frac{1}{\tau}\left[f_i(\mathbf{x},t) - f_i^{eq}(\rho(\mathbf{x}),u(\mathbf{x}),t) \right] \tag{12.4}$$

where τ is a relaxation parameter that determines how quickly the system evolves to equilibrium; τ is related to a lattice kinematic viscosity, ν, as (Qian et al., 1992)

$$\nu = \frac{c_s^2\left[\tau - 0.5\right]}{6} \tag{12.5}$$

where c_s is the speed of sound (commonly $1/\sqrt{3}$ for the D2Q9 and D3Q19 models, but $\sqrt{2}$ for D3Q15).

The equilibrium distribution is parameterized such that mass, momentum and energy are conserved (Qian et al., 1992; Martys and Chen, 1996):

$$f_0^{eq}(\mathbf{x}) = \omega_0\rho(\mathbf{x})\left(1 - \frac{3}{2}\mathbf{u}(\mathbf{x})^2\right) \tag{12.6}$$

$$f_i^{eq}(\mathbf{x}) = \omega_i\rho(\mathbf{x})\left(1 + 3\mathbf{e}_i \cdot \mathbf{u}(\mathbf{x}) + \frac{3}{2}\left[3(\mathbf{e}_i \cdot \mathbf{u}(\mathbf{x}))^2 - \mathbf{u}(\mathbf{x})^2\right]\right) \tag{12.7}$$

where $\omega_0 = 1/3$, $\omega_i = 1/18$ for links of type II, and $\omega_i = 1/36$ for links of type III. For the sake of brevity, the equations for streaming and collision operators are often condensed into

$$f_i(\mathbf{x} + \mathbf{e}_i, t + 1) = f_i(\mathbf{x},t) - \frac{1}{\tau}\left[f_i(\mathbf{x},t) - f_i^{eq}(\rho(\mathbf{x}),\mathbf{u}(\mathbf{x}),t)\right] \tag{12.8}$$

While streaming, calculation of mass and velocity, and collision form the core of LB models, additional rules and boundary conditions are necessary to implement a real-world system in a LBM simulation. For simulation of fluid flow through porous media it is necessary to deal with fluid-solid boundaries and to define a means to generate flow.

Solid-phase geometry can be easily implemented by imposing the "bounce back" rule near fluid–solid boundaries (Figure 12.3). Without this rule the streaming operator would inadvertently cause mass loss by streaming particle distribution members into the solid phase. At the same time fluid nodes near the solid phase boundary would have undefined links that point away from the solid phase boundary. The bounce back rule simply states that mass entering the solid phase is bounced back to the originating node along the same path but opposite in direction. No mass is lost and fluid nodes just

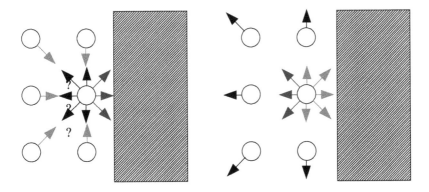

FIGURE 12.3
An illustration of the bounce back rule. Links that would be streamed into the solid return to the originating node with an opposite direction (gray), thus defining links that would otherwise be unknown as indicated by question marks in the left-hand image.

outside the solid phase have their entire particle distribution defined. The bounce back rule also causes a reduction in macroscopic velocity profiles near solid surfaces and near-parabolic velocity profiles in cylindrical channels (for more information see Chen and Doolen, 1998). Improved schemes are available (e.g., Pan et al., 2006) but such methods are computationally more intensive.

12.2.3 Pressure, Gravity, and Flow

With the equations in the previous section it is possible to simulate fluid movement inside a porous medium. Fluid movement can be induced by using boundary conditions that impose a pressure gradient between the inlet and outlet. Such methods have been used with success by Maier et al. (1996) and Zou and He (1997) among many others. Such pressure gradients can be imposed by slightly varying densities at the inlet and outlet boundaries. Another method is to use an artificial gravity term, or body force, that modifies the momentum of all fluid nodes (Martys and Chen, 1996)

$$\rho(\mathbf{x})\mathbf{u}^+(\mathbf{x}) = \rho(\mathbf{x})[\mathbf{u}(\mathbf{x}) + \mathbf{g}_L] \qquad (12.9)$$

where $\mathbf{g}_L$ is the gravitational force in lattice units and $\mathbf{u}^+(\mathbf{x})$ is the modified velocity that is subsequently used in the computation of the equilibrium distribution.

12.2.4 Phase Separation

Phase separation can be obtained by introducing an interaction between masses in adjacent nodes. Several approaches are possible, including those that define interactions between different components (i.e., different chemicals), such as the Shan and Chen (1993, 1994) and subsequent Martys and Chen (1996) approach (MC) which was used by Schaap et al. (2007) to simulate water–air and water–soltrol systems. The Martys and Chen (1996) method requires two or more distinct particle distributions to be present at each node, one for each component.

Fluid–fluid and fluid–solid interaction effects are implemented in the MC model by modifying the momentum of the components:

$$\rho_\alpha(\mathbf{x})\mathbf{u}^+(\mathbf{x}) = \rho_\alpha(\mathbf{x})\mathbf{u}(\mathbf{x}) + \tau_\alpha[\mathbf{F}_{c,\alpha}(\mathbf{x}) + \mathbf{F}_{a,\alpha}(\mathbf{x})] \qquad (12.10)$$

where $\mathbf{u}^+(\mathbf{x})$ is the modified velocity, τ_α is the relaxation parameter of component a, and $\mathbf{F}_c$ and $\mathbf{F}_a$ are the cohesion (interaction between fluid components), and adhesion forces (component to solid adhesion), respectively. Note that the velocity vector $\mathbf{u}$ is not subscripted by α because we are effectively dealing with the mixture of two components, even if there is fluid separation (cf. Schaap et al., 2007). The effective ensemble velocity of the mixture is therefore (Martys and Chen, 1996):

$$\mathbf{u}(\mathbf{x}) = \frac{\sum_{\alpha=1}^{2} \rho_\alpha(\mathbf{x}) \mathbf{u}_\alpha(\mathbf{x}) / \tau_\alpha}{\sum_{\alpha=1}^{S} \rho_\alpha(\mathbf{x}) / \tau_\alpha} \tag{12.11}$$

The effective mass is simply the sum over the component masses present at each node

$$\rho(\mathbf{x}) = \rho_1(\mathbf{x}) + \rho_2(\mathbf{x}) \tag{12.12}$$

while the ensemble kinematic viscosity is

$$\nu_L = c_s^2 \frac{\sum_{\alpha=1}^{2} (\rho_\alpha / \rho) \tau_\alpha - 0.5}{6} \tag{12.13}$$

Fluid cohesion forces causing interfacial tension are defined by

$$\mathbf{F}_{c,\alpha}(\mathbf{x}) = -\rho_\alpha(\mathbf{x}) \sum_i G_{c,\alpha,\alpha',i} \rho_{\alpha'}(\mathbf{x} + \mathbf{e}_i) \mathbf{e}_i, \quad \alpha \neq \alpha' \tag{12.14}$$

The variables α and α' denote the two *different* fluid components: in the MC two-component model there is no interaction between components of the same type, nor interaction between different components at the same node. Fluid miscibility is set by the interaction strength parameter $G_{c,\alpha\alpha',i}$ that is equal to $2G_c$ for type II links, but G_c for more distant type III links (Martys and Chen, 1996). The value of G_c should be positive and identical for both fluid components. Increasing G_c beyond a critical value will lead to progressively purer component mixtures.

Adhesive forces between solid phase and a fluid component are determined by the presence of solids in neighboring nodes that surround a fluid node:

$$\mathbf{F}_{a,\alpha}(\mathbf{x}) = -\rho_\alpha(\mathbf{x}) \sum_i G_{a,\alpha} I(\mathbf{x} + \mathbf{e}_i) \mathbf{e}_i \tag{12.15}$$

here $I(\mathbf{x} + \mathbf{e}_i)$ is an indicator function that is equal to 1 for a solid node and 0 for a pore node. G_a controls the interaction strength and wetting properties of the fluids and $G_{a,\alpha}$ is set to $2G_a$ for type II links and equal to G_a for type III links. To simulate systems of variable wettability a positive (negative) interaction strength would be used for nonwetting (wetting) fluid. By changing the interaction strength a different contact angle can be obtained (Huang et al., 2007).

With the MC method it is not possible to create large density differences between the two fluids (less than 0.5:1, cf. Schaap et al., 2007), formally limiting the method to fluid–fluid systems rather than fluid–vapor, or fluid–air systems. It is, however, possible to create reasonably large density ratios (1:500) between a fluid and its dissolved phase in the other fluid (Schaap et al., 2007). Using dimensional analysis Schaap et al. (2007) and Porter et al. (2009) showed that the model can successfully represent water–air systems as long as the capillary force acting upon the wetting fluids dominates over inertial, viscous and gravitational forces. In such a case the presence of high-density nonwetting phase can simply be considered to be a "ghost phase" that is needed to create correct wetting phase behavior. As a result of this analysis, Schaap et al. (2007) showed that accurate representations of observed capillary pressure and saturation were possible (Figure 12.4). In this case the drainage branch of the pressure–saturation graph represents the removal of the wetting phase from a water-saturated glass-bead porous medium (i.e., entry of air). The

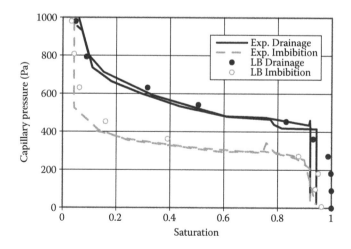

FIGURE 12.4
Observed (lines) and simulated relation between capillary pressure and saturation. (Reproduced and modified from Schaap, M.G. et al. 2007, *Water Resour. Res.* 43:W12S06, doi:10.1029/2006WR005730. With permission of American Geophysical Union.)

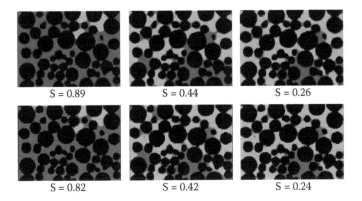

FIGURE 12.5

Observed (top row) and simulated patterns of wetting (water, dark gray) and nonwetting (air, light gray) fluid. Black is the solid phase, in this case glass beads. "S" denotes the saturation fraction, note that the values on the top and bottom row do not completely correspond. (Adapted from Porter, M.L., M.G. Schaap, and D. Wildenschild. 2009, *Adv. Water Resour.* 32(11):1632–1640, doi:10.1016/j.advwatres.2009.08.009. With permission.)

imbibition branch, however, represents wicking of wetting phase into an initially dry medium.

In a more extensive analysis, Porter et al. (2009) showed that the model was actually able to accurately capture observed patterns of wetting fluid distribution (Figure 12.5) as well as provide a reasonable estimate of observed wetting–nonwetting phase interfacial area (see Porter et al., 2009 for details). With the previous results this leads to great confidence that multiphase LB simulations are indeed capable of simulating physical processes under certain constraints. It is, however, desirable to investigate whether previous assumptions of dominance of capillary forces over other forces (Schaap et al., 2007) can be relaxed, such that a wider range of physical conditions (e.g., densities and temperatures) can be simulated.

12.2.5 Extending the Shan–Chen Model with van der Waals Equations of State

In this section, we will focus on true phase separation in the sense that liquid and vapor phases of single chemical can coexist provided that the temperature and density are under the critical temperature. The method that we follow here is based on Shan and Chen (1993, 1994) and recently modified by Yuan and Schaefer (2006). The method below differs from the MC method used earlier by Schaap et al. (2007) and Porter et al. (2009) in that it uses only one component per node to accomplish phase separation, which immediately reduces the computational requirements by 50%.

An cohesive interaction force between masses at different neighboring nodes can be introduced as defined in Shan and Chen (1993, 1994) and Yuan and Schaefer (2006):

$$\mathbf{F} = -\psi(\mathbf{x}) \sum_i G_c \alpha_i \psi(\mathbf{x} + \mathbf{e}_i) \mathbf{e}_i \qquad (12.16)$$

where α_i equals 0, 2, and 0.5 for type I, II, III links, respectively. $\psi(\mathbf{x})$ is the "effective mass" that depends on the local density, $\rho(\mathbf{x})$; while G_c is the cohesive interaction strength, similar as in the MC model. Pressure, p can be defined as follows:

$$p(\mathbf{x}) = c_s^2 \rho(\mathbf{x}) + \frac{c_0}{2} G_c \left[\psi(\mathbf{x}) \right]^2 \qquad (12.17)$$

where c_s is the speed of sound $(1/\sqrt{3})$ while c_0 is 6 for the D2Q9 and D3Q19 models. From this equation it becomes clear that ψ modifies the linear relation between pressure and density (first term in RHS). Suitable parameterizations of ψ allow phase separation to occur. Shan and Chen (1993, 1994) defined ψ as

$$\psi(\mathbf{x}) = \rho_0 [1 - \exp(-\rho(\mathbf{x})/\rho_0)] \qquad (12.18)$$

where ρ_0 is a constant that can be related to a critical temperature. Yuan and Schaefer (2006), however, demonstrated that while this equation is simple, it is also of limited practical value because it allows for only limited density contrast between the fluid and vapor phase. Instead, they proposed to use van der Waal models. Such formalisms have several advantages, such as a better match between modeled and experimentally derived relations among pressure, temperature, and density (the so-called equation of state, or EOS). Yuan and Schaefer (2006) discussed several alternative van der Waals-type EOSs, including the original model, Redlich–Kwong, Redlich–Kwong–Soave, Peng–Robinson, and Carnahan–Starling (see Yuan and Schaefer, 2006 for details). Here we focus on the Carnahan–Starling (CS) EOS because of its simplicity and because it yields stable simulations with large density contrasts and relatively large temperature ranges. The CS EOS is defined as follows:

$$p = \rho RT \frac{1 + b\rho/4 + (b\rho/4)^2 - (b\rho/4)^3}{1 - (b\rho)^4} - a\rho^2 \qquad (12.19)$$

where R is the gas constant, T the temperature in Kelvin, $a = 0.4963 \, R^2 T_c^2 / p_c$ and $b = 0.18727 RT_c/p_c$. T_c, p_c, and ρ_c are the temperature, pressure, and density at the critical point, respectively (here p_c does not correspond to the concept of capillary pressure; we will later use the capital P_c for capillary pressure). At the critical point, $T_c = 0.0943$, $p_c = 0.00442$, and $\rho_c = 0.122$. From Equation 12.18 it follows that ψ for CS is defined as

$$\psi(\rho) = \sqrt{\frac{2(p - c_s^2)}{c_0 G_c}} \tag{12.20}$$

The coexistence graph for the CS model as well as for the water–vapor system (IAPWS, NISTIR-5078) is shown in Figure 12.6. The horizontal axis represents the reduced temperature, $T_r = T/T_c$, while the logarithmic vertical axis represents the reduced density, $\rho_r = \rho/\rho_c$, the critical point is therefore located at (1, 1). Phase separation to the left of the critical point is clearly visible with an increasingly dense liquid phase and less dense vapor phase with decreasing reduced temperature. Liquid and vapor can coexist when the average volume density of a domain is between the liquid and vapor branches. In this region equilibrium densities are pre-scribed: for example, even if we would initialize a volume with and aver-age density $\rho_r = 0.1$ at $T_r = 0.7$, this system would spontaneously separate into liquid water with a higher density ($\rho_r = 2.75$) and a vapor phase with a lower density ($\rho_r = 0.016$). It is impossible to have a stable liquid phase at

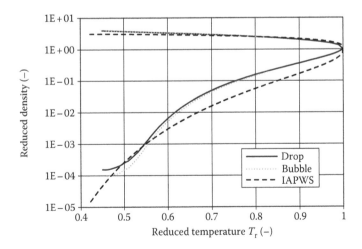

FIGURE 12.6
Coexistence curve for liquid–vapor branches in the Carnahan–Starling (CS) model. The horizontal axis represents the reduce temperature ($T_r = T/T_c$), and the vertical axis gives the reduced density ($\rho_r = \rho/\rho_c$). The coordinate (1,1) thus represents the critical point. For each $T_r < 1$ there are two densities: a higher density fluid phase and a lower-density vapor phase. The dashed line gives the coexistence curve for water (IAPWS-95, NISTIR-5078), the solid and dotted lines give the LB simulation for a drop and bubble, respectively. See text for more information.

any given $T_r < 1$ when the mean density is below that of the vapor branch. Likewise, when the mean density is higher than that of the liquid branch no stable vapor phase will be possible. For $T_r > 1$, no meaningful distinction can be made between liquid and gas phases: the substance is then in a super critical state.

From Figure 12.6, it is clear that the (IAPWS, NISTIR-5078) curve for the water–vapor system differs from that of the LB simulations with the CS EOS. The correspondence between the simulated and experimental liquid branch could qualitatively be described as "reasonable," though the CS EOS tends to overestimate liquid density at $T_r < 0.7$. There are significant differences in the vapor branch however, the CS EOS overestimates the vapor density in the range $0.55 < T_r < 1$, but generally underestimates the density for $T_r < 0.55$. It should be noted that below $T_r < 0.55$ the LB results no longer qualitatively track the IAPWS data. This is probably caused by increasing spurious currents as noted by Yuan and Schaefer (2006), which increase significantly at lower T_r and ultimately make the LB simulation unstable around $T_r \approx 0.45$. We expect that a better quantitative correspondence between the LB results and experimental data can be obtained if the a and b parameters in the CS EOS (Equation 12.19) would be adjusted.

Figure 12.6 also indicated the difference in results for LB simulations for a liquid drop surrounded by vapor (solid line) and a vapor bubble embedded in liquid (dotted line). Although it is hard to see because of the logarithmic vertical scale, these systems exhibit slightly different liquid branches, with differences in density of $\approx 1\%$ at $T_r < 0.7$. Much larger relative differences are visible for the vapor branch, especially where $T_r < 0.7$. Here the vapor densities around the liquid drop are higher than the vapor densities in the bubble. The relative differences in vapor density between drop and bubble is qualitatively understandable: relative to a flat interface, liquid–vapor interfaces in the physical world exhibit lower vapor pressures in gas bubbles and higher over liquid drops.

Figure 12.7 shows the density ratio of the vapor–liquid branch against reduced temperature. Ratios of more than 10,000 can be reached if T_r is lowered far enough, albeit at the cost of the already noted increased instabilities and spurious currents near the interfaces (cf. Yuan and Schaefer, 2006). This graph demonstrates the real advantage of the van der Waals type of EOS over the previously used MC implementation (Schaap et al., 2007; Porter et al., 2009), which typically are unable to reach *component* density ratios of more than 500. Consistent with Figure 12.6, the density ratios for the drop are lower than for the bubble, while the LB simulations yields a different curve shape than the IAPWS data.

Figure 12.8 shows the surface tension (in lattice units) for the CS EOS over a range of T_r as derived from the pressure difference between the fluid and vapor phase and the interfacial area of a drop (solid line). Also shown is the surface tension of pure water (σ_p, in physical units) as a function of temperature as described from the IAPWS surface tension equation:

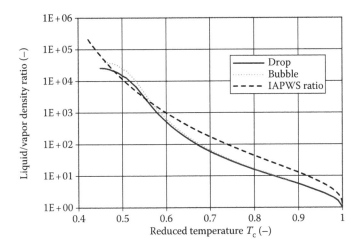

FIGURE 12.7
Density ratios for the vapor–liquid systems displayed in Figure 12.6.

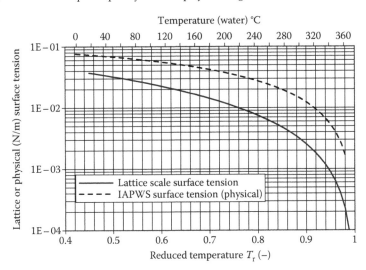

FIGURE 12.8
Simulated surface tension for the CS model as a function of T_r, in lattice units. Also shown is the experimental surface tension for the water–vapor system (IAPWS). Because of the difference in units the curves do not overlap. The horizontal axis on top displays the temperature in centigrade for the water–vapor system.

$$\sigma = 235.8\gamma^{1.256}(1 - 0.625\gamma) \qquad (12.21)$$

where $\gamma = 1 - T_r$.

It is obvious that the two curves do not match, but that is understandable because the units are different (lattice units for the simulation, physical units for the IAPWS data). However it would be convenient if the general shape of the two curves is similar. Such shape similarity make it possible to simulate fluids where temperature varies in space and/or over time, which is currently not easily possible in most LB approaches, such as the MC approach. Using Schaap et al. (2007) it is possible to relate the two curves through the capillary pressure, P_c, that is, the difference between the pressures in the liquid and vapor phase on either side of an interface:

$$P_{c,p} = \frac{\sigma_p P_{c,L}}{h_p \sigma_L} \tag{12.22}$$

Where $P_{c,p}$ and $P_{c,l}$ are the physical and lattice capillary pressure, respectively. Likewise, σ_p and σ_l are the physical and lattice surface tension and h_p is the size of a LB node in physical units. Although the shape similarity between the two curves in Figure 12.8 is not perfect, it does make it possible use the CS EOS where the temperature varies within limited temperature ranges.

Figure 12.8 also shows that the CS EOS can be applied down to a T_r of 0.46, which corresponds to a temperature physical temperature $\approx 25°C$ (T_c for the water–vapor system is 373.946°C). This may presently be disappointing because most environmental applications are within the 0–50°C range. However, it is possible to use other EOS (e.g., Yuan and Schaefer, 2006) and we think it is possible to modify the current CS EOS such that it performs better at low T_r. The use of such modified van der Waals-type equations of state within a LB framework has significant advantages over the MC approach, including: more physical realism, explicitly defined temperatures and higher liquid vapor density ratios and lower computational demands.

12.2.6 Implementation of Solid Phase

So far we have used the CS EOS in a liquid–vapor system without solid phase. To allow for wicking of fluid into a porous medium it is necessary to define a fluid–solid interaction term. Analogous to Equations 12.15 and 12.16 we define:

$$F_a = -\psi^2(x) \sum_i G_a \alpha_i I(x + e_i) e_i \tag{12.23}$$

where G_a is the fluid–solid interaction parameter similar to G_a, while $I(x + e_i)$ is an indicator function that denotes the presence of a solid phase. Note that $\psi(x)$ occurs as a square in this equation to be consistent with Equation 12.15. With the above formulation the fluid–solid interaction only depends on the fluid effective mass $\psi(x)$, the value of G_a, and the presence of solid phase.

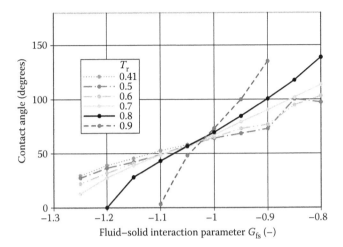

FIGURE 12.9
Contact angle as a function of G_{fs} and T_r

With fluid–solid interaction, we also introduce the concept of contact angle, that is, the angle of intersection of the curved meniscus with a straight solid phase plane (the angle is measured through the liquid phase). Figure 12.9 shows the dependence of the contact angle on G_a, as well as T_r for a range of simulations carried out in a straight capillary. Our main objective in the following is to maintain a zero contact angle for a T_r of 0.8, which means that G_a should be –1.2 (Figure 12.9). Other contact angles at this temperature can be obtained by choosing larger G_a. Note that the contact angle is more sensitive (i.e., varies strongly over small ranges of G_a) at high T_r, and less sensitive at low T_r, which is a consequence of the surface tension between the liquid and vapor phases. We also note that, in our simulations, the maximum contact angle reached was about 140°; no stable simulations were possible for larger contact angles.

12.3 Wicking with the CS Equation of State

12.3.1 Wicking into Straight Capillaries

Wicking into straight capillaries (i.e., capillary rise) is the simplest system that can be imagined for evaluating the performance of the CS EOS against a physically verifiable situation. Capillary rise, Δz, for a 2D capillary with straight walls is defined as follows:

$$\Delta z = \frac{\sigma \cos(\beta)}{\rho g R_{cap}}$$ (12.24)

Where σ is the surface tension (lattice or physical), β is the contact angle (0 for a prefect wetting fluid), ρ is the density (of water in this case), and R_{cap} the radius of the capillary. Note that this equation is missing a prefactor of 2 because we are dealing with a 2D capillary (a channel) and not a 3D cylindrical capillary.

In the remainder of this chapter we choose $T_r = 0.8$. This corresponds to a temperature of 244.5°C for water-vapor system, in which case the density of water is 806.2 and 18.3 kg/m³ for the liquid and vapor phase, respectively; the surface tension of water at this temperature is 0.027 N/m. The reduced lattice density for the CS-EOS is 2.50 and 0.15 for the liquid and vapor phases, while the lattice surface tension at this temperature is 0.00747 (all in lattice units). The above conditions are, of course, quite different from an environmental standpoint of view, in which case temperature typically range between 0°C and 50°C. Although some of the previous results indicated we could use the CS-EOS to about $T_r = 0.45$, we also observed instabilities in the simulation once solid phase was introduced (also see Yuan and Schaefer, 2006). To avoid the instabilities we arbitrarily chose $T_r = 0.8$ for the purposes of this chapter but similar results would be obtained for different T_r. We are currently investigating how to reduce instabilities for lower environmental-range applications.

The lattice scale can be linked through the gravitational attraction (9.81 m/s² for the physical scale) with the following equation:

$$g_L = \frac{\sigma_L \rho_p g_p h_p^2}{\sigma_L \rho_L}. \tag{12.25}$$

where the subscript L and p refer to the lattice and physical scale respectively and h_p is the size of a LB unit. Suitable choices for h and g_L are 6.809×10^{-5} (meters, or 68 microns per node) and 3.3×10^{-5} (lattice units). Other values are also possible, in fact Equation 12.25 merely indicates that g_L is proportional to the square of h. Note that that the choices above presume that the CS EOS is indeed a good representation of the water–vapor system, even though previous results have demonstrated some discrepancies.

Figure 12.10 shows the capillary rise in a bundle of capillaries with radii, R_{cap}, of 4.5, 5.5, 6.5, 7.5, 8.5, 9.5, 10.5, 11.5, 12.5, 13.5, 14.5, and 17.5 lattice units (dark gray is liquid, light gray is vapor, the capillary walls are black). The white line in this graph and the strip immediately to the right of the bundle of capillaries indicates the flat liquid water level if no capillaries are present. This level gives us the reference level from which to measure capillary rise. The final rise was determined for each capillary and plotted against $1/R_{cap}$ on the right-hand side of Figure 12.10 (symbols). While the observed capillary rise data exhibit a linear response as expected it is also evident that the observed response is stronger (capillary rise is higher) than what is expected from Equation 12.24 (solid line). We do not present simulation results for

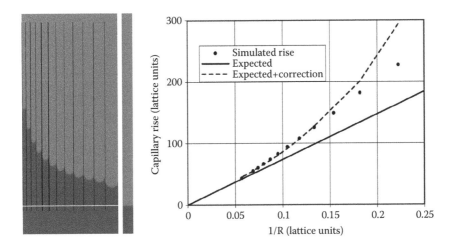

FIGURE 12.10
Wetting phase rise in a bundle of 12 straight 2D capillaries (left), with the fluid level of the conditions without capillaries immediately to the right. The simulated capillary rise (symbols), expected rise (Equation 12.25) and the expected rise corrected for hydrostatic equilibrium are shown in the diagram at the right.

$R_{cap} > 17.5$ because for such large radii gravity starts to deform the capillary into an interface with a nonuniform curvature.

The explanation for this discrepancy is threefold. First, the expression given in Equation 12.24 does not account for hydrostatic equilibrium in the vapor phase: in common physical situations the atmospheric pressure upon each interface is virtually the same. This assumption is possible because the vapor density is low and the distances involved are in the order of centimeters for the capillary rise and kilometers for the thickness of the atmosphere. Second, from Figure 12.11 it can be seen that both the vapor phase *and* the liquid phase are compressible. The density in the liquid phase ranges from 0.31 to 0.29 (lattice units) for the capillary with a radius of 4.5 lattice units over a distance of 290 lattice units (i.e., $290 \times 68.09 = 463$ μm). This makes the compressibility of the liquid phase in the LB model much stronger than is reasonable for physical situations, but also affects the capillary rise (Equation 12.24) because the lattice density is not constant. Third, the solid–fluid attraction tends to accumulate fluid at the capillary walls with a higher density than the free fluid outside the capillaries. This is visible in Figure 12.11 where there is a divergence in the center-line vapor-phase density among the three capillaries. For example, the increase in vapor-phase density downward is much stronger for $R_{cap} = 4.5$ than for $R_{cap} = 17.5$ (lattice units) whereas density profiles in the physical world would overlap. The increase itself is another consequence of hydrostatic equilibrium and compressibility in the vapor phase, especially at this relatively high T_c (high vapor densities).

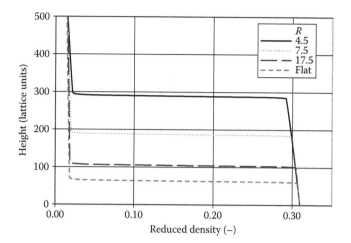

FIGURE 12.11
Density profiles in three capillaries ($R = 4.5$, 7.5, and 17.5) and for the situation with no capillaries (flat interface).

When we correct for hydrostatic equilibrium and density gradients we obtain a fairly accurate representation of the simulated capillary rise (dashed line in the diagram of Figure 12.10). This correction indicates that up to about $1/R_{cap} \approx 0.15$ (i.e., $R_{cap} \approx 7$) the simulation is reasonable if hydrostatic equilibrium is taken into account. Beyond $1/R_{cap} = 0.15$ it is likely that the narrow capillaries increasingly affect the contact angle and/or the surface tension. At present we ignore the narrow-capillary effect, but note that an improved interaction with the solid phase is in order.

12.3.2 Wicking into a Random Porous Medium

Finally we conduct a simulation with the same conditions as the bundle of capillaries with gravity, but now for a periodic random packing of overlapping polydisperse discs with sufficient density to allow a permeable medium of 256×256 lattice units. If we use the same scaling conversions as before for the bundle of capillaries, this system represents a porous medium of 1.7×1.7 cm. Using Schaap et al. (2007) we can determine the time scaling as

$$\Delta t_p = \rho_w \upsilon_L \frac{h_p^2}{\mu_w} \qquad (12.26)$$

which for a density, ρ_w, of 806.2 kg/m³, and a dynamic viscosity, μ_w, of 1.068×104 Pa s for water at 244.5°C yield a time step of 5.8×10^{-3} seconds

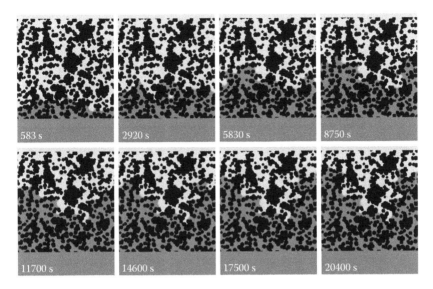

FIGURE 12.12
Liquid (dark gray) wicking into a porous medium of overlapping discs, the text in each image indicates the physical time that has elapsed.

for a resolution $h = 6.809 \times 10^{-5}$ m and a lattice kinematic viscosity $\upsilon_L = 0.1667$ (Equation 12.5 with $\tau = 1$). As can be seen in Figure 12.12, the wicking against gravity now leads to an irregular invasion front with flow upward against gravity. The invasion is rapid at first, but slows later on in the simulation around 8720 s. True equilibrium sets in a few million time steps after the last image. It is interesting to note that occasionally enclosed vapor phase is found. These bubbles are not vapor phase that was left behind by the advancing wetting front, but bubbles that form for brief periods as the main front advances rapidly to fill a large pore. This condition leads to a temporary pressure decrease within the liquid phase, making it possible for vapor to appear at $t = 2920$ s.

It is also easy to conduct the same simulations under different gravitational conditions. The top left image in Figure 12.13 shows the wicking at 5830 s with gravity pointing downward as before. The middle and rightmost image on the top row show zero-gravity and gravity upward making it progressively easier for the liquid to enter the porous medium, that is, the fluid invades faster and even leaks out from the top in the negative gravity case (top right). The bottom row duplicates the 1g case at 1 million time steps (bottom left), but now shows the zero and negative gravity cases for equivalent mass invasion, which occurs at 2330 and 1170 s for the zero and negative gravity case, respectively. These images indicate that different front geometries are obtained under different gravitational conditions.

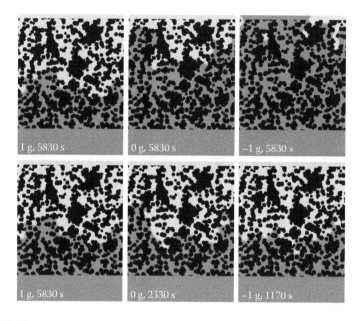

FIGURE 12.13
Wicking into a porous medium under different gravitational conditions, the text in each image indicates the gravitational conditions (positive is down) and the elapsed time.

The wicking in each of the three cases is driven by the sorptivity of the medium and the direction and strength of gravity. The cumulative infiltration, I (i.e., cumulative wicking amount) is given by (Philip, 1969)

$$I = S\sqrt{(t)} + a^*t \qquad (12.27)$$

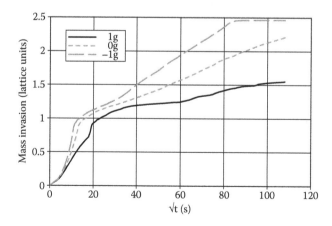

FIGURE 12.14
Cumulative mass invasion (wicking) versus the square toot of the number of iterations, for the 1g, 0g, and –1g case.

where t is time, S the sorptivity, and a is a parameter that is related to the hydraulic conductivity of the medium and the gravitational acceleration. Figure 12.14 plots the cumulative mass invasion (lattice units) versus the square root of time. It is clear that for the 1g case (black line) there is no linear relation, indicating that the mass invasion is dominated by gravity and hydraulic conductivity, that is, the second term on the right hand size of Equation 12.27. However, the 0g and −1g cases do exhibit a linear relationship, after the mass invasion reaches a value of 1.5 at which case the sorptivity controls the mass invasion. The flat horizontal line for −1g indicates that the fluid has reached the other end of the porous medium and is leaking out at the top.

12.4 Conclusion

The application of the LB method to simulations of flow and transport in micro-scale porous media is a relatively new branch to computational fluid dynamics with a very active research and development community. There are many aspects that we have not touched upon, such as numerical and parallelization efforts (the method generally is very computationally intensive) and more efficient and accurate numerical schemes, such as multiple relaxation time methods (d'Humieres et al., 2001, Ginzburg, 2007). The work we presented in this chapter emphasized the application of the LB method to observables, thus demonstrating that the method can provide reasonable representations of reality, even after making simplifying assumptions (Schaap et al., 2007). A further development of this research allowed Porter et al. (2009) to also validate thermodynamic pressure–saturation–interfacial area theories by Hassanizadeh and Gray (1993) and Gray et al. (2002). Implementing realistic equations of state (EOS) into a multi-phase LB framework is very attractive for a number of reasons, not in the least because fewer simplifying assumptions are needed, but also because it becomes possible to use physically more realistic relations among pressure, density and temperature. Using work by Yuan and Schaefer (2006) we have shown that it is possible to reach much higher liquid–vapor density ratios than previously possible. We demonstrated that a wide range of the subcritical water–vapor can be covered, though it is not presently possible to reach the actual triple point of water (0°C) with the equations of state we have investigated. We expect that by modifying existing EOS specifically for simulating the water–vapor system in LB models we may be able to reach a more physically realistic match than presented here. In addition to the water–vapor system we are also working on testing EOS to sub- and supercritical CO_2 to investigate pore-scale aspects of geological carbon sequestration in deep saline aquifers.

Acknowledgment

Dr. Schaap would like to acknowledge partial support from DOE-BES grant DE-FG02-11ER16278.

Nomenclature

Δz	Capillary rise in physical units (m) or lattice units
α, α'	Indicator of fluid type or phase in the MC model
β	Contact angle
γ	$1-T_r$
ρ	Density, subscripts "p" and "L" denote physical units (kg m^{-3}) or lattice units, where needed
ρ_c	Critical density (kg m^{-3} or lattice units)
ρ_r	Reduced density (no units)
σ_p, σ_L	Physical and lattice surface (or interfacial) tension (N m^{-1} or lattice units), respectively
τ	Relaxation parameter, related to the fluid viscosity (lattice units)
υ	Lattice kinematic viscosity
ω_0, ω_i	Prefactors for the type I, II, and III distribution members i in f^{eq}
ψ	Effective density (lattice units)
a,b	Coefficients in the Carnahan–Stirling and other van der Waals-type models
c_0	Coordination number, equal to 6 for the D2Q9 and D3Q19 models
c_s	Speed of sound ($\sqrt{3}$, lattice units)
$\mathbf{e}$	2D or 3D matrix to indicate nearest and next nearest nodes
$\mathbf{F}_c, \mathbf{F}_a$	Gravitational, cohesive, adhesive forces, respectively for the MC model (in N or lattice units)
$f^{eq}(\rho(\mathbf{x}),\mathbf{u}(\mathbf{x}),t)$	Equilibrium particle distribution for ρ and $\mathbf{u}$ at grid point $\mathbf{x}$ and time t (lattice units)
$f(\mathbf{x},t)$	Particle distribution at grid point $\mathbf{x}$ and time t (lattice units)
$G_c, G_a,$	Interaction strength between fluid phases and fluid and solid, respectively (lattice units)
g_p, g_L	Gravitational acceleration (m s^{-2} or lattice units)
h_p	Physical size of a LB node (m)
I	Infiltration (m)

i	Index for particle distribution member in $\mathbf{f}$, $\mathbf{f}^{eq}$, and $\mathbf{e}$
$I()$	Indicator function
IAPWS	International Association for the Properties of Water and Steam
NIST	National Institute of Standards
p	Pressure (Pa, or lattice units)
p_c and ρ_c	Critical pressure (Pa or lattice units) and density (kg m^{-3}), respectively
$P_{c,p}$, $P_{c,L}$	Physical and lattice capillary pressure (Pa, or lattice units)
p_r and ρ_r	Reduced pressure (no units)
R	Gas constant
R_{cap}	Capillary radius in meters or lattice units
S	Sorptivity (ms$^{1/2}$)
T_c	Critical temperature (Kelvin or lattice units)
T_r	Reduced temperature (no units)
t	Time (seconds or lattice units)
$\mathbf{u}$	Velocity (2D or 3D vector, lattice units)
$\mathbf{u}^+$	Velocity modified by external forcing (2D or 3D vector, lattice units)
$\mathbf{x}$	Grid coordinate, a 2 or 3 member vector indicating integer position in x, y, z

---

References

Adler, P.M., C.G. Jacquin, and J.A. Quiblier. 1990, Flow in simulated porous media, *Int. J. Multiphase Flow* 16(4):691–712.

Chen, S. and G. D. Doolen. 1998, Lattice Boltzmann method for fluid flows, *Ann. Rev. Fluid Mech.* 30:329–364.

d'Humieres, D., I. Ginzburg, M. Krafczyk, P. Lallemand, and L.-S. Luo. 2001, Multiple–relaxation–time lattice Boltzmann models in three dimensions, *Philosophical Transactions of the Royal Society of London. Series A: Mathematical, Physical and Engineering Sciences*, 360(1792), 437, doi:10.1098/rsta.2001.0955.

Ferréol, B. and D.H. Rothman. 1995, Lattice Boltzmann simulations of flow through Fontainebleau sandstone, *Transp. Porous Media* 20:3–20.

Ginzburg, I. 2007, Lattice Boltzmann modeling with discontinuous collision components: Hydrodynamic and advection—diffusion equations, *J. Stat. Phys.* 126(1), 532:157–206, doi:10.1007/s10955–006–9234–4.

Gunstensen, A.K. and D.H. Rothman. 1991, Lattice Boltzmann model for immiscible fluids, *Phys. Rev. A.* 43:4320–4327.

Gray, W.G., A.F.B, Tompson, and W.E. Soll. 2002, Closure conditions for two-fluid flow in porous media, *Transp. Porous Media* 47:29–65.

Hassanizadeh, S.M. and W.G. Gray. 1993, Thermodynamic basis of capillary pressure in porous media, *Water Resour. Res.* 29:3389–3405.

He, X., S. Chen, and G.D. Doolen. 1998, A novel thermal model for the lattice Boltzmann method in incompressible limit, *J. Comput. Phys.* 146:282–300.

Heijs, A.W.J. and C.P. Lowe. 1995, Numerical evaluation of the permeability and the Kozeny constant for two types of porous media, *Phys. Rev. E* 61(5):4346–4352.

Huang, H., D.T. Thorne, M.G. Schaap, and M.C. Sukop. 2007, *A priori* determination of contact angles in Shan-and- Chen-type multi-component multiphase lattice Boltzmann models, *Phys. Rev. E* 76:066701.

IAPWS. 1996, International Association for the Properties of Water and Steam http://www.iapws.org/relguide/IAPWS-95.htm

Inamuro, T., T. Ogata, S. Tajima, and N. Konishi. 2004a, A lattice Boltzmann method for incompressible two-phase flows with large density differences, *J. Comp. Phys.* 198:628–644.

Inamuro, T., T. Ogata, and F. Ogino. 2004b, Numerical simulation of bubble flow by the lattice Boltzmann method, *Future Gener. Comp. Syst.* 20:959–964.

Luo, L. 1998, Unified theory of lattice Boltzmann models for nonideal gases, *Phys. Rev. E* 81:1618–1621.

Maier, R.S., R.S. Bernard, and D. W. Grunau. 1996, Boundary conditions for the lattice Boltzmann method, *Phys. Fluids* 8(9):1788–1801.

Martys, N.S. and H. Chen. 1996, Simulation of multicomponent fluids in complex three-dimensional geometries by the lattice Boltzmann method, *Phys. Rev. E* 53:743–750.

Masad E., B. Muhunthan, and N. Martys. 2000, Simulation of fluid flow and permeability in cohesionless soils, *Water Resour. Res.* 36(4):851–864.

NISTIR-5078. 1998, National Institute of Standards http://www.nist.gov/srd/upload/NISTIR5078.htm

O'Connor, R.M. and J.T. Fredrich. 1999, Multiscale flow modling in geologica materials, 1999, *Physics and Chemistry of the Earth, Part A: Solid Earth and Geodesy*, 24(7):611–616, doi: 10.1016/S1464–1895(99)00088–5.

Pan, C., M. Hilpert, and C.T. Miller. 2001, Pore-scale modeling of saturated permeabilities in random sphere packings, *Phys. Rev. E* 64:6702.

Pan, C., L.-S. Luo, and C.T. Miller. 2006, An evaluation of lattice Boltzmann schemes for porous medium flow simulation, *Comput. Fluids* 35(2):898–909.

Philip, J.R. 1969, Theory of infiltration, *Adv. Hydrosci.* 5:215–296.

Porter, M.L., M.G. Schaap, and D. Wildenschild. 2009, Lattice-Boltzmann simulations of the capillary pressure–saturation–interfacial area relationship for porous media, *Adv. Water Resour.* 32(11):1632–1640, doi:10.1016/j.advwatres.2009.08.009.

Qian, Y.H., D. d'Humieres, and P. Lallemand. 1992, Lattice BGK models for Navier–Stokes Equation, *Europhysics Lett.* 17:479–484.

Quiblier, J.A. 1984, A new 3-dimensional modeling technique for studying porous media, *J. Colloid Interface Sci.* 98(1):84–102, doi:10.1016/0021–9797(84)90481–8.

Schaap, M.G, M.L. Porter, B.S.B. Christensen, and D. Wildenschild. 2007, Comparison of pressure-saturation characteristics derived from computed tomography and lattice Boltzmann simulations, *Water Resour. Res.* 43:W12S06, doi:10.1029/2006WR005730.

Shan, X. and H. Chen. 1993, Lattice Boltzmann model for simulating flows with multiple phases and components, *Phys. Rev. E* 47:1815–1819.

Shan, X. and H. Chen. 1994, Simulations of non ideal gases and liquids-gas phase transitions by the Lattice Boltzmann equation, *Phys. Rev. E* 49:2941–2948.

Succi, S. 2001, *The Lattice Boltzmann Equation for Fluid Flow and Beyond*, Oxford Scientific Publications, Oxford, Great Britain.

Swift, M.R., W.R. Osborn, and J.M. Yeomans. 1995, Lattice Boltzmann simulation of nonideal fluids, *Phys. Rev. Lett.* 75:830–833.

Swift, M.R., E. Orlandini, W.R. Osborn, and J.M. Yeomans. 1996, Lattice Boltzmann simulations of liquid-gas and binary fluid systems, *Phys. Rev. E* 54:5041–5052.

Yuan, P. and L. Schaefer. 2006, Equations of state in a lattice Boltzmann model, *Phys. Fluids* 18, 042101, doi: 10.1063/1.2187070.

Zhang, D., R. Zhang, S. Chen, and W.E. Soll. 2000, Pore scale study of flow in porous media: Scale dependency, REV and statistical REV. *Geophys. Res. Lett.* 27:1195–1198.

Zhang, J., B. Li, and D.Y. Kwok. 2004, Mean-field free-energy approach to the lattice Boltzmann method for liquid–vapor and solid–fluid interfaces, *Phys. Rev. E* 69:032602.

Zhang, J. and D.Y. Kwok. 2009, A mean-field free energy lattice Boltzmann model for multicomponent fluids. *Eur. Phys. J. Special Top.* 171:45–53.

Zheng, H.W., C. Shu, and Y.T. Chew. 2006, A lattice Boltzmann model for multiphase flows with large density ratio, *J. Comput. Phys.* 218:353–371, doi: 10.1016/j.jcp.2006.02.015.

Zou, Q. and X. He. 1997, On pressure and velocity boundary conditions for the lattice Boltzmann BGK model, *Phys. Fluids* 9(6):1591–1599.

Index